AF677523

10 Years Plant Molecular Biology

10 Years Plant Molecular Biology

Edited by

Robbert A. Schilperoort and Leon Dure

Reprinted from Plant Molecular Biology, Vol. 19 (1992)

Kluwer Academic Publishers
Dordrecht / Boston / London

ISBN 0-7923-1480-8

Published by Kluwer Academic Publishers,
P.O. Box 17, 3300 AA Dordrecht, The Netherlands.

Kluwer Academic Publishers incorporates
the publishing programmes of
D. Reidel, Martinus Nijhoff, Dr W. Junk and MTP Press.

Sold and distributed in the U.S.A. and Canada
by Kluwer Academic Publishers,
101 Philip Drive, Norwell, MA 02061, U.S.A.

In all other countries, sold and distributed
by Kluwer Academic Publishers Group,
P.O. Box 322, 3300 AH Dordrecht, The Netherlands.

Printed on acid-free paper

All Rights Reserved
© 1992 Kluwer Academic Publishers
No part of the material protected by this copyright notice may be reproduced or
utilized in any form or by any means, electronic or mechanical,
including photocopying, recording or by any information storage and
retrieval system, without written permission from the copyright owner.

Printed in the Netherlands

Contents

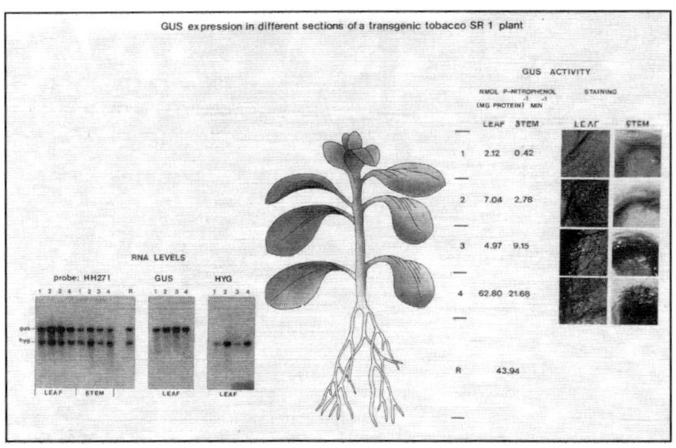

Front cover: The expression of a mannopine synthase *gusA* fusion in a transgenic F_1 tobacco plant. Histochemical and biochemical assays show higher GUS activities in the old parts. RNA blot analysis shows comparable steady state levels of GUS mRNA in all tissues. Hensgens et al (Plant Molecular Biology, submitted) concluded that GUS activity accumulates in time and is not a straight reflection of promoter activity.

Plant Molecular Biology **19**: vii, 1992.
R. A. Schilperoort & L. Dure (eds.), 10 Years Plant Molecular Biology.

Preface

It is with great pleasure that Leon Dure and I, together with the publisher of *Plant Molecular Biology*, present to you this special publication which appears as a special issue of *PMB* as well as a book available to a wider audience.

The publication of this issue is special to us because it represents a landmark in the history of the journal and, to a certain extent, in the underlying scientific discipline. It is being published as a celebration of the 10 years' existence of the journal and of 10 years of publishing high-quality scientific papers. After starting from scratch, the result of 10 years' hard work is that the journal now has a leading place in the area of plant molecular biology. This could only have happened through the constant support and advice of the scientific community. Hundreds of fine papers that have shaped the field have been published during the 10 years of the journal's existence and fascinating and sometimes astonishing progress has been made in almost all areas of plant research.

It is very clear nowadays that plants offer several opportunities for basic studies, e.g., on development and embryogenesis, and that the fundamental principles layed open contribute to the development of new tools for plant breeding.

Within the scope of the present publication, the editors have had to make a difficult choice from the many important subjects that have contributed to the remarkable progress of our molecular biological understanding of complex biological problems. This has resulted in review papers showing the present state of the art in genetic engineering, gene expression and its manipulation, microbe and insect interactions with plants, transposable elements and gene tagging, plant and organ development, the function and structure of the genome of chloroplasts, and lipid biosynthesis.

All papers have been written in such a way that they are also useful for non-experts interested in a particular field, as well as for students following courses in plant molecular biology. Besides presenting the state of the art, each paper gives some historical background to the developments in the field as well as perspectives for further basic research and applications. Because of the latter, scientists and students engaged in plant breeding will also profit from this publication.

The editors would like to express their gratitude to all the prominent scientists who have contributed excellent papers and their latest results to "Ten Years of Plant Molecular Biology". We all know how much time this takes. We would also like to thank the reviewers who have given their time critically appraising the papers.

ROB A. SCHILPEROORT
The Editor

Plant Molecular Biology **19**: 1–14, 1992.
© 1992 *Kluwer Academic Publishers. Printed in Belgium.*

Transcriptional and post-transcriptional regulation of gene expression in plants

Cris Kuhlemeier
Institute of Plant Physiology, University of Berne, Altenbergrain 21, CH-3013 Berne, Switzerland

Key words: plant genes, transcription, RNA stability, translation, signal transduction

Introduction

In the past few years virtually every aspect of plant gene expression has been covered by thoughtful, in-depth reviews. In this article, it is not my goal to repeat, combine or update those reviews. Rather, I will attempt to give as general as possible an overview of plant gene expression. In order to reach this goal I have selected a small number of examples to illustrate what, in my opinion, are important concepts in plant gene expression.

Many concepts in plant gene expression derive from animal, yeast and bacterial systems. This interaction between plant and non-plant research has been and continues to be extremely rewarding. I will emphasize these general concepts, but at the same time highlight where plant gene expression is different or our understanding more advanced.

Plant nuclear genes are like other eukaryotic genes

Over the past years a great number of plant genes have been cloned, and if one general conclusion can be drawn from the accumulated data it is that plant genes are very much like animal or yeast genes. Plant genes use the same genetic code, are split by introns, and use regulatory mechanisms that are similar in principle. However, that is not to say that genes can be shuffled at will between plants and animals. Introns in plant nuclear genes, for instance, are removed by the same lariat-type mechanism as in animals, but the sequence requirements for splice site recognition are subtly different. Similar things can be said about plant promoters. There are many variations on the general theme but the basics remain the same.

Not too surprisingly, the first genes to be cloned were those that are highly expressed. What could one learn from such a cloned cDNA? Let us take the case of the small subunit of Rubisco, a very abundant well-characterized enzyme. Cloning provided the amino acid sequence not only of the mature protein, but also of the N-terminal transit sequence, which targets the protein to the proper compartment, the chloroplast stroma. Comparison of many such transit sequences can shed light on what makes a transit sequence specific for the chloroplast and not for the mitochondrial compartment. What is important for chloroplast targeting is apparently not so much a particular sequence but rather the overall structure, or in this case 'random coil': i.e. the absence of any structure [36].

Southern blotting and genomic cloning showed that many proteins are encoded in the plant chromosomes by multiple gene copies. In pea the small subunit of Rubisco is encoded by five genes, each with two introns [17]. In all plants analysed to date there are 4–12 gene members per genome. In pea these *rbcS* genes code for identical mature proteins, but in other organisms the corresponding proteins have small sequence variations. The *rbcS* sequence variations are minor and there is no evidence that they are correlated with different functions of the proteins. Of course, in many instances isozymes with different functions are well-known, a good example being enzymes such as glyceraldehyde 3-phosphate dehydrogenase (GAPDH) for which both cytoplasmic and chloroplast forms exist [10, 61, 81]. The cytoplasmic form is involved in glycolysis, whereas the chloroplast enzyme catalyses the reverse reaction in the Calvin cycle. This picture of two isozymes has been greatly refined with the aid of the cloned genes. In maize the chloroplast enzyme is a heterodimer encoded by the *gapA* and *gapB* genes. The cytoplasmic enzyme is homomeric and encoded by several *gapC* genes. In cases such as GAPDH, but also many others, the relatively straightforward standard methods of molecular biology have enabled physiologists and biochemists to obtain highly precise information not only on the gene families but more importantly on the peculiarities of the encoded isozymes. The individual enzymes can often be overexpressed in *Escherichia coli* or yeast and subsequently purified and analyzed for structure and enzymatic functions.

Thus, gene cloning methods have provided a wealth of data on proteins that would have been impossible to obtain with classical biochemical or physiological methods.

Nuclear genes are primarily regulated at the level of transcription

To understand the function of a protein it is necessary not only to characterize enzymatic activities, it is equally important to know where and when the protein is present. Let us turn again to the example of the small subunit of Rubisco [22, 26, 46, 94, 96]. The major Rubisco activity is present in green leaves and *in vitro* translation/immunoprecipitation showed a correlation between the prevalence of translatable mRNA and Rubisco protein. Subsequent studies using *rbcS* cDNA showed that mRNA levels in etiolated seedlings rise after a red light pulse. The red light effect is far-red-reversible, establishing the involvment of the photoreceptor phytochrome in the modulation of mRNA levels. So-called nuclear run-off experiments indicated that these modulations of mRNA levels arise primarily from modulations of the rate of transcription and not from post-transcriptional processes such as differential mRNA stability.

The *rbcS* genes within a given plant show minor sequence divergence. Yet the expression of individual genes varies considerably [22, 46, 96]. In petunia expression varies only quantitatively, i.e. there are highly expressed genes and lowly expressed genes, but all genes appear to be expressed in the same organs at the same relative levels and at the same time in development. In tomato there are both differences in the transcript levels *and* in the patterns of expression. Such differential gene expression can be observed in many gene families, often more dramatically than in the *rbcS* family. Another example are again the GAPDH encoding genes. The maize *gapA* and *gapB* genes which code for the chloroplastic forms are induced by light whereas *gapC* genes for cytosolic GAPDH are not light-induced. Genes for glycolytic enzymes are usually induced during anaerobiosis, because the low energy efficiency of fermentation requires an increased flux through the glycolytic pathway. It is interesting that of the cytosolic GAPDH genes only *gapC1* is anaerobically induced, whereas *gapC2* mRNA levels remain constant [61, 81].

The *cis*-acting elements

In the previous section we have seen that even closely related genes may have very different patterns of expression. What makes a gene expressed the way it is? This question can be answered by mutational analysis. In complex eukaryotes such as plants this can only be done by starting with the cloned gene, mutating it *in vitro* and returning the mutated gene to the plant. The required gene transfer techniques are described in an accompanying paper. Suffice it to say here that basically there are two approaches. One relies on the introduction of DNA into protoplasts and the assay of mRNA or protein within a few days [85]. These systems are fast and semi-quantitative, but have the drawback that some characteristics, such as tissue specificity, cannot be scored. More recently, novel techniques such as particle bombardment [28, 70] and microinjection (G. Neuhaus, pers. comm.) have been developed to obtain transient expression in differentiated tissues. The second approach utilizes transgenic plants, plants that are identical to wild-type plants except for the fact that they have a mutated gene integrated into the genome. Transgenic plants take more time to raise and analyse but have the advantage that they allow us to study the gene in its natural environment: the intact plant [1, 46].

Mutational analyses of the type described above have defined two classes of DNA sequence important for transcription of a gene (Fig. 1). First there is the TATA box, or a functionally related sequence that binds the RNA polymerase complex and determines at what site transcription will start, about 30 bases downstream. Mutations in the TATA box interfere with proper transcription initiation. Constructs consisting of a TATA box fused to a reporter gene usually give low to undetectable transcript levels. It must be noted that very little is known about plant TATA boxes or genes without TATA boxes. This is in strong contrast with the flood of publications on the upstream elements. There is some evidence that the TATA box, or sequences in the close vicinity, may be important for the light-regulated expression of pea *rbcS* genes [47, 49, 66].

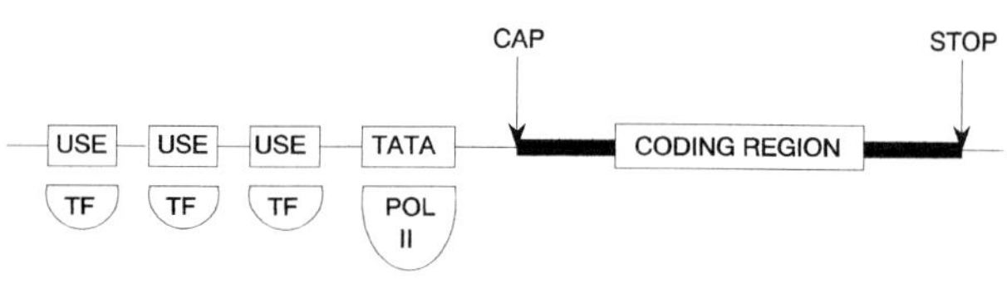

Fig. 1. Schematic view of a plant nuclear gene. The 'coding region' is the DNA sequence between the ATG translational initiation codon and the TAA, TGA or TAG translational stop codons. The coding region may be interrupted by introns, sequences that are present in the DNA and in the primary RNA transcript, but are removed by splicing and therefore absent in the mature cytoplasmic mRNA. The beginning and the end of the DNA region that is transcribted into RNA are indicated by CAP and STOP. The thick black lines represent the 5′ leader and the 3′ tail of the mRNA. Upstream of the CAP site is the TATA box, which is the binding site for RNA polymerase II and associated factors. The upstream sequence elements (USEs) can bind a variety of transcription factors (TFs). The drawing is not to scale!

The second class of DNA elements includes the binding sites for proteins that can interact with the RNA polymerase complex. Such *cis*-acting elements can function at variable distances from the TATA box. Even if their orientation is reversed they may still work. Often these elements are regulatory, i.e. they only enhance (or repress) transcription under specified cellular or environmental conditions. A classical example is the heat-shock element, which, when fused upstream of a TATA box/reporter gene, increases transcription only at high temperatures [76].

One of the best studied plant promoters is the 35S promoter. This very strong viral promoter produces the cauliflower mosaic virus (CaMV) 35S genomic RNA. In early experiments approximately 1000 bp of promoter DNA including a few basepairs beyond the transcription start site were fused to various reporter genes. Analysis in protoplasts, transformed calli and in transgenic plants demonstrated that the reporter genes were always expressed and at high levels, and were insensitive to various endogenous and environmental cues such as hormones, heat shock or light [43, 69, 72]. This so-called constitutive expression made the 35S promoter popular as a control for experiments analysing other, regulated promoters. Deletion analysis of the 35S promoter showed that the 350 bp adjacent to the TATA box were sufficient for high expression. The upstream 300 bp from about −350 to −50 (relative to the transcription start site and thus not including the TATA box) can be inverted and even placed 3′ of the reporter gene without loss of function [3]. A startling observation was made when the 35S promoter was deleted to −105 or −90: expression became organ-specific. No expression could be found in leaves or stems but in roots there were considerable levels of the CAT reporter gene mRNA and enzyme activity [79]. This finding naturally led to an important question. Is a constitutive promoter a simple promoter that contains one or more copies of a simple *cis*-

acting element that confers constitutive expression? Or is perhaps a constitutive promoter not simple at all but rather a complex array of various regulatory *cis*-acting elements, and is it the sum of all these specialized *cis*-elements that results in a constitutive, non-specialized promoter?

The latter model had proven correct for the SV40 promoter, a highly expressed constitutive animal viral promoter [71, 88]. Detailed studies on the 35S promoter have now been performed which demonstrate that subsegments of the 35S promoter confer widely varying patterns of gene expression upon the GUS reporter gene supporting the combinatorial model of promoter function (for review see [3]).

The *trans*-acting factors

The TATA box is the binding site for RNA polymerase II and its associated factors. The other *cis*-acting regulatory elements can bind a wide variety of DNA binding proteins. These proteins must interact with the RNA polymerase complex either directly, or via so-called bridging proteins which have no affinity for DNA themselves but are thought to have contact sites for both RNA polymerase and upstream DNA-binding proteins [54].

Some of the upstream binding proteins are probably general transcription factors, present in all or at least most cell types and active under most if not all conditions. Other factors may be more specialized. However, it should be kept in mind that the transcription rate and its regulation are very likely determined not just by the intrinsic properties of a transcription factor and its cognate binding site but rather by a complicated interplay of multiple factors and multiple binding sites (e.g. [26]). One factor may have different affinities to multiple sites and it may bind cooperatively. A factor may compete with another factor for a single or overlapping binding sites resulting in changed interactions with the RNA polymerase complex. Post-translational modification of transcription factors may influence all the above.

The above concepts derive mainly from experiments with bacteria, yeast and animals. However, also our knowledge of plant transcription factors is increasing rapidly. Here I review some of the information on plant transcription factors. A list of well-characterized factors is given in Table 1.

RNA polymerase and associated proteins

RNA polymerases will synthesize RNA when provided with a DNA template, Mg^{2+} ions and the four ribonucleoside triphosphates. Of the three RNA polymerases present in eukaryotes RNA polymerase II transcribes the nuclear protein-encoding genes. Plant RNA polymerase II has been isolated from a number of monocot and dicot species and displays similar subunit structure [32]. There are two large subunits with M_r 180–220 and M_r 140 and eight small subunits with M_r between 16 and 40. The largest subunit contains 35–40 tandem copies of the heptapeptide PTSPSYS at its carboxy-terminus. Similar repeats are also present in other eukaryotic RNA polymerases. Labelling with ^{32}P-phosphate indicated extensive phosphorylation probably at the threonine, serine and tyrosine residues of the heptapeptide repeat. Phosphorylation/dephosphorylation of an RNA polymerase may be important for interactions with histones or other transcription factors.

Of the accessory proteins TFIIA, B, D, E, F, known from HeLa cells, only TFIIA and TFIID have been characterized in plants. TFIIA appears to be very similar to its animal counterpart [11]. TFIID, which has some homology to bacterial sigma factor, binds to the TATA box and thus may be a key determinant of the transcription initiation site. Screening of an *Arabidopsis thaliana* cDNA library with heterologous probes revealed that *Arabidopsis* contains two distinct TFIID genes [25]. Whether these two genes code for functionally distinct proteins is an interesting speculation at present.

Table 1. Plant transcription factors.

Factor	Class	Target sequence	References
TFIIA	general	–	11
TFIID-1	general	TATA box	25
TFIID-2	general	TATA box	25
AT-1		AT-rich	16, 87
3AF-1	zinc-finger	AT-rich	49
no name		AT-rich	12
ASF-2		GATA	48
GA-1		GATA	19, 87
GC-1	Sp1-like?	GC-rich	87
GT-1		GTGG	29, 31, 87
GT-2		GTGG	18
Knotted-1	homeobox		96
Athb-1	HD-ZIP		80
Athb-2	HD-ZIP		80
HSF8	heat shock	GAAnnTTC	86
HSF24	heat shock	GAAnnTTC	86
HSF30	heat shock	GAAnnTTC	86
TGA1a + b	bZIP	TGACG	41, 42, 98
GBF		G/C ACGTG	20, 87
OCSTF	bZIP	GACGTA	89a
TAF1	bZIP	ACGTG	70
O-2	bZIP		34, 55, 89
HBP-1 a + b	bZIP	GACGTG	91, 92
EmBP-1	bZIP	G/C ACGTG	33
CPRF-1, 2, 3	bZIP	CACGTG	97
Deficiens	MADS		90
Agamous	MADS		14
TM3-TM8	MADS		78
AGL1-AG26	MADS		59
B1	HLH	CAGGTGC	28
Myb-like	HLH		38
Lc	HLH		56, 57
C1	HLH		27, 75
floricaula			15
viviparus-1			64

Abbreviations: HD-ZIP, homeodomain-leucine zipper; bZIP, basic domain-leucine zipper; MADS, MCM1-*Agamous-Deficiens*-SRF1 family. HLH, helix-loop-helix.

Specific DNA-binding proteins and (putative) transcription factors

The initial characterization of proteins binding to well-characterized *cis*-acting elements was made by gel retardation and footprinting assays [30]. In a gel retardation assay a labelled DNA fragment is incubated with a nuclear extract and then run on a non-denaturing gel. A protein-DNA complex will migrate slower compared to free DNA. The specificity of the interaction can be monitored by adding excess of unlabelled DNA to the binding reaction. DNA with a sequence related to the *cis*-acting element will compete for binding, unrelated DNA will not.

DNA footprinting techniques rely on the principle that proteins will protect their DNA-binding sequences from attack by nucleases or chemical agents. Footprinting can thus identify the protein-binding sites on a piece of DNA with high resolution. More recently, methods have been developed to clone the genes for DNA-binding proteins. As yet no plant transcription factor has been purified directly from nuclear extracts. However, methods have been devised to clone the genes for DNA-binding proteins.

Most successfully, radioactively labelled oligonucleotides comprising well-defined *cis*-acting elements have been used as probes to screen expression libraries. The resulting cDNAs in most cases have been shown to code for proteins with characteristics of animal and yeast transcription factors. From the accumulated data an interesting yet somewhat confusing picture is emerging. The probes used for the library screens corresponded to very diverse *cis*-acting regulatory elements. Yet many of the genes isolated so far appear to be structurally related (Table 1). Many fall into the class of the so-called bZip proteins, putative transcription factors that contain a leucine zipper dimerization motif and a basic DNA-binding domain. In particular, over the basic DNA binding domain, there is a high degree of similarity. This is unexpected since the factors were isolated using *cis*-acting elements from genes regulated by cues as different as light and abscisic acid. Close inspection of the *cis*-acting elements, however, reveals that all have a CACGTG palindromic core motif or closely related sequence. Although no exhaustive analysis has been made, in several cases it could indeed be shown that the putative transcription factors could bind to more sites than only their cognate *cis*-element. For instance, factor TAF-1 binds not only to the cognate *cis*-element in the ABA-regulated rice *rab16* gene, but also to G-box motifs found in various light regulated genes [70]. How to explain this? One possibility is that all these factors are general factors that are only indirectly involved in regulation of gene expression. Other as yet unidentified factors may interact with the general factors to bring about regulated gene expression. A second possibility is that the binding affinities *in vitro* do not reflect the *in vivo* reality. Gel retardation assays measure only binding affinities and are not necessarily a good indication of transcription rates, the biologically relevant parameter. It should be pointed out that only in a few cases has evidence been presented that these binding proteins can actually modulate transcription. TGA-1 stimulates transcription in HeLa cell and plant *in vitro* systems [42, 99] and TAF-1, when introduced into whole plants as a cDNA, can increase expression of a reporter gene carrying copies of the cognate *cis*-acting element [70].

A completely different approach has also led to the cloning of genes coding for (putative) transcription factors. Since the beginning of the century a considerable number of regulatory mutations have been described. In maize, mutants regulating anthocyanin biosynthesis or storage protein production have been well characterized. In *Antirrhinum majus*, pea and *Arabidopsis* so-called homeotic mutants are known that alter the identity of an organ. Great progress has been made recently in cloning the genes defined by these genetic defects. Virtually all of such genes seem to code for transcription factors. I take as an example the *opaque-2* mutation in maize. Maize homozygous for the *o–2* mutation has a reduced content of the 22 kDa zein storage proteins and a protein called b32, whereas the 19 kDa zeins are relatively unaffected. The lack of b32 and the 22 kDa zeins appears to result from a lack of the corresponding mRNAs. The *o–2* mutation maps to the short arm of chromosome 7 and is unlinked to known 22 kDa zein genes. An *o–2* mutant allele caused by insertion of transposon Spm1 was cloned using the transposon as a probe [34, 55, 89]. The wild-type *o–2* gene could then be isolated from a wild-type maize library. Sequence analysis shows that the *o–2* gene encodes yet another bZIP transcription factor. Indeed, the *o–2* protein binds to *cis*-elements in the b-32 target gene and transient expression studies show that it can activate a reporter gene preceded by b-32 *cis*-acting elements [34, 55]. Therefore the *o–2* gene, genetically defined as a specific regulator of a specific subset of storage protein genes, belongs to a family of ubiquitous transcription factors. Homeotic genes from *Antirrhinum* and *Arabidopsis* defined by the *deficiens* and agamous floral mutations were cloned using similar strategies as for the maize *o–2* gene. The deduced proteins have sequence homology over the DNA-binding domain to yeast and human transcription factors. They are now collectively named MADS box proteins [14, 91]. The *def* and *agamous* proteins seem to be very precise regulators of steps in the pathway of floral development, although it is not known yet what their target genes are. On the other hand, it has been found in *Arabidopsis* and tomato that the MADS box genes are members of multigene families, some of which appear to be expressed ubiquitously [78, 59].

In summary, a growing number of transcription factors are being characterized. Many are structurally related. Detailed knowledge about their *in vitro* binding specificity is accumulating. The challenge is now to understand how these factors bring about the very diverse and very precise regulation of target genes.

Table 1 gives an overview of cloned or at least well-characterized (putative) plant transcription factors. Clearly many of these factors are structurally related and bind to very similar DNA sequences.

Cis-acting elements for post-transcriptional regulation?

The majority of the *cis*-acting elements have been found in the 5′ upstream regions of plant genes. In most cases it has been proven, or at least assumed, that these elements are involved in the modulation of transcription rates. However, it must also be pointed out that in most cases a search was made exclusively for such upstream transcriptional elements.

Possible *cis*-acting elements downstream of the TATA box, modulating either transcription or post-transcriptional processes are often not considered in experimental designs and could easily be overlooked. *Cis*-acting elements in 'unusual' places have been described in several genes. The first intron of the maize ADH gene is required for high transcript levels, a phenomenon that is not clearly defined as purely transcriptional [58]. The ABA-responsive *Em* gene from wheat has an upstream regulatory element that mediates ABA responsiveness. Then there is a second element encompassing the 5′ non-translated leader that enhances reporter gene expression 10-fold [37, 60]. It is easy to imagine that this second element does not function at the DNA level, but rather influences stability or translation of the mRNA. In the pea gene coding for ferredoxin the only light-regulatory elements encountered are in the protein coding region. Again, although effects on transcription cannot be ruled out, a role in mRNA stability may be more likely [21, 95].

Research focused on mRNA stability determinants has only just begun. Detailed information on the *cis*-acting RNA sequences and the proteins that interact with them should become available in the near future.

The formation of 3′ ends of mRNAs appears to be different between plant and mammalian genes. The conserved hexanucleotide AATAAA found in most mammalian genes 10–30 basepairs before the 3′ end is absent in many plant genes. The requirement for such a site may be less stringent. No sequences downstream of the polyadenylation site appear to be necessary, but further upstream elements have been found [65, 83]. An interesting problem is posed by the termination and polyadenylation of the CaMV 35S RNA [84]. This RNA is transcribed from the circular CaMV DNA genome as a terminally redundant RNA, i.e. transcription goes all around the circle, passes the transcription start site and stops some 200 nucleotides beyond. These last 200 bp are sufficient for correct termination of reporter gene constructs and the question is why transcription does not terminate during the first passage over the termination site. With a number of constructs having increasing length of DNA between transcription start and termination sites it could be shown that a minimal distance between the two is required for proper termination.

Translational regulation can be important too

An example of very well documented translational regulation of nuclear gene expression is provided by the *Amaranthus rbcS* genes [5, 6]. The light-responsive expression of the *rbcS* genes is one of the paradigms of transcriptional regulation in plants. Studies by Klessig and coworkers make it clear that there exists a second tier of regulation. With *Amaranthus* seedlings these authors could show that after transfer from light to dark, mRNA levels for both *rbcS* and the chloroplast-encoded *rbcL* subunits remain unchanged for at least 6 h. On the other hand, incorporation of ^{35}S-methionine in the encoded proteins ceases completely within 2 h. Subsequent experiments demonstrated that the mRNA remains bound to polysomes, implicating regulation at the level of translation elongation. Based on the animal literature a possible involvement of elongation factor EF-2 can be surmised [82]. In contrast, when seedlings were transferred from dark to light, recruitment of *rbcS* mRNA into polysomes was observed, indicating regulation at the translation initiation step. Thus one of the workhorses for transcription studies is also extremely useful for research on translation.

In *Volvox* cultures synchronized by a light-dark cycle, the major events in the juvenile-to-adult-transition are light-dependent. The effect of light is not exerted at the transcriptional but at the translational level [44].

The promoter for the CaMV 35S RNA has been studied by several groups in great depth and with exciting results (see before). The translation of the proteins encoded by this mRNA is at least as interesting. Translation of the genome-size RNA is thought to give rise to at least five proteins. The existence of such polycistronic mRNAs in eukaryotes has been in doubt for many years and only relatively recently was it shown unambiguously for poliovirus RNA that downstream open reading frames can be translated through internal initiation [77]. A number of dicistronic reporter gene constructs were prepared and transient expression in host protoplasts measured. The conclusion from these experiments was that always only the first open reading frame in a dicistronic construct is translated. Expression of downstream cistrons, however, could be observed when the viral ORF VI gene was co-transfected. Effects on splicing, nuclear-cytoplasmic transport or mRNA stability could be ruled out. Thus the ORF VI gene product acts as a trans-activator to stimulate translation from downstream open reading frames in polycistronic mRNAs [7].

The efficiency of translation of ORF VII, the first gene in the 35S mRNA, is severely impaired by sequences in the 600 nt leader sequence preceding ORF VII. Within these 600 nt, mutational analysis identified a mosaic of inhibitory and stimulatory elements. None of the mutations influenced steady state mRNA levels to any great extent and thus, the effects again, must be at the level of translation [23]. Translational enhancer sequences have been described for the 5′ leaders of several plant viruses [24, 40].

In all branches of molecular biology the interest in the mechanisms of translation declined dramatically in the 1980s. This may be due to the fact that most often regulation of gene expression is at the level of transcription. Translational regulation of the GCN4 in yeast and of the ferritin gene in animal cells are two of the rare genes in which gene-specific translational regulation has been demonstrated and studied in great detail [67, 13]. And in these two genes the traditional translation initiation and elongation factors do not seem to be centrally involved. Yet, translation initiation factors are likely to play crucial roles in cellular responses. For instance, the gene for translation initiation factor eIF-4E, the cap-binding protein, has recently been shown to act as an oncogene when overexpressed in mammalian cells [52]. Translation initiation factor eIF-4A, a putative RNA helicase, is thought to unwind secondary RNA structure in the 5′ leader of mRNAs to enable the scanning ribosome to reach the initiator AUG. Injection of purified eIF-4A into *Xenopus* oocytes can activate dormant mRNAs [2].

Plant translation initiation factors have been fairly well-characterized from wheat germ extracts [39, 51]. They are very much like the factors in rabbit reticulocytes. This is not surprising since wheat germ and rabbit reticulocyte cell-free extracts are both standard systems for *in vitro* translation of mRNAs and the differences between the two systems are minor. We have recently isolated genes for plant eIF-4A and found a multigene family of highly divergent genes [73]. This is in contrast to yeast and mouse where duplicate genes code for identical or highly similar proteins, respectively. The divergence of the plant eIF-4A genes suggests that they may have dissimilar functions, for example they could perhaps translate various mRNAs with different efficiencies.

Signal transduction

Molecular-biological experiments have provided ample evidence that internal and external signals can modulate the expression of specific genes. A major question remaining pertains to the intermediary steps. In the case of light: how does light succeed in turning transcription on or off? The first step is relatively well defined, at least for red/far-red reversible reactions. The light is perceived by the photoreceptor phytochrome. A large body of data documents how red light can change the physical conformation of the $M_r = 120\,000$ chromoprotein [94]. Far-red light can reverse this conformational change. But, what comes afterwards? The idea that the activated photoreceptor could bind directly to *cis*-acting regulatory elements – as is the case in glucocorticoid-induced gene expression in mammalian systems – has been abandoned. Thus, there must be intermediary steps to transduce the signal from the activated photoreceptor to the transcriptional apparatus. Research into the nature of these intermediaries has so far mostly followed along the lines set out for non-plant systems. Evidence implicating protein kinases, Ca^{2+} and calmodulin, G-proteins, phosphoinositides has been obtained in various systems. A good example is again the phytochrome mediated response. The approach usually taken is to find compounds that can interfere with the signal transduction chain with the goal of eliciting the response in the absence of the natural stimulus. Clearly, a multicellular plant is less suitable as an experimental system and single cell systems have been sought that retain phytochrome responsiveness. Wheat protoplasts respond to red light treatment by increasing in volume and this red-light-induced swelling is far-red-reversible. The red-induced swelling requires Ca^{2+}, and importantly, swelling can occur in the dark when the protoplasts are incubated in the presence of Ca^{2+} and Ca^{2+} ionophores. The results are interpreted to mean that phytochrome induces the opening of Ca^{2+} channels in the plasma membrane. Subsequent experiments with phorbol esters and GTP/GDP analogues indicate the involvement of a membrane-bound GTP-binding protein [89]. Evidently, it will be interesting to compare these results with results obtained in other phytochrome-mediated systems.

Signal transduction during development: the events upstream and downstream

The signal pathway leading to the activation of light-regulated genes is likely to involve more than just a linear amplification of the signal. Some light-regulated genes are turned on faster or at lower fluence rates than others; some are turned off by light. Also the light pathway must interact with other pathways that determine cell specificity or hormone responsiveness. To understand how signals cross-react and network to induce highly specific patterns of gene expressions is a challenge for the future.

In the case of light regulation we know very well at least what is at the end of the signal transduction chain. The *rbcS* gene is transcribed, the transcript is translated, the protein transported into the chloroplast and combined with the chloroplastic *rbcL* gene product to form the Rubisco enzyme. Finally, there is a wealth of data about the enzymatic activities of the protein. Thus the steps after transcription initiation are known in detail.

What about developmental pathways? In some cases we know, or think we know, the signals and we may have some ideas about how they modulate gene expression. Nodule formation on leguminous roots can be initiated by an oligosaccharide secreted by the infecting *Rhizobium* [53; T. Bisseling's paper elsewhere in this volume]. An early signal in flower development is florigen, which is not so well-characterized but appears to be produced by leaves and transported to the vegetative shoot apex where it is thought to initiate the floral transition [4]. At the end of the signal transduction chain is the flower, a complex structure, very distinct from the vegetative organs. Between florigen and flower must be many steps of which we know only two: the homeotic genes and the flower-specific genes.

Mutations in homeotic genes drastically alter the identity of organs. Thus in the *Antirrhinum majus deficiens* mutant petals are changed into sepals and carpels form instead of stamens [14]. Best known are the homeotic mutants that alter flower development. However, homeotic mutations in vegetative organs have also been described [62]. The *deficiens* gene and several other floral homeotic genes have been cloned and sequenced and been shown to have strong homology to known transcription factors, in particular over the DNA-binding domains [14]. Homeotic genes that determine organ identity are well known from *Drosophila* and many of them also code for transcription factors. In flies, genes have been described that act even earlier and specify the body plan [68]. In the context of this review it is interesting that one of these very early genes codes not for a transcription factor but for a putative RNA helicase and thus may act at the post-transcriptional level [35, 50]. Very early developmental mutations have recently also been described in *Arabidopsis*, and their characterization at the molecular level is eagerly awaited [63].

It is attractive to think of homeotic genes as central switches, reacting to positional, developmental and environmental cues, and determining a cascade of subsequent events, finally leading to organ formation.

The signal transduction pathways that turn the central switches (i.e. lead to expression of homeotic genes) are not known in detail. The chemical nature of florigen remains elusive despite intensive efforts. On the other hand, what is the result of the expression of the homeotic genes? Since most appear to code for transcription factors it is reasonable to assume that they will activate target genes downstream in the pathway. Genes that are expressed only in petals, in stamens, in the tapetum layer of the stamen etcetera have been isolated and their spatial and temporal expression determined in great detail [45]. The question to be answered in the near future is how the *cis*-acting elements of these target genes interact with the homeotic-type transcription factors.

Conclusion

Ten years ago a small number of plant genes had been cloned and sequenced. Today not only have more genes been sequenced, we have also learned a great deal about how they are expressed. Small *cis*-acting elements have been delineated, mostly in the upstream sequences, that can confer correct regulation upon reporter genes. More recently genes have been isolated coding for proteins that bind to these *cis*-acting elements. A major object of research in the near future will be to understand how the often ubiquitous transcription factors cooperate with one another, with as yet undiscovered factors, and with the *cis*-acting elements, to bring about the finely tuned regulation of individual genes. In summary, the molecular cloning of plant genes has allowed for an unprecedented level of detail in the characterization of the gene products. We are beginning to understand how genes are regulated.

Acknowledgements

Numerous colleagues contributed to this review by making (p)reprints available. Drs Urs Feller, Andrew Fleming, Susan Flores and Gunther Neuhaus were very helpful by critically reading the manuscript. I am extremely grateful to Ms L. Häusermann, M. Zeder and R. Hintermann for their expert secretarial assistance and for their patience.

References

1. An G, Ebert PR, Mitra A, Ha SB: Binary vectors. In: Plant Molecular Biology Manual, pp. A3/1–A3/19. Kluwer Academic Publishers, Dordrecht (1988).
2. Audet RG, Goodchild J, Richter JD: Eukaryotic initiation factor 4A stimulates translation in microinjected *Xenopus* oocytes. Devel Biol 121: 58–68 (1987).
3. Benfey PN, Chua NH: The cauliflower mosaic virus 35S promoter: Combinatorial regulation of transcription in plants. Science 250: 959–966 (1990).
4. Bernier G: The Control of floral evocation and morphogenesis. Annu Rev Plant Physiol Plant Mol Biol 39: 175–219 (1988).
5. Berry JO, Breiding DE, Klessig DF: Light-mediated control of translational initiation of ribulose-1,5-bisphosphate carboxylase in amaranth cotyledons. Plant Cell 2: 795–803 (1990).
6. Berry JO, Carr JP, Klessig DF: mRNAs encoding ribulose-1,5-bisphosphate carboxylase remain bound to polysomes but are not translated in amaranth seedlings transferred to darkness. Proc Natl Acad Sci USA 85: 4190–4194 (1988).
7. Bonneville JM, Sanfaçon H, Fütterer J, Hohn T: Post-transcriptional *trans*-activation in cauliflower mosaic virus. Cell 59: 1135–1143 (1989).
8. Bossen ME, Dassen HHD, Kendrick RE, Vredenberg WJ: The role of calcium ions in phytochrome-controlled swelling of etiolated wheat (*Triticum aestivum* L.) protoplasts. Planta 174: 94–100 (1988).
9. Bossen ME, Kendrick RE, Vredenberg WJ: The involvement of a G-protein in phytochrome-regulated Ca^{2+}-dependent swelling of etiolated wheat protoplasts. Physiol Plant 80: 55–62 (1990).
10. Brinkmann H, Martinez P, Quigley F, Martin W, Cerff R: Endosymbiotic origin and codon bias of the nuclear gene of chloroplast glyceraldehyde-3-phosphate dehydrogenase from maize. J Mol Evol 26: 320–328 (1987).
11. Burke C, Yu XB, Marchitelli L, Davis EA, Ackermann S: Transcription factor IIA of wheat and human function similarly with plant and animal viral promoters. Nucl Acids Res 18: 3611–3620 (1990).
12. Bustos MM, Guiltinan MJ, Jordano J, Begum D, Kalkan FA, Hall TC: Regulation of β-glucuronidase expression in transgenic tobacco plants by an A/T-rich, *cis*-acting sequence found upstream of a french bean β-phaseolin gene. Plant Cell 1: 839–853 (1989).
13. Casey JL, Hentze MW, Koeller DM, Caughman SW, Rouault TA, Klausner RD, Harford JB: Iron-responsive elements: Regulatory RNA sequences that control mRNA levels and translation. Science 240: 924–928 (1988).
14. Coen ES, Meyerowitz EM: The war of the whorls: genetic interactions controlling flower development. Nature 353: 31–37 (1991).
15. Coen ES, Romero JM, Doyle S, Elliot R, Murphy G, Carpenter R: *Floricaula*: A homeotic gene required for flower development in *Antirrhinum majus*. Cell 63: 1311–1322 (1990).
16. Datta N, Cashmore AR: Binding of a pea nuclear protein to promoters of certain photoregulated genes is modulated by phosphorylation. Plant Cell 1: 1069–1077 (1989).
17. Dean C, Pichersky E, Dunsmuir P: Structure, evolution and regulation of RbcS genes in higher plants. Annu Rev Plant Physiol. Plant Mol Biol 40: 415–439 (1989).
18. Dehesh K, Bruce WB, Quail PH: A trans-acting factor that binds to a GT-motif in a phytochrome gene promoter. Science 250: 1397–1399 (1990).
19. Donald RGK, Cashmore AR: Mutation of either G box or I box sequences profoundly affects expression from the *Arabidopsis rbcS-1A* promoter. EMBO J 9: 1717–1726 (1990).
20. Donald RGK, Schindler U, Batschauer A, Cashmore AR: The plant G box promoter sequence activates transcription in *Saccharomyces cerevisiae* and is bound *in vitro* by a yeast activity similar to GBF, the plant G box binding factor. EMBO J 9: 1727–1735 (1990).
21. Elliot RC, Dickey LF, White MJ, Thompson WF: Cis-acting elements for light regulation of pea ferredoxin I gene expression are located within transcribed sequences. Plant Cell 1: 691–698 (1989).
22. Fluhr R, Kuhlemeier C, Nagy F, Chua NH: Organ-specific and light-induced expression of plant genes. Science 232: 1106–1112 (1986).
23. Fütterer J, Gordon K, Sanfaçon H, Bonneville JM, Hohn T: Positive and negative control of translation by the leader sequence of cauliflower mosaic virus pregenomic 35S RNA. EMBO J 9: 1697–1707 (1990).
24. Gallie DR, Kado CI: A translational enhancer derived

from tobacco mosaic virus is functionally equivalent to a Shine-Dalgarno sequence. Proc Natl Acad Sci USA 86: 129–132 (1989).

25. Gasch A, Hoffmann A, Horikoshi M, Roeder RG, Chua NH: *Arabidopsis thaliana* contains two genes for TFIID. Nature 346: 390–394 (1990).
26. Gilmartin PM, Sarokin L, Memelink J, Chua NH: Molecular light switches for plant genes. Plant Cell 2: 369–378 (1990).
27. Goff SA, Cone KC, Fromm ME: Identification of functional domains in the maize transcriptional activator C1: comparison of wild-type and dominant inhibitor proteins. Genes Devel 5: 298–309 (1991).
28. Goff S, Klein TM, Roth BA, Fromm ME, Cone KC, Radicella JP, Chandler VL: Transactivation of anthocyanin biosynthetic genes following transfer of *B* regulatory genes into maize tissues. EMBO J 9: 2517–2522 (1990).
29. Green PJ, Kay SA, Chua NH: Sequence-specific interactions of a pea nuclear factor with light-responsive elements upstream of the *rbcS-3A* gene. EMBO J 6: 2543–2549 (1987).
30. Green PJ, Kay SA, Lam E, Chua NH: *In vitro* DNA footprinting. In: Plant Molecular Biology Manual, pp. B11/1–B11/22. Kluwer Academic Publishers, Dordrecht (1989).
31. Green PJ, Yong MH, Cuozzo M, Kano-Murakami Y, Silverstein P, Chua NH: Binding site requirements for pea nuclear protein factor GT-1 correlate with sequences required for light-dependent transcriptional activation of the *rbcS-3A* gene. EMBO J 7: 4035–4044 (1988).
32. Guilfoyle TJ, Dietrich MA, Prenger JP, Hagen G: Phosphorylation/dephosphorylation of the carboxyl terminal domain in the largest subunit of RNA polymerase II. Curr Top Plant Biochem Physiol 9: 299–312 (1990).
33. Guiltinan MJ, Marcotte WR Jr, Quatrano RS: A plant leucine zipper protein that recognizes an abscisic acid response element. Science 250: 267–271 (1990).
34. Hartings H, Maddaloni M, Lazzaroni N, Di Fonzo N, Motto M, Salamini F, Thompson R: The O_2 gene which regulates zein deposition in maize endosperm encodes a protein with structural homologies to transcriptional activators. EMBO J 10: 2795–2801 (1989).
35. Hay B, Yeh Jan L, Nung Jan Y: A protein component of drosophila polar granules is encoded by *vasa* and has extensive sequence similarity to ATP-dependent helicases. Cell 55: 577–587 (1988).
36. von Heijne G, Nishikawa K: Chloroplast transit peptides. The perfect random coil? FEBS Lett 278: 1–3 (1991).
37. Hetherington AM, Quatrano RS: Mechanisms of action of abscisic acid at the cellular level. New Phytol, in press.
38. Jackson D, Culianez-Macia F, Prescott AG, Roberts K, Martin C: Expression patterns of *myb* genes from *Antirrhinum* flowers. Plant Cell 3: 115–125 (1991).
39. Jaramillo M, Browning K, Dever TE, Blum S, Trachsel H, Merrick WC, Ravel JM, Sonenberg N: Translation initiation factors that function as RNA helicases from mammals, plants and yeast. Biochim Biophys Acta 1050: 134–139 (1990).
40. Jobling SA, Gehrke L: Enhanced translation of chimaeric messenger RNAs containing a plant viral untranslated leader sequence. Nature 325: 622–625 (1987).
41. Katagiri F, Lam E, Chua N-H: Two tobacco DNA binding proteins with homology to the nuclear factor CREB. Nature 340: 727–730 (1989).
42. Katagiri F, Yamazaki K, Hrikoshi M, Roeder RG, Chua NH: A plant DNA-binding protein increases the number of active preinitiation complexes in a human *in vitro* transcription system. Genes Devel 4: 1899–1909 (1990).
43. Kay R, Chan A, Daly M, McPherson J: Duplication of CaMV 35S promoter sequences creates a strong enhancer for plant genes. Science 236: 1299–1302 (1987).
44. Kirk MM, Kirk DL: Translational regulation of protein synthesis, in response to light, at a critical stage of volvox development. Cell 41: 419–428 (1985).
45. Koltunow AM, Truettner J, Cox KH, Wallroth M, Goldberg RB: Different temporal and spatial gene expression patterns occur during anther development. Plant Cell 2: 1201–1224 (1990).
46. Kuhlemeier C, Green PJ, Chua NH: Regulation of gene expression in higher plants. Annu Rev Plant Physiol 38: 221–257 (1987).
47. Kuhlemeier C, Strittmatter G, Ward K, Chua NH: The pea *rbcS-3A* promoter mediates light responsiveness but not organ specificity. Plant Cell 1: 471–478 (1989).
48. Lam E, Chua N-H. ASF2: A factor that binds to the cauliflower mosaic virus 35S promoter and a conserved GATA motif in cab promoters. Plant Cell 1: 1147–1156 (1989).
49. Lam E, Kano-Murakami Y, Gilmartin P, Niner B, Chua NH: A metal-dependent DNA-binding protein interacts with a constitutive element of a light-responsive promoter. Plant Cell 2: 857–866 (1990).
50. Lasko F, Ashburner M: The product of the *Drosophila* gene *vasa* is very similar to eukaryotic initiation factor-4A. Nature 335: 611–617 (1988).
51. Lax SR, Lauer SJ, Browning KS, Ravel JM: Purification and properties of protein synthesis initiation and elongation factors from wheat germ. Meth Enzymol 118: 109–128 (1986).
52. Lazaris-Karatzas A, Montine KS, Sonenberg N: Malignant transformation by a eukaryotic initiation factor subunit that binds to mRNA 5′ cap. Nature 345: 544–547 (1990).
53. Lerouge P, Roche P, Faucher C, Maillet F, Truchet G, Promé JC, Dénarié J: Symbiotic host-specificity of *Rhizobium meliloti* is determined by a sulphated and acylated glucosamine oligosaccharide signal. Nature 344: 781–784 (1990).

54. Lewin B: Commitment and Activation at Pol II Promoters: A tail of protein-protein interactions. Cell 61: 1161–1164 (1990).
55. Lohmer S, Maddaloni M, Motto M, DiFonzo N, Hartings H, Salamini F, Thompson RD: The maize regulatory locus *Opaque-2* encodes a DNA-binding protein which activates the transcription of the *b-32* gene. EMBO J 10: 617–624 (1991).
56. Ludwig SR, Habera LF, Dellaporta SL, Wessler SR: *Lc*, a member of the maize *R* gene family responsible for tissue-specific anthocyanin production, encodes a protein similar to transcriptional activators and contains the *myc*-homology region. Proc Natl Acad Sci USA 86: 7092–7096 (1989).
57. Ludwig SR, Wessler SR: Maize *R* gene family tissue-specific helix-loop-helix proteins. Cell 62: 849–851 (1990).
58. Luehrsen KR, Walbot V: Intron enhancement of gene expression and the splicing efficiency of introns in maize cells. Mol Gen Genet 225: 81–93 (1991).
59. Ma H, Yanofsky MF, Meyerowitz EM: *AGL1-AGL6*, an *Arabidopsis* gene family with similarity to floral homeotic and transcription factor genes. Genes Devel 5: 484–495 (1991).
60. Marcotte WR, Russell SH, Quatrano RS: Abscisic acid-responsive sequences from the *Em* gene of wheat. Plant Cell 1: 969–976 (1989).
61. Martinez P, Martin W, Cerff R: Structure, evolution and anaerobic regulation of a nuclear gene encoding cytosolic glyceraldehyde-3-phosphate dehydrogenase from maize. J Mol Biol 208: 551–565 (1989).
62. Marx GA: Developmental mutants in some annual seed plants. Annu Rev Plant Physiol 34: 389–417 (1983).
63. Mayer U, Torres RA, Berleth T, Miséra S, Jürgens G: Mutations affecting body organization in the *Arabidopsis* embryo. Nature 353: 402–353 (1991).
64. McCarthy DR, Hattori T, Carson CB, Vasil V, Lazar M, Vasil IK: The viviparous-1 developmental gene of maize encodes a novel transcriptional activator. Cell 66: 895–905 (1991).
65. Mogen BD, MacDonald MH, Graybosch R, Hunt AG: Upstream sequences other than AAUAAA are required for efficient messenger RNA 3′-end formation in plants. Plant Cell 2: 1261–1272 (1990).
66. Morelli G, Nagy F, Fraley RT, Rogers SG, Chua NH: A short conserved sequence is involved in the light-inducibility of a gene encoding ribulose 1,5-bisphosphate carboxylase small subunit of pea. Nature 315: 200–204 (1985).
67. Muller PP, Trachsel H: Translation and regulation of translation in the yeast *Saccharomyces cerevisiae*. Eur J Biochem 191: 257–261 (1990).
68. Nüsslein-Volhard C, Frohnhöfer HG, Lehmann R: Determination of anteroposterior polarity in *Drosophila*. Science 238: 1675–1681 (1987).
69. Odell JT, Nagy F, Chua NH: Identification of DNA sequences required for activity of the cauliflower mosaic virus 35S promoter. Nature 313: 810–812 (1985).
70. Oeda K, Salinas J, Chua NH: A tobacco bZip transcription activator (TAF-1) binds to a G-box-like motif conserved in plant genes. EMBO J 10: 1793–1802 (1991).
71. Ondek B, Shepard A, Herr W: Discrete elements within the SV40 enhancer region display different cell-specific enhancer activities. EMBO J 6: 1017–1025 (1987).
72. Ow DA, Jacobs JD, Howell SH: Functional regions of the cauliflower mosaic virus 35S RNA promoter determined by use of the firefly luciferase gene as a reporter of promoter activity. Proc Natl Acad Sci USA 84: 4870–4874 (1987).
73. Owttrim GW, Hofmann S, Kuhlemeier C: Divergent genes for translation initiation factor eIF-4A are coordinately expressed in tobacco. Nucl Acids Res 19: 5491–5496 (1991).
74. Paz-Ares J, Ghosal D, Wienand U, Peterson PA, Saedler H: The regulatory *c1* locus of *Zea mays* encodes a protein with homology to *myb* proto-oncogene products and with structural similarities to transcriptional activators. EMBO J 6: 3553–3558 (1987).
75. Paz-Ares J, Ghosal D, Saedler H: Molecular analysis of the C1-I allele from *Zea mays*: a dominant mutant of the regulatory c1 locus. EMBO J 9: 315–321 (1990).
76. Pelham HRB, Bienz M: A synthetic heat-shock promoter element confers heat-indudibility on the herpes simplex virus thymidine kinase gene. EMBO J 1: 1473–1477 (1982).
77. Pelletier J, Sonenberg N: Internal initiation of translation of eukaryotic mRNA directed by a sequence derived from poliovirus RNA. Nature 334: 320–325 (1988).
78. Pnueli L, Abu-Abeid M, Zamir D, Nacken W, Schwarz-Sommer Z, Lifschitz E: The MADS box gene family in tomato: temporal expression during floral development, conserved secondary structures and homology with homeotic genes from *Antirrhinum* and *Arabidopsis*. Plant J 1: 255–266 (1991).
79. Poulsen C, Chua NH: Dissection of 5′ upstream sequences for selective expression of the *Nicotiana plumbaginifolia rbcS-8B* gene. Mol Gen Genet 214: 16–22 (1988).
80. Ruberti I, Sessa G, Lucchetti S, Morelli G: A novel class of plant proteins containing a homeodomain with a closely linked leucine zipper motif. EMBO J 10: 1789–1791 (1991).
81. Russel DA, Martin MS: Differential expression and sequence analysis of the maize glyceraldehyde-3-phosphate dehydrogenase gene family. Plant Cell 1: 793–803 (1989).
82. Ryazanov AG, Shestakova EA, Natapov PG: Phosphorylation of elongation factor 2 by EF-2 kinase affects rate of translation. Nature 334: 170–173 (1988).
83. Sanfaçon H, Brodmann P, Hohn Th: A dissection of the

cauliflower mosaic virus polyadenylation signal. Genes Devel 5: 141–149 (1991).

84. Sanfaçon H, Hohn Th: Proximity to the promoter inhibits recognition of cauliflower mosaic virus polyadenylation signal. Nature 346: 81–84 (1990).
85. Saul MW, Shillito RD, Negrutiu I: Direct DNA transfer to protoplasts with and without electroporation. In: Plant Molecular Biology Manual, pp. A1/1–A1/16. Kluwer Academic Publishers, Dordrecht (1988).
86. Scharf KD, Rose S, Zott W, Schöff F, Nover L: Three tomato genes code for heat stress transcription factors with a region of remarkable homology to the DNA binding domain of the yeast HSF. EMBO J 9: 4495–4501 (1990).
87. Schindler U, Cashmore AR: Photoregulated gene expression may involve ubiquitous DNA binding proteins. EMBO J 9: 3415–3427 (1990).
88. Schirm S, Jiricny J, Schaffner W: The SV40 enhancer can be dissected into multiple segments, each with a different cell type specificity. Genes Devel 1: 65–74 (1987).
89. Schmidt RJ, Frances AB, Burr B: Transposon tagging and molecular analysis of the maize regulatory locus *opaque-2*. Science 238: 960–963 (1987).
90. Singh K, Dennis ES, Ellis JG, Llewellyn DJ, Tokuhisa JG, Wahleithner JA, Peacock WJ: OCSBF-1, a maize Ocs enhancer binding factor: Isolation and expression during development. Plant Cell 2: 891–903 (1990).
91. Sommer H, Beltran JP, Huijser P, Heike P, Lönnig WE, Saedler H, Schwarz-Sommer Z: *Deficiens*, a homeotic gene involved in the control of flower morphogenesis in *Antirrhinum majus*: the protein shows homology to transcription factors. EMBO J 9: 605–613 (1990).
92. Tabata T, Takase H, Takayama S, Mikami K, Nakatsuka A, Kawata T, Nakayama T, Iwabuchi M: A protein that binds to a cis-acting element of wheat histone genes has a leucine zipper motif. Plant Cell 2: 891–903 (1990).
93. Tabata T, Nakayama T, Mikami K, Iwabuchi M: HBP-1a and HBP-1b: leucine zipper-type transcription factors of wheat. EMBO J 10: 1459–1467 (1991).
94. Thompson WF, White MJ: Physiological and molecular studies of light-regulated nuclear genes in higher plants. Annu Rev Plant Physiol Plant Mol Biol 42: 423–66 (1991).
95. Thompson WF, Elliott RC, Dickey LF, Gallo M, Pedersen TJ, Sowinski DA: Unusual Features of the light response system regulating ferredoxin gene expression. In: Thomas B, Johnson CB, (eds) Phytochrome Properties and Biological Action. NATO ASI Series, vol. H50, pp. 201–216. Springer-Verlag, Berlin (1988).
96. Tobin EM, Silverthorne J: Light regulation of gene expression in higher plants. Annu Rev Plant Physiol 36: 569–93 (1985).
97. Vollbrecht E, Veit B, Sinha N, Hake S: The developmental gene *Knotted-1* is a member of a maize homeobox gene family. Nature 350: 241–243 (1991).
98. Weisshaar B, Armstrong GA, Block A, da Costa e Silva O, Hahlbrock K: Light-inducible and constitutively expressed DNA-binding proteins recognizing a plant promoter element with functional relevance in light responsiveness. EMBO J 10: 1777–1786.
99. Yamazaki KI, Katagiri F, Imaseki H, Chua NH: TGA1a, a tobacco DNA-binding protein, increases the rate of initiation in a plant *in vitro* transcription system. Proc Natl Acad Sci USA 87: 7035–7039 (1990).

Plant Molecular Biology **19**: 15–38, 1992.
© 1992 *Kluwer Academic Publishers. Printed in Belgium.*

Agrobacterium and plant genetic engineering

Paul J.J. Hooykaas and Rob A. Schilperoort*
Clusius Laboratory, Institute of Molecular Plant Sciences, Leiden University, Wassenaarseweg 64, 2333 AL Leiden, Netherlands (author for correspondence)*

Key words: Agrobacterium tumefaciens, plant transformation, plant tumours, T-DNA, Ti plasmid, transgenic plants

Plant tumour induction

The Ti plasmid

More than eighty years ago now Smith and Townsend [141] published an article in which they presented evidence that the bacterium which is now called *Agrobacterium tumefaciens* is the causative agent of the widespread neoplastic plant disease crown gall (Fig. 1). Since then a large number of scientists throughout the world have focused their research on this organism in an effort to analyse the molecular mechanism underlying the process of crown gall induction in detail. This was driven by the hope that this would lead to a better understanding of oncogenesis in general, and to the development of remedies for such diseases. After a period of diminished interest in the system *Agrobacterium* and crown gall research revived when it became apparent that oncogenic gene transfer from *Agrobacterium* to plants might form the molecular basis of crown gall induction, and thus the transfer system might be exploited for the genetic engineering of plants.

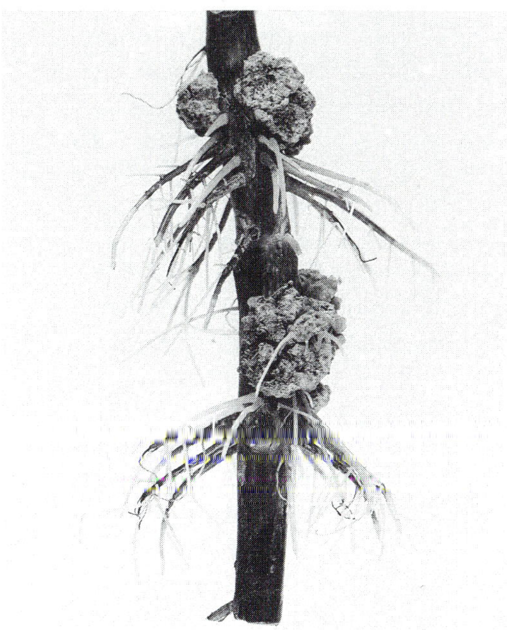

Fig. 1. Crown gall tumours induced by *Agrobacterium tumefaciens.*

A key discovery made some fifteen years ago now was the finding that virulent strains of *A. tumefaciens* contain a large extrachromosomal element, harbouring genes involved in crown gall induction [195]. Although researchers initially were thinking of a replicative form of a tumorigenic lysogenic virus (bacteriophage) in *Agrobacterium* with similarity to the oncogenic viruses that had been discovered in animal systems by then, in fact the extrachromosomal element that was found turned out to be a plasmid of an exceptionally large size (more than 200 kb). Because of its role in plant tumour induction this plasmid was called the Ti (tumour-inducing) plasmid [171]. The introduction of the Ti plasmid into related bacterial species such as the root nodule-inducing bacterium *Rhizobium trifolii* [71] or the leaf nodule inducer *Phyllobacterium myrsinacearum* [172] led to tumour-inducing strains, stressing the importance of the virulence determinants on the plasmid for the tumorigenicity of their bacterial hosts. However, introduction and maintenance in more distantly related bacteria such as *Escherichia coli* or *Pseudomonas aeruginosa* did not result in tumour-inducing strains [66], indicating that other factors – most likely chromosomally determined – were also important.

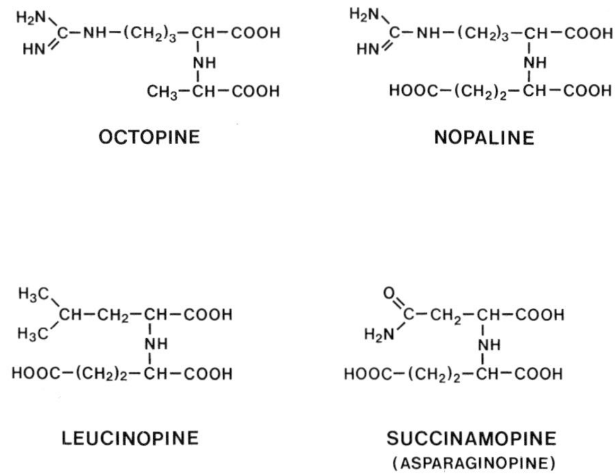

Fig. 2. Structural formulae of four characteristic opines.

T-DNA structure and function

Crown gall cells are tumorous, i.e. they proliferate autonomously in the absence of the phytohormones (auxins and cytokinins) that are needed for the growth of normal plant cells [23]. Because of this property grafting of aseptic (*Agrobacterium*-free) crown gall tissue onto a normal plant results in tumour formation. In *in vitro* culture crown gall cells grow and form a callus even when the growth stimulating phytohormones are absent from the culture medium. Another feature of crown gall cells is that they produce and excrete amino acid and sugar derivatives that are not formed by normal plant cells [158]. One of the first of such compounds that was characterized was octopine, a product formed by condensation of arginine with pyruvic acid that was formerly only known from *Octopus*, hence its name. Now such plant tumour-specific compounds are generally referred to as opines (Fig. 2). The type of opines formed by crown gall cells depends on the infecting *Agrobacterium* strain [19, 123]. Thus *Agrobacterium* strains can be classified according to the typical opines present in tumours as octopine, nopaline, leucinopine and succinamopine type strains.

The fact that crown gall cells differ from normal plant cells in the two properties mentioned motivated a search for the presence of agrobacterial DNA in crown gall cells. This search led to conflicting results and was unsuccessful until restriction enzymes became available to dissect the genome into a number of discrete fragments that could be separated by gel electrophoresis. Using such isolated fragments Chilton *et al.* [32] were the first to demonstrate the presence of about 20 copies of a segment of the octopine Ti plasmid for which a physical map had been established in the meantime [33] in an aseptic octopine crown gall line. Apparently, this piece of DNA had been introduced into plant cells by *Agrobacterium* during crown gall induction. Whether this segment of DNA had indeed oncogenic properties was however doubted soon after its discovery, when Koekman *et al.* [84] reported that in fact the deletion of this DNA stretch from the octopine Ti plasmid did not lead to a loss of oncogenicity by the host bacterium. Moreover, Ledeboer [91] found that poly-adenylated transcripts homologous to another, adjacent segment of the Ti plasmid were present in plant tumour lines. The development of a new, more sensitive technique for the detection of specific, short gene sequences in large genomes by Southern [144] helped to reconcile these older, apparently conflicting data. Using the method of Southern it was found in a number of laboratories that strains with octopine Ti plasmids are exceptional in that they have two segments of Ti plasmid DNA that are independently transferred to plant cells during tumour induction [17, 43, 161]. One of these two stretches of DNA turned out to be oncogenic to plant cells and was called the left-transferred DNA (T_L-DNA). The other segment of the octopine Ti plasmid that can be transferred was found to have no oncogenic properties and was called the right-transferred DNA (T_R-DNA). Octopine crown gall tumour lines always contain the T_L-DNA and sometimes also the T_R-DNA. In fact, the line studied initially by Chilton *et al.* [32] was exceptional in that it had a large number of copies of the T_R-DNA segment. Later on it was demonstrated that this tumour line contained the oncogenic T_L-DNA as well [161]. Tumour lines induced by nopaline, succinamopine and leucinopine strains of *Agrobacterium* contain a segment of oncogenic T-DNA that is at least partially homologous to the T_L-DNA transferred by octopine strains [17]. In all tumour lines analysed the T-DNA was invariably found to be integrated in the nuclear genome of plant cells, and to be absent from organelles [34].

By comparison of the T-DNA structure in a large number of independent tumour lines it was found that the T-DNA corresponds to a rather precisely defined segment of the Ti plasmid, and that no permutations occur during its integration into the plant genome. This latter finding suggested that the T-DNA is delivered into plant cells as a linear stretch of DNA. Sequencing of the T regions in different Ti plasmids showed that these regions are surrounded by a conserved 24 bp direct repeat [190]. Since tumour lines do not contain Ti plasmid sequences originally located outside of the (T-)region as defined by this repeat, it was logical to assume that this direct repeat functions as a recognition signal for the transfer apparatus. The copy number of the T-DNA in transformed plant lines is usually low varying from one to a few copies, although lines with up to a dozen copies have also rarely been found. If more than one copy is present, these may be located at different loci in the plant genome, or at the same locus where they occur in a direct or inverted orientation towards each other [120, 147].

The T-DNA contains a number of genes that are expressed in the transformed plant cells. Transcript maps have been made for the octopine Ti T_L- and T_R-DNAs (Fig. 3) as well as for T-DNAs from some other types of Ti plasmids. The bacterially derived T-DNA genes are apparently surrounded by expression signals that are recognized by the transcriptional factors of the plant. Sequence analysis of the T-DNA showed in fact that the T-DNA genes have well-known 5′ and 3′ eukaryotic expression signals such as the TATA box for transcription initiation and the AATAAA box involved in transcription termination and poly-adenylation [10]. Besides these common sequences T-DNA genes have plant-specific regulatory sequences in which they differ from each other and which make the level of their expression controllable by tissue types or signal compounds such as phytohormones [89]. Although the T-DNA genes have their own expression signals, expression is still influenced by the neighbouring chromosomal sequences. This position effect can even lead to complete silencing of the genes. The molecular principle underlying this

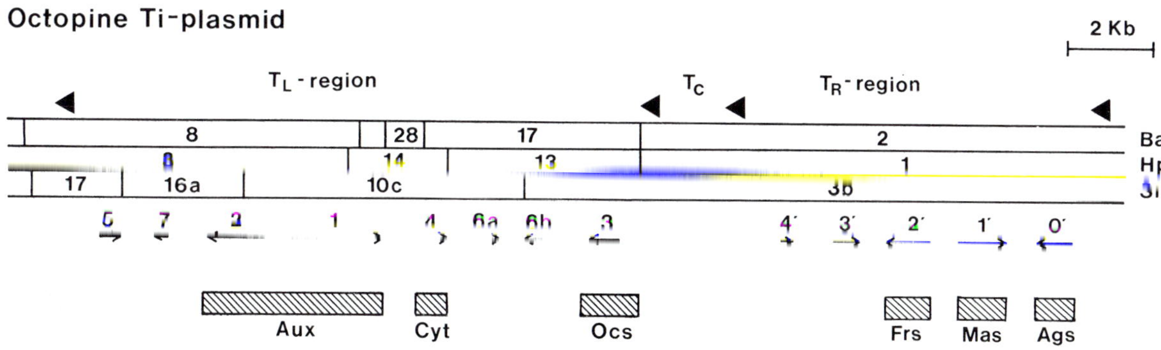

Fig. 3. Map of the transcripts encoded by the octopine Ti T-DNAs. Triangles indicate border repeats. The loci for Aux, Cyt, Ocs, Frs, Mas and Ags contain genes for IAA synthesis, isopentenyl transferase, octopine synthase, fructopine synthase, mannopine synthase and agropine synthase, respectively [53, 186].

phenomenon is unknown, but is thought to have to do with the chromatin structure at the insertion site. For some tumour lines it has been found that the expression of a few or all of the T-DNA genes is affected by DNA methylation. Treatment of such lines with the demethylating agent 5-azacytidine leads to the reappearance of T-DNA gene expression [5, 61, 176].

The T-DNA does not integrate at specific positions in the nuclear genome. Seven Ti T-DNA inserts were mapped on five different chromosomes of tomato [38], while for *Crepis capillaris* T-DNA was found to integrate into any of its three chromosomes [6]. The selection for expression of the oncogenic properties of the T-DNA will of course restrict the apparent integration sites to those regions that are transcriptionally active in the course of tumour induction and development. The DNA sequences of wild-type and T-DNA-tagged genomic loci were compared in order to find out whether integration occurs at certain preferred DNA sequences [54, 98, 100]. The results from the analysis of 17 independent insertion events showed that T-DNA integration occurs via illegitimate recombination on short stretches of DNA homology and is accompanied by short (29–73 bp) target site deletions.

T-DNA genes are responsible for the ability of crown gall cells to grow *in vitro* in the absence of phytohormones and to produce opines. Mutagenesis studies of the octopine Ti plasmid revealed that genes in the T_R region are responsible for production of the opines agropine and mannopine in transformed plant cells [85], while a gene in the T_L region is necessary for octopine formation [93]. This latter gene was cloned and expressed in *E. coli* and then found to code for the enzyme octopine synthase, which is able to convert arginine and pyruvic acid into octopine. The mutagenesis of the T_L region of the octopine Ti plasmid revealed that mutations at three loci led to changes in oncogenicity of the host bacterium. Such mutants were non-oncogenic on tomato, and formed tumours with an aberrant morphology on tobacco and kalanchoe [52, 111]. Since the supplementation of either auxin or cytokinin restored oncogenicity of such mutants on tomato, it was concluded that the mutations had inactivated genes which cause either an auxin or a cytokinin effect in plants [111]. Therefore, two of the T-DNA genes involved were called *aux* genes and the third the *cyt* gene. The *aux* mutants induced shooty tumours on tobacco and kalanchoe, whereas the *cyt* mutants formed rooty tumours on these plant species. Because of this the *aux* genes later on were also called *tms* (tumour morphology shoot) or *shi* (shoot inhibition) genes, and the *cyt* gene the *tmr* (tumour morphology root) or *roi* (root inhibition) gene [52, 93]. These tumour phenotypes on tobacco correspond to the response of tobacco tissue to an excess of auxin or cytokinin, respectively, in *in vitro* tissue culture media [138]. Expression of the *cyt* gene in *E. coli* showed that the protein encoded by this gene was an isopentenyl-transferase capable of catalysing the formation of the cytokinin isopentenyl-AMP from isopentenyl-pyrophosphate and AMP [11]. Therefore, the *cyt* gene is often called *ipt* gene now. Similarly, the expression of each of the two *aux* genes in *E. coli* revealed that they together mediate a pathway for synthesis of the auxin indole-acetic acid (IAA). The protein encoded by the *aux-1* gene turned out to be a mono-oxygenase capable of converting tryptophan into indole acetamide (IAM), and hence the *aux-1* gene is now called *iaaM* [159]. The enzyme determined by the *aux-2* gene had hydrolase activity, and was capable of converting IAM into IAA. The *aux-2* gene is, therefore, now called *iaaH* [132]. It has to be noted here that the pathway of IAA synthesis via the intermediate IAM does not occur normally in plants, but proceeds via indole pyruvic acid as an intermediate. The (over)production of an auxin and a cytokinin via the T-DNA explains why crown gall cells proliferate even in the absence of externally applied phytohormones. Crown gall cells apparently are 'autocrine.' Production of opines, the second characteristic property of crown gall cells, is similarly explained by the finding that the T-DNAs have genes coding for opine synthases.

Besides the genes mentioned above, the octopine T_L-DNA contains some genes with a still unknown function. Inactivation of these genes did not affect oncogenicity of the host bacterium. For one of these genes (a gene named *6*[b]) it was shown that it had an oncogenic effect for certain plant species even in the absence of the other T-DNA genes [70]. For another one (a gene called *5*) it was recently found that it encodes an enzyme capable of forming indole-lactic acid, an inhibitor of the auxin response and thus a modulator of the effects brought about by an excess of auxin [89]. Although these genes are not of prime importance for tumour formation on most plant species, it may be that they are necessary for oncogenicity on certain specific host plants.

The T regions of nopaline, succinamopine and leucinopine Ti plasmids embrace *ipt*, *iaaH* and *iaaM* genes that are closely related to those of the octopine Ti T_L region. It is interesting to know that these same genes are also present in some other species of phytopathogenic bacteria such as *Pseudomonas syringae* pv. *savastanoi* that produce auxin and cytokinin [191]. Most *Agrobacterium* strains induce tumours on a wide range of dicotyledonous plant species. However, strains isolated from grapevine often have a limited host range for tumour induction [115]. These limited host range (LHR) strains are clearly of the octopine type, but have strongly rearranged T_L and T_R regions in the Ti plasmid [26]. Many of these LHR strains lack a functional *ipt* gene in their T_L region, and it has been demonstrated that this is one of the reasons for their limited host range. Reintroduction of the *ipt* gene from a wide host range strain resulted in an extension of the host range of the LHR strain AG57 [68]. The absence of the *ipt* gene from the T-DNA may be one of the reasons that the LHR strains are efficient tumour inducers on grapevine. A wide host range strain was found to become able to induce tumours on certain grapevine varieties only after the inactivation of its *ipt* gene [60]. Evidence is accumulating that in LHR strains the *6*[b] gene is a very important determinant of oncogenicity [164].

Since tumour formation had not been observed on monocotyledonous plant species [44], it was generally assumed that T-DNA transfer does not occur to such plants. However, tumour formation is the end-result of a complex process in which a large number of discrete steps are involved including recognition of plant (target) cells by *Agrobacterium*, attachment of the bacterium to the plant cells, T-DNA transfer, T-DNA expression, T-DNA integration into the genome, and symptom expression by the transformed plant cells. It can be imagined that T-DNA transfer is not always accompanied by symptom formation. Indeed, when transformed plant cells cannot be stimulated to divide by the phytohormones that are overproduced, or when the genes for production of these phytohormones are not expressed at all, transformation would not lead to tumour formation. Since the T-DNA contains genes for opine production, opines and opine synthases can be used as alternative markers for the detection of T-DNA transfer. On the basis of these considerations in 1983 we set out to find evidence for T-DNA transfer to plant species that do not form tumours in response to infection with *Agrobacterium*, viz. the monocots *Chlorophytum capense* and *Narcissus* cv. Paperwhite [72]. These plants were infected with different types of *Agrobacterium* strains and small swellings different from typical tumours were observed some weeks after infection. In order to avoid misinterpretation due to earlier observed artefacts or to the presence of compounds unique to these plant species [37], we avoided substrate feeding and used extracts of these swellings directly in an enzyme assay that was developed earlier to detect opine synthase activity [113]. In this way we obtained evidence that not only specific opines, but also the specific opine synthases were present in tissues that had been infected with *Agrobacterium*. As expected, octopine and octopine synthase activity were only present in tissues that had been infected with octopine strains and not in those infected with nopaline strains, while nopaline was found only in tissues infected with nopaline strains. Thus, unequivocal evidence was found for the – perhaps unexpected – transfer of T-DNA to the 'non-host' – with regard to tumour development – monocotyledonous plant species [72]. Our findings were corroborated by subsequent similar findings in which other monocotyledonous plant species such as *Asparagus officinalis* [62],

Dioscorea bulbifera (yam) [133], *Zea mays* [55] and *Triticum aestivum* (wheat) [182] were used. Later on the even more sensitive reporter system of agroinfection was developed by Grimsley *et al.* [56] in which a plant virus is transmitted to plant cells at the infection sites concomitantly with the T-DNA. Transformed plants are then recognized by the symptoms of viral infection. Using this reporter system further clear evidence was obtained for T-DNA transfer to gramineous species such as *Zea mays* [57], *Triticum aestivum* and *Hordeum vulgare* (barley) [20]. From the data obtained so far it is obvious that the T-DNA transfer system may be exploited for the introduction of DNA into an extremely wide variety of plant species, although the efficiency with which this occurs may differ from one species to the other.

Genetic colonization

The process of crown gall induction consists of a large number of discrete, essential steps. First, wounding of the plant is necessary [22] to allow entrance of the bacteria and to make available compounds that induce its virulence system (see below). The bacteria multiply in the wound sap and attach to the walls of plant cells in the wound [95, 131]. Subsequently, the T-DNA is transferred and expressed in the plant cells even before integration [75]. After integration T-DNA expression is maintained at a particular stable level depending on the position of integration. After some time tumours develop due to cell divisions triggered by the continuous production of auxin and cytokinin via T-DNA encoded enzymes. The resulting tumours consist of a mixture of transformed (T-DNA-containing) and normal plant cells [177]. The T-DNA-containing cells produce and excrete opines that are consumed specifically by the infecting agrobacteria. Octopine strains can utilize octopine but not nopaline, while nopaline strains catabolize nopaline, but not octopine [19, 123]. The genes for opine catabolism are located on the Ti plasmid (Fig. 4). An opine may act not only as an inducer of its catabolic genes, but also as an aphrodisiac and activate the conjugative

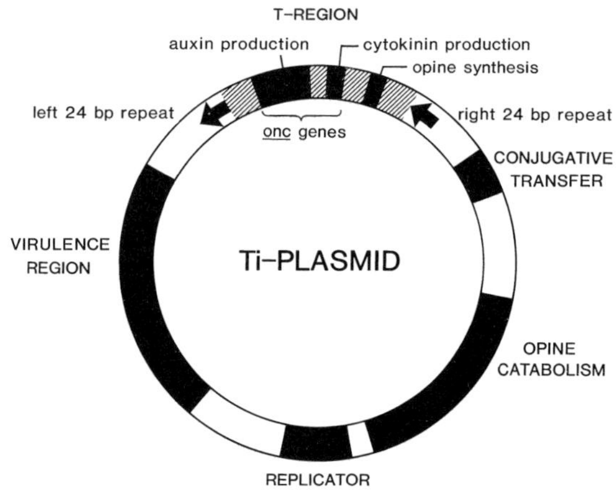

Fig. 4. Genetic map of an octopine Ti plasmid.

transfer system of the Ti plasmid [82, 158]. This is the reason that the Ti plasmid is widely spread through the bacterial population in plant tumours [81]. Since agrobacteria exploit plant cells by inducing these to produce compounds that are of specific use only to agrobacteria and do this by way of genetic engineering, the process has been called genetic colonization.

Molecular mechanism of T-DNA transfer

Virulence genes

By genetic analysis of *Agrobacterium* it was shown that besides the *onc* genes (*ipt*, *iaaM*, *iaaH*) present in the T-DNA a large number of other genes involved in tumorigenicity are present either on the Ti plasmid in a segment of 40 kb called the virulence region (*vir* genes) or on the chromosome (*chv* genes). By introducing T-DNA into plant cells *in vitro* via direct gene transfer, it was found that T-DNA by itself is sufficient for provoking the transformation of normal plant cells into tumour cells [87]. This was the first evidence that the *vir* and *chv* genes do not have an essential oncogenic function, but rather determine the apparatus necessary for *in vivo* transfer of the T-DNA from *Agrobacterium* to plant cells. Also molecular genetic experiments performed by Leemans *et al.* [93] and Hille *et al.* [67] showed that none of the T-DNA genes is required for T-DNA transfer. Even when all the genes that are naturally present in the T region are inactivated or replaced with other genes the transfer of T-DNA still occurs provided that the border repeat remains intact.

The *chvA* and *chvB* genes are necessary for the attachment of *Agrobacterium* to plant cell walls [46]. It was found that the *chvB* gene codes for a 235 kDa protein involved in the formation of a cyclic β-1,2 glucan [200], while there is evidence that the *chvA* gene determines a transport protein located in the bacterial inner membrane necessary for the transport of the β-1,2 glucan into the periplasm [29]. Mutations in another chromosomal virulence gene, which is called *pscA* or *exoC*, also lead to bacteria which do not produce β-1,2 glucan [160]. This points to a possible role of β-1,2 glucan in the attachment of agrobacteria to plant cell walls. The addition of β-1,2 glucan to *chv* mutants in plant infection experiments, however, did not result in tumour formation. More recently it was found that *chv* mutants lack a protein that was called rhicadhesin and that might be involved in attachment [139]. Addition of this protein to *chv* mutants during plant infection led to a partial restoration of the ability to induce tumours. The 40 kb *vir* region of the octopine Ti plasmid embraces 24 genes involved in virulence. These genes are present in 8 operons called *virA-virH*, which are co-regulated and thus form a regulon (see below). Most of the operons contain several genes (Fig. 5). The complete nucleotide sequence of the *vir* region of a nopaline Ti plasmid [128] and an octopine Ti plasmid [15] have been established.

Regulation of the virulence genes

With the exception of the *virA* and *virG* genes, the *vir* operons are not transcribed during normal vegetative growth [150]. Therefore, one of the early steps in plant tumour induction concerns the coordinate activation of the virulence system, when the bacteria are present near (wounded) plant tissues and sense plant cell exudate factors [150]. While the known *chv* genes are constitutively expressed, the *vir* genes are silent until they become induced by certain plant factors. Stachel *et al.* [148] identified these plant factors from tobacco as being the phenolic compounds acetosyringone and α-hydroxyacetosyringone (Fig. 6). These compounds are released from plant tissue, especially after wounding, which has been long known to be a prerequisite for plant tumorigenesis via *Agrobacterium*. Further work by several groups including our own showed that besides (α-hydroxy)acetosyringone several other phenolic compounds can act as *vir* inducers including well-known lignin precursors such as coniferyl alcohol and sinapinic acid [103, 143, 145]. Recent evidence suggests that also some flavonoids known as *nod* inducers in *Rhizobium* may act as *vir* inducers [197].

For different plant species or plant tissues different substituted phenols may be responsible for induction. For instance, although (α-hydroxy)-acetosyringone was found to be the most prominent inducer in solanaceous plants such as tobacco, tomato and potato, other species rather tended to have derivatives of benzoic acid or cinnamic acid as inducers [143, 146]. For wheat suspension cells for instance the *vir* inducer re-

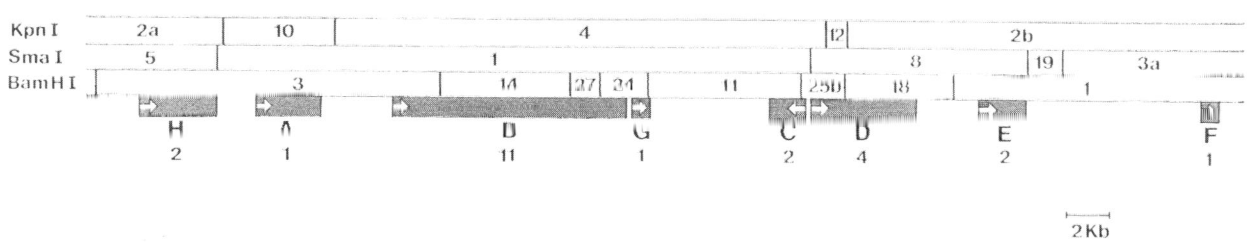

Fig. 5. Loci present in the virulence region of the octopine Ti plasmid. Letters refer to names of the *vir* operons, numbers to the number of genes in these operons. The direction of transcription is indicated.

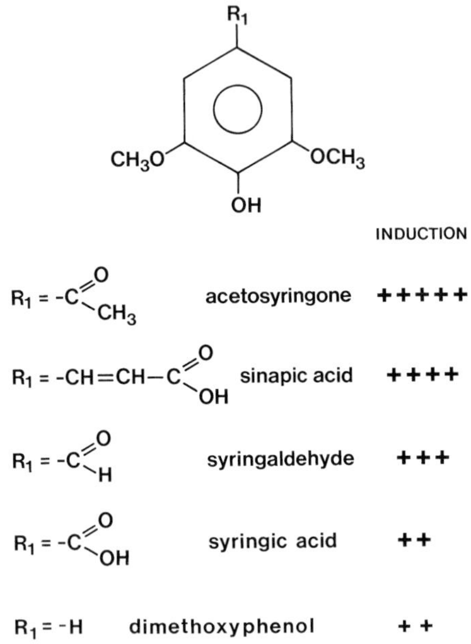

Fig. 6. Induction of the virulence genes by acetosyringone and some other phenolic compounds.

leased was found to be ethyl ferulate [107]. Plants may also excrete compounds that are inhibitory to *vir* induction. Certain maize varieties have an inhibitor in their root exudate that turned out to be 2,4-dihydroxy-7-methoxy-1,4-benzoxazin-3-one (DIMBOA) [130]. In a few cases the amounts of *vir*-inducing compounds were found not to be optimal for tumour induction to occur. In such cases the addition of acetosyringone during infection stimulated tumour induction. This was the case for leaf disc infection of *Arabidopsis thaliana* [135] and tuber disc infection of the monocot *Dioscorea bulbifera* [133].

The induction of the *vir* system can be monitored most easily for indicator strains carrying a *vir* gene promoter linked to a reporter gene such as the *lacZ* gene (encodes the enzyme β-galactosidase) of *E. coli*. The presence of β-galactosidase is easily measured with substrates such as 0–nitrophenyl-β-d-galactopyranoside (ONPG) or 5-bromo-4-chloro-3–indolyl-β-d-galactopyranoside (X-gal) that release coloured compounds after β-galactosidase action. Using cultures of such indicator strains it was found that particular conditions have to be met – even in the presence of an inducer – in order to obtain optimal *vir* induction [4, 104, 150, 167]: (1) the pH of the medium must be between 5 and 6, (2) the temperature must be between 20 and 30 °C, (3) the presence of yeast extract in the medium must be avoided, (4) a high sugar content must be present in the medium. Recently, it was found that some specificity exists in the sugars that are required for optimal *vir* induction [28, 136, 142]. In 'conditioned' plant medium the presence of a high level of inositol enhances *vir* expression [142]. Also some non-catabolizable sugars such as 2-deoxy-*d*-glucose and 6-deoxy-*d*-glucose have such a stimulatory effect [7, 136]. This sugar effect is seen most clearly when phenolic inducers are limiting. Although the *vir* systems of different Ti plasmids are similar and are inducible in similar induction media, there are small, but significant differences in their prerequisites for optimal induction [167]. For instance octopine and leucinopine Ti strains require a lower pH for optimal induction than nopaline and succinamopine Ti strains. This probably reflects small differences in the regulatory proteins that control the *vir* system. The maximal level to which the *vir* system can be induced also varies for different strains. While the *vir* system in LHR strains is only inducible by acetosyringone to a low level, it is possible to induce the supervirulent leucinopine Ti strain Bo542 to a much higher level than octopine or nopaline Ti strains under the same conditions. There are some indications that the level to which the *vir* system can be induced correlates with virulence on certain host plants [127, 187].

Two proteins encoded by the virulence region, VirA and VirG, mediate the activation of the other *vir* genes in the presence of phenolic inducers [150]. Sequence analysis revealed that the *virA* [94, 105] and *virG* [106, 188] genes resemble genes of other two-component regulatory systems such as *envZ-ompR*, *ntrB-ntrC*, *dctB-dctD* [108]. In such two-component systems the VirA-like proteins are thought to be sensors for a specific signal (phenolic compounds in the case of VirA), while the VirG-like proteins are thought to be DNA-binding activator proteins. Recent data obtained with the NtrB-NtrC [80], EnvZ-OmpR

[2] and CheA-CheY [21] systems have shown that the sensor protein can become phosphorylated and then can act as a specific protein phosphorylase for the accompanying activator protein. *In vitro* binding of the OmpR protein to the *omp* promoters, on which this protein acts, turned out to be efficient if the protein preparation was from cells that had been grown in a high-osmolarity medium, but inefficient if the preparation was from cells grown in a low-osmolarity medium [50]. Osmolarity is what is being sensed by the EnvZ protein which acts on OmpR. For the *vir* system we and others have shown that the VirA protein is present in the bacterial inner membrane [94, 105] and therefore is theoretically in the proper position to sense phenolic compounds directly. The topology of the VirA protein was analysed in more detail by making fusions with the *E. coli* protein PhoA (for alkaline phosphatase) devoid of its signal sequence [104, 189]. Since the alkaline phosphatase protein only becomes active after transport across the bacterial inner membrane, it can be used as a tool to probe the topology of predicted transported proteins [96]. Since only PhoA-VirA fusions in the predicted periplasmic domain of VirA resulted in alkaline phosphatase activity, these experiments support the model in which the VirA protein has a cytoplasmic N-terminus, a first hydrophobic transmembrane α-helix (TM1), a periplasmic domain, a second hydrophobic transmembrane α-helix (TMZ) and a large C-terminal cytoplasmic domain (Fig. 7). With this topology VirA resembles many other sensor proteins (e.g. EnvZ, Tar, Tsr). In order to find out where the receptor function for acetosyringone was located in VirA, we made hybrids between VirA and the *E. coli* Tar protein (sensor for aspartate and maltose) and assayed the function of such hybrid proteins in a *virA* mutant background [104]. Our results showed unexpectedly that the periplasmic domain does not contain the receptor for acetosyringone, but point to the possibility that an acetosyringone receptor domain is located in the TM2 region or in a neighbouring cytoplasmic portion of VirA [104]. Acetosyringone is a lipophilic compound that easily would accumulate in the bacterial inner

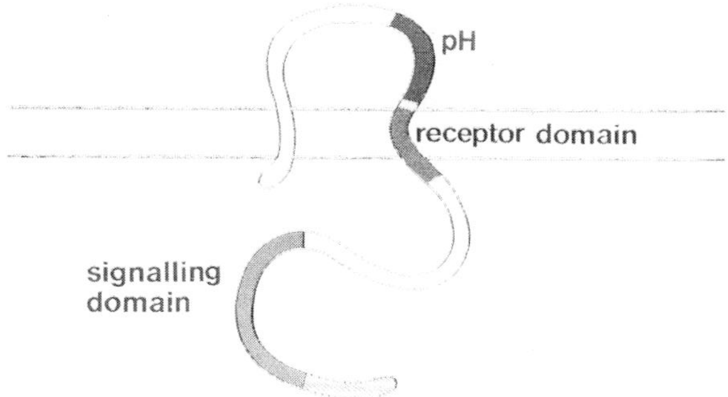

Fig. 7. Topology of the VirA protein. Positions of the receptor domain and signalling (kinase) domain are indicated; pH refers to an area influencing pH sensitivity of induction.

membrane or pass through it. Certain mammalian receptors such as the adrenergic receptors are known to have their receptor domain (for phenolic compounds such as dopamine) formed by a number of transmembrane helices [51, 154]. It is rather striking that serine and cysteine residues that form part of the binding pocket of the adrenergic receptors are also present and conserved in the TM2 domains of VirA proteins from various Ti plasmids. Another interesting feature of the VirA TM2 domain is its potential to form a leucine zipper structure. It may therefore be that dimerization of VirA plays a role in the signal transduction via VirA.

Part of the cytoplasmic domain of VirA is homologous to other sensor proteins. It was found that this domain can act as an autokinase phosphorylating itself on a conserved histidine residue [74, 78]. The phosphorylated VirA protein has the capacity to transfer its phosphate to a conserved aspartate residue in the VirG protein *in vitro* [77]. It is likely that phosphorylation modulates the DNA-binding VirG protein in a way that it stimulates *vir* gene transcription (Fig. 8). The VirG protein binds to specific areas of *vir* promoters probably as a dimer or multimer [118, 133]. The structure of the N-terminal part of the VirG protein has been predicted on the basis of the crystal structure of the homologous CheY protein of *E. coli* which revealed that the VirG protein has an acidic pocket similar to that of CheY where phosphorylation occurs [125]. The

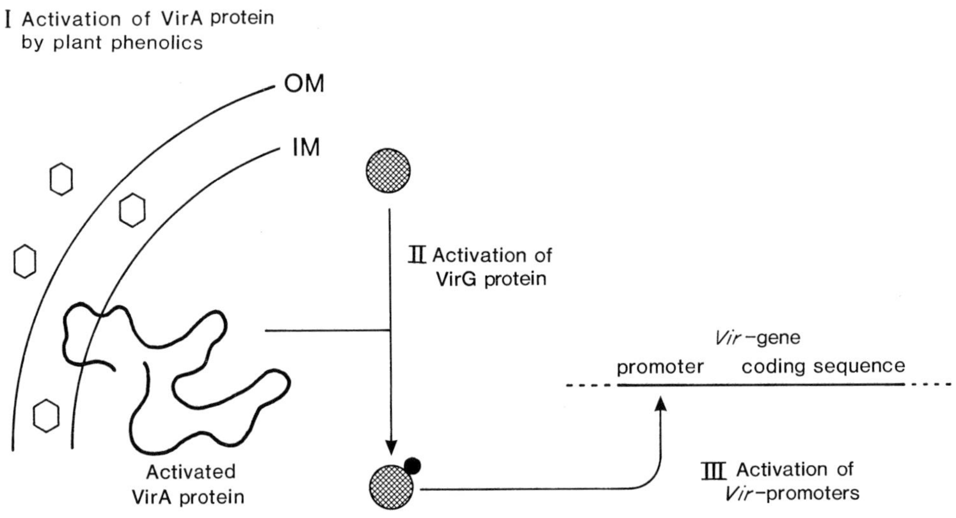

Fig. 8. Activation of the *vir* genes via VirA and VirG.

C-terminal part of VirG has been shown to have DNA-binding activity [118].

It is remarkable that the C-terminal 110 amino acids of VirA have similarity to the N-terminal (phosphorylation) domain of VirG. The function of this domain is unknown, but one could speculate about a regulatory (modulating) role in the process of *vir* regulation. Deletion of part of this region strongly reduced the activity of the VirA protein, but the deletion of the complete 110 amino acid domain led to a VirA mutant protein that was in *in vitro* assays almost as active as the wild-type VirA protein [166, 187].

T-DNA processing and transfer

After induction of the *vir* system single-stranded (ss) molecules (called T-strands) that represent the bottom strand of the T-region can be detected in *Agrobacterium* [149]. However, with lower frequency also double-stranded (ds) T-molecules have been detected by physical or genetic analysis [47, 151, 163], and therefore it is still a matter of debate in what form the T region segment is transferred to plant cells. The formation of both ss- and ds-T molecules is dependent on the activity of two proteins called VirD1 and VirD2 that are encoded by the *virD* operon of the *vir* region [194]. These proteins together determine an endonuclease activity capable of nicking (introducing ss breaks) the border repeats at a precise site, which is of course in agreement with their suspected role as recognition signals for the transfer system. Since nicking occurs at a precise site which is conserved in all border repeats sequenced so far, it was a surprise to find that mutagenesis of this nick site in the right-border repeat of the T-region (conversion of 3′G-T5′ into 3′A-T5′) did not lead to avirulence of the host bacterium [169].

It is likely that the nick sites act as starting points for DNA synthesis in the 5′ → 3′ direction. T-strands will then be released by displacement (Fig. 9). Alternatively, such nicks may be used by recombination systems to dissociate the T-region in a ds form from the Ti plasmid at a low frequency [163]. That the nick sites in the border repeats define the DNA segment that is transferred to plant cells became evident from the fact that from transformed plant cells 3 bp at most from the right-border repeat and 21 bp from the left-border repeat can be recovered [9]. It had been observed earlier that the deletion of the right-border repeat almost completely abolished T-DNA transfer [112, 134], whereas the deletion or mutation of the left repeat only led to a slightly lower frequency of transfer [67]. This observed

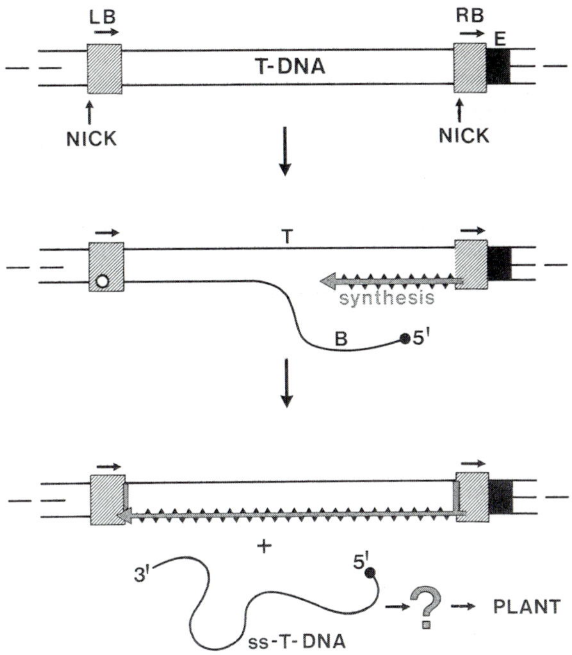

Fig. 9. Model for formation of T-strands in *Agrobacterium tumefaciens*. Black dot at the 5′ end of the T-strand indicates the covalently bound VirD2 protein.

polarity of the system is of course in agreement with the importance of T-strands as intermediates since these are probably formed after 5′ → 3′ displacement synthesis from right border to left border repeat. The absence of a right border repeat would be lethal to such a system since T-strand synthesis would not start at all, while it can be imagined that termination of the process might occur – albeit with lower efficiency – even when a left border repeat is absent. For the formation of ds T-molecules via recombination left and right-border repeat would be equally important. Therefore, recombination is apparently not involved in formation of T-DNA intermediates. However, one can still speculate that ds intermediates are formed in alternative ways.

A question that remains is why the DNA segment to the left of the left border repeat is not transferred with high efficiency to plant cells. Of course transfer of this area would not lead to tumour formation, but the same is true for the T_R region of the octopine Ti plasmid which is detected regularly in tumour lines. The reason for this turned out to be the fact that left border areas are much less efficient in acting as starting sites for DNA synthesis than right borders [122, 168]. This difference is not due to the subtle differences in the nucleotide sequences of the left and right 24 bp border repeats, but rather is caused by the presence of an enhancer next to the right sequence [121]. This T-DNA transfer enhancer, also called 'overdrive' by Peralta and Ream [121], strongly enhances T-strand formation in *Agrobacterium* [170]. It is called an enhancer because it functions in both orientations and at different positions and distances from the border repeat sequence [170]. The deletion of the enhancer sequence leads to a diminished virulence of the host bacterium [122, 168]. Toro *et al.* [165] have found that the VirC1 protein specifically binds to the overdrive sequence. Mutations in the *virC* operon result in an attenuation of virulence. In order to find out how plant cells would deal with ssDNA, we have introduced ssDNA into tobacco protoplasts via electroporation [126]. Results indicate that ssDNA is quickly converted into dsDNA in plant cells. In spite of this, ssDNA turned out to be a more effective vehicle for stable plant cell transformation (somewhat higher transformation frequencies) than dsDNA [126]. If *Agrobacterium* indeed introduced ssDNA into plant cells, this result may at least partially explain why *Agrobacterium* is so efficient in transforming plant cells. One should note that it is likely, however, that *Agrobacterium* does not introduce naked DNA molecules into plant cells. Recent evidence indicates that T-strands retain the VirD2 protein covalently attached to the 5′ terminus [63, 183]. The presence of VirD2 makes the 5′ end of the T-strand less vulnerable to an attack by exonucleases [47]. Besides the VirD2 protein may act as a pilot to direct the T strand to the nucleus of the transformed plant cell, since it contains nuclear targeting sequences [64]. The 69 kDa VirE2 protein encoded by the second open reading frame (orf) of the *virE* operon is a ssDNA-binding protein, which is able to coat the T-strands by cooperative binding leading to long, thin nucleo-protein filaments [39]. However, the presence of such nucleoprotein complexes (T-complexes) in transformed plant cells has not been demonstrated yet.

The introduction of nicks at border repeats by the VirD system followed by ss T-strand formation is reminiscent of what occurs in the initial steps of bacterial conjugation. In this latter process a specific nick is made at a sequence called origin of transfer (*oriT*) via Mob- or Tra-proteins, which is followed by the formation of single-stranded molecules via (rolling circle) displacement synthesis [185]. In bacteria ssDNA is transferred from donor to recipient via a conjugative pore (encoded by Tra proteins) that is formed after the bacteria have been brought into close contact via the sex pilus (encoded by other Tra proteins) [185]. Recently, Pansegrau and Lanka [116] observed that there was not only homology between the border repeat sequences of Ti plasmids and the *oriT* of *incP* plasmids, but also between the nicking enzymes VirD2 of Ti and TraI in *incP* plasmids. In fact, the *virD* operon of Ti plasmids, which contains four genes called *virD1*, *virD2*, *virD3* and *virD4*, seems analogous to the mobilization operon of the *incP* plasmids containing the genes *traJ*, *traI*, *traH* and *traG*, although only *traG* and *virD4* share strong DNA homology [199]. It is interesting to note that the *traK* gene of *incP* plasmids, which does not bind to the nick site of *oriT* but to a neighbouring enhancing sequence, shares a high proline content with the product of the *virC1* gene, which binds to the T-DNA transfer enhancer.

The recent data described above make clear that the initial steps of T-DNA transfer and bacterial conjugative transfer are similar. They are in line with the initially surprising finding of Buchanan-Wollaston *et al.* [25] that *incQ* plasmids are transferred to plant cells from agrobacteria harbouring the Ti virulence genes provided that the *oriT* sequence and the *mob* genes of the *incQ* plasmids are functional. These results point to a strong relationship between T-DNA transfer from *Agrobacterium* to plant cells and conjugative DNA transfer between bacteria. Since in the latter case ssDNA is transferred from donor to recipient, this might be taken as an extra argument in favour of ssDNA being the material that is introduced by *Agrobacterium* into plant cells.

Above the roles played by VirD, VirC and VirE proteins in T-complex formation are described in detail, as well as the way *vir* expression is regulated via VirA and VirG. The remaining Vir-proteins are not involved in regulation of expression or T-strand formation; only those encoded by the *virB* operon are essential for virulence. Sequencing of the octopine [162, 184a] (and nopaline [88, 128]) Ti *virB* locus showed that it contains a complex operon consisting of 11 genes (Fig. 10). Most of the proteins predicted for the *virB* operon are located in the membrane, and we and others have therefore suggested that these proteins together may form a structure (conjugal pore or pilus) through which the T-DNA is delivered into the plant cell [162, 184]. The *virB11* gene has an ATP binding site [162], and more recently the protein was found indeed to have ATPase activity [35]. It may therefore be involved in delivering energy required for T-DNA transfer. Remarkably, the *virB11* gene has clear DNA homology with the *comG* gene of *Bacillus subtilis* that is involved in ss*DNA* uptake by competent cells of this bacterium [3]. The VirB10 protein was found to form aggregates sticking from the inner

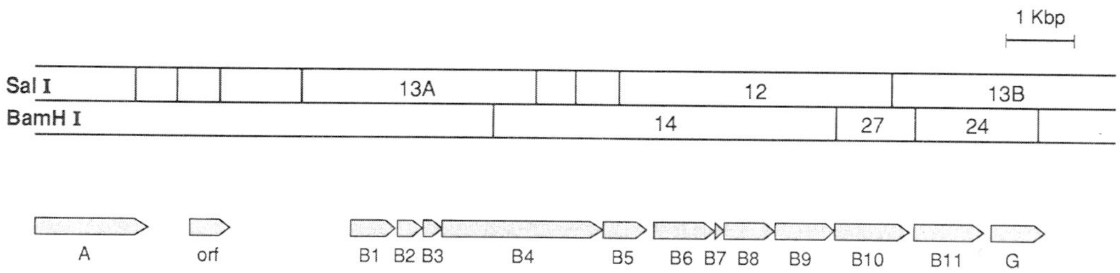

Fig. 10. Structure of the *virB* operon as determined by nucleotide sequence analysis [88, 128, 162, 184].

membrane into the periplasm [184]. Via the *phoA* system (described above for determining the VirA topology in detail) further evidence for export or membrane location was obtained for the *virB1*, *virB2*, *virB5*, *virB6*, *virB7* and *virB10* gene products [14]. In this way it was also demonstrated that the small open reading frame corresponding to the *virB7* gene indeed encoded a protein that is exported over the inner membrane and may have an outer membrane location [14].

Recently, it was found that the conjugative transfer system of *incP* plasmids can be used to introduce DNA into yeast cells [59, 137]. Apparently the *incP* type conjugal pore can be formed even between such widely diverse organisms as yeasts and bacteria. Since it might be that the *virB* operon determines a transfer apparatus similar to that of conjugative plasmids, we tried to find some experimental evidence for this. The approach we took was to investigate whether indeed the *vir* system could replace the *tra* system of conjugative plasmids in the mobilization of the non-conjugative wide host range *incQ* plasmids between bacteria. Hereby, we speculated that the transfer apparatus determined by the *virB* system would not be specific to bridge bacterial cells with plant cells but would also be able to bring together bacterial donors and recipients. Of course we used an octopine Ti plasmid from which the (octopine-inducible) conjugative transfer genes had been deleted in these experiments [13]. In full agreement with our hypothesis we found that the *vir* system was able to mobilize *incQ* plasmids into recipient *A. tumefaciens* and *E. coli* cells. As expected the system only was operative after induction with acetosyringone. Mutagenesis experiments showed that the mutation of *virA*, *virG*, *virB* or *virD4* led to a complete loss of *incQ* transfer ability [13]. This corroborates with the proposed role of the VirB and VirD4 proteins in determining a transfer apparatus similar to that of conjugative plasmids.

The octopine and nopaline Ti plasmids have a few accessory *vir*-genes that are specific for these plasmids and affect the host range for tumour formation. In the octopine Ti plasmid these are *virF* and *virH*, in the nopaline Ti plasmid a gene called *tzs*. The *tzs* gene codes for an enzyme that is similar to that determined by the T-DNA gene *ipt* and is involved in cytokinin production that is excreted from the cells as *trans*-zeatin [12]. The presence of this gene might result in enhanced tumorigenicity on certain host plants [198]. The *virH* operon consists of two genes that code for proteins that show some similarity to cytochrome P450 enzymes [79]. These proteins may therefore have a role in the detoxification of certain plant compounds that might otherwise adversely affect the growth of *Agrobacterium*. Enhanced tumorigenicity was observed for bacteria having the *virH* genes as compared to those lacking these on certain hosts [79]. The *virF* operon encodes one 23 kDa protein which shows no obvious homology to any of the proteins for which sequences are available in data banks [102]. Presence of the *virF* gene in octopine Ti strains makes these vastly superior to nopaline strains in transferring DNA to *Nicotiana glauca* and some other plant species. Using reporter genes we recently found that *virF* plays a role in T-DNA delivery rather than symptom formation [124]. A striking feature of *virF* is that it like *virE* shows '*trans*-complementation', i.e. bacteria lacking *virF* can be complemented for tumour formation by coinfection with bacteria lacking a T region but having *virF*. Cell exudates or cell extracts of *virF* $^{+}$ cells do not give *trans*-complementation [124]. Therefore, it may not be a product made via *virF* that is needed for complementation but rather the VirF protein itself. Indeed, *trans*-complementation only works if the complementing bacterium carries a complete *vir* system. Localization experiments showed that the *virF* gene product has at least partially a membrane location, but evidence for secretion was not found [124]. All these data point to the possibility that the VirF protein is delivered into plant cells via the *vir* system and functions there. In order to test this we made transgenic *N. glauca* plants in which the *virF* gene is expressed from the CaMV 35S promoter. Such engineered *N. glauca* plants are now equally good hosts for *virF* $^{+}$ as for *virF* $^{-}$ strains, showing that indeed the VirF protein can exert its function when present in plant cells [124]. Together our results indicate that proteins are delivered into plant cells via the *vir*-system even in situations when there is no T-DNA transfer. In view of the similarities between T-DNA transfer and conjugative plasmid transfer the same may be true for the latter process.

Applications

Vector systems

Although besides the T-DNA no other parts of the Ti plasmid become integrated into the genome of plant cells [17], it has long been debated whether the entire Ti plasmid or just the T-DNA segment was introduced into plant cells via *Agrobacterium*. Experiments in which the T-region was separated from the rest of the Ti plasmid [45, 69]. Genetic experiments showed that these two parts were maintained on independent replicons indeed, and did not form a cointegrate again [69]. This firmly established that no physical linkage between the T-region and the rest of the Ti plasmid was necessary for T-DNA transfer to occur. As described above the transfer system is determined by the *vir* and *chv* genes, while the 24 bp direct repeat which flanks the T-region is essential as recognition signal for the transfer apparatus (Fig. 11). On the basis of these results vector systems for the transformation of plants have been developed (Fig. 12). These can be distinguished into two types: (1) *cis* systems in which new genes are introduced via homologous recombination into an artificial T-DNA already present on the Ti plasmid [196], (2) binary systems in which new genes are cloned into plasmids containing an artificial T-DNA, which are subsequently introduced into an *Agrobacterium* strain harbouring a Ti plasmid with an intact *vir* region, but lacking the T region [16, 45, 69].

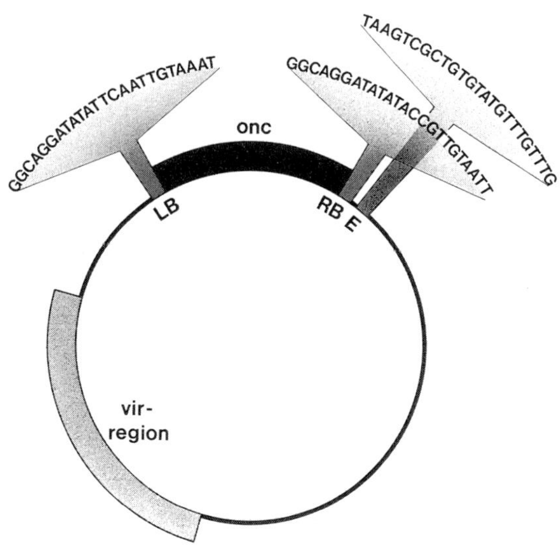

Fig. 11. Regions of the Ti plasmid important for tumorigenicity: *vir* region, border repeats (RB, LB) and enhancer (E) are involved in T-DNA transfer, and the T-DNA with *onc* genes brings about symptoms on plants.

Transgenic plant cells carrying a wild-type (oncogenic) Ti T-DNA are tumorous and cannot be regenerated into plants. However, plant cells transformed with disarmed, i.e. non-oncogenic T-DNA behave in the same way as untransformed plant cells of the same species in tissue culture and during regeneration.

After use of *Agrobacterium* for the delivery of disarmed T-DNA, mature transformed plants are being obtained for an ever increasing list of plant species including crops such as tobacco [73], potato [153], rapeseed [31] and asparagus [27]. Such transgenic plants were indistinguishable from untransformed plants, although sometimes aberrations were observed due to somaclonal variation occurring during tissue culture. In order to be able to detect or select transformed plant cells new markers have been developed. Selection markers are based on the sensitivity of plant cells to antibiotics and herbicides. It was found that

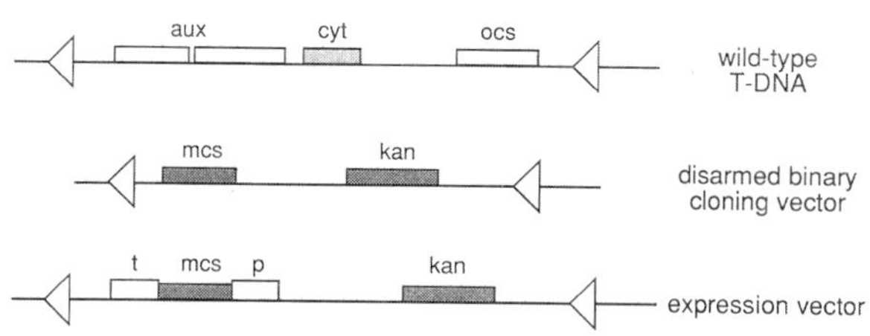

Fig. 12. Construction of plant vectors in which the *onc* genes in the T region are replaced by genes that do not disturb plant development. mcs, multiple cloning site; kan, kanamycin resistance gene for plants; p, promoter; t, terminator.

expression of (bacterial) genes coding for enzymes capable of detoxifying such compounds in plant cells can make these resistant. To this end chimaeric genes were constructed in which the (bacterial) sequence coding for the detoxifying enzyme was surrounded by plant expression signals that were obtained from the cauliflower mosaic virus (19S, 35S promoter), T-DNA genes (e.g. for octopine or nopaline synthase) or endogenous plant genes. Vectors are now available which allow selection for instance for kanamycin resistance [18] via the neomycin phosphotransferase (NPTII) gene from the bacterial transposon Tn*5*, hygromycin resistance [173] via the hygromycin phosphotransferase (HPT) gene from *E. coli*, methotrexate resistance [48] via the dihydrofolate reductase (DHFR) gene from mouse, or bialaphos (a herbicide) resistance [42] via the *bar* gene from *Streptomyces hygroscopicus*. In certain instances herbicides may be preferable because these can be sprayed and are well taken up by plants. Screening for transformation can be done by using the genes for opine synthase activities. Relatively new plant reporter genes include those coding for the enzymes luciferase [114], which gives light emission, β-galactosidase [157] and β-glucuronidase [76]. Because of the presence of endogenous β-galactosidase activity in many plant tissues, the use of β-glucuronidase is usually preferred. Reporter enzyme activity can be measured quantitatively using umbelliferyl derivatives, which release umbelliferone after enzymatic activity that can be measured fluorometrically. Histological staining for the reporter enzymes can be done using 5-bromo-4-chloro-3-indolyl derivatives, which release a compound after enzymatic activity that is quickly converted into indigo (blue) with oxygen. In order to avoid expression of β-glucuronidase by *Agrobacterium*, gene constructs were made in which the gene lacked a bacterial ribosome-binding site [75] or contained an intron in its coding sequence [178]. Such constructs are being used for an early detection of transformation via *Agrobacterium* [75].

The *Agrobacterium* vector system is being used extensively now for the transfer of various traits to (crop) plants as well as for the study of gene function in plants. Applications include the transfer of genes affecting such widely diverse traits as: resistance to viruses [1], herbicide tolerance [42], altered flower colour [175], altered shelf life of tomato [140], male sterility [97], cold tolerance [65], altered source-sink relationships [180], altered starch composition [179], starch derivatization to cyclodextrin [109], and resistance to pathogenic bacteria [8]. Although none of the transgenic crops produced is ready for marketing, field tests have been performed for quite a few of such modified crops and it is likely therefore that we shall begin to see these on the market in years to come.

Although the introduction of new traits into plants via the *Agrobacterium* system is now a common practice, there are still shortcomings in the system. The first is that it seems sometimes difficult to transform those cells in a tissue that are able to regenerate. It might be that these are in layers too deep to be reached by *Agrobacterium*, or simply are not targets for T-DNA transfer. Recently, an alternative system was developed for plant transformation with which it is – in contrast with previous alternatives such as Ca^{2+}/PEG coprecipitation, electroporation and microinjection – possible to introduce DNA sequences directly into cells rather than into protoplasts (cell walls removed) that have to be used with most of these alternative methods. In this novel procedure a particle gun is used with which small tungsten or gold microprojectiles that are coated with DNA are shot into plant tissues [83]. When a microprojectile reaches the nucleus, the DNA segment that it brought along is able to integrate into the genome and express its genes [192]. Such transformed plant cells can be regenerated into fertile, mature plants even for difficult species such as soybean [101] and rice [36, 129]. Although this has not been studied in detail, delivery of DNA via the particle gun may have the same disadvantages as other naked DNA transformation methods, i.e. scrambling of DNA copies and integration of multiple DNA copies that may be prone to recombination, rearrangement or silencing. The *Agrobacterium* system does not have such disadvantages, which is probably due to the structure with which the T-complex is delivered into plant cells and the activity of Vir proteins in the plant cells. It may therefore be a good idea to use the particle gun to deliver DNA-protein complexes into plant cells that are similar to the T-complex known from *Agrobacterium*.

Perspectives

Although genes that are introduced into plant cells are usually expressed there may be large variations in the levels at which the genes are expressed. Such 'position' effects can affect even genes closely linked on one T-DNA in a different manner. The reason for this is unknown, but it may have to do with the chromatin structure at the integration site. Also general regulatory systems including those that act via the methylation of DNA may be involved in this [99]. The copy number of T-DNA often does not correlate with the expression level [119]. It has been observed that the introduction of extra T-DNA copies even can lead to gene inactivation, a phenomenon for which there is as yet no clear explanation and that has been called co-suppression [99]. For mammalian cells it has been found that transformation with genes that are surrounded by matrix attachment regions (MARs), sequences that form the contact points for chromatin proteins, leads to expression that is independent of the integration position [152]. For such genes there was found to be a direct correlation between copy number and expression level [152]. Thus the addition of MARs, which have been isolated from plants as well in the meantime [58], to genes that are to be delivered into plants, may help to avoid variations in the level of expression after transformation.

An alternative way to avoid position effects would be to target the genes to predetermined sites in the genome where expression is guaranteed. This may be accomplished by using either site-directed or homologous recombination systems. To this end the bacterial Cre-*loxP* [41] system was introduced into plant cells. Systems for gene targeting via homologous recombination would have the additional advantage that they could also be used for the replacement or modification of genes endogenous to the plant genome. Unfortunately, in contrast to lower eukaryotes such as yeasts, fungi and protists, where integration occurs preferentially via homologous recombination, plants like mammalian cells integrate new segments of DNA only efficiently via illegitimate recombination. However, recent results show that homologous recombination can occur with a low frequency between an incoming DNA segment and a homologous copy endogenous to the plant genome. Whether the DNA was introduced via direct DNA transfer [117] or via the *Agrobacterium* vector system [92, 110] did not make much difference for the frequency with which recombination was observed, i.e. in 1 out of 10^4 to 10^5 transformants. From our results in this area we obtained unequivocal evidence that gene targeting, i.e. the modification of a locus in the plant genome, can occur in plant cells after the introduction of a homologous repair construct via *A. tumefaciens* [110]. In mammalian cells initially similar low frequencies for gene targeting were obtained. However, extensive further research led to the identification of variables that affect the frequency of gene targeting [30] and now gene targeting is used as a standard tool for the modification of the mammalian genome and the analysis of gene function. Therefore, one may hope that a similar development will be possible for plants if the process of homologous recombination is studied more carefully in these organisms.

The *Agrobacterium* vector system has also been used to tag and therefore identify plant genes influencing plant morphology (plant height, flower morphology, trichome formation). This approach has been especially successful for *Arabidopsis thaliana* for which Feldmann developed a simple seed transformation protocol with which large numbers of independent T-DNA-tagged mutants were obtained [49]. Using a T-DNA-tagged homeotic mutant, the *Arabidopsis* gene *agamous* was identified, which was found to encode a transcriptional regulator necessary for flower development [193].

Also special purpose T-DNA vectors have been developed for the identification of particular plant genes. These include vectors that have a promoterless resistance or indicator gene located close to the border repeat [156]. Activation of expression can occur after integration into a transcriptionally active area. Unexpectedly, it was found that such gene activation occurs with an extremely high frequency, i.e. 30–50% of the plant cells transformed expressed the promoterless reporter or resistance gene [86]. This was similar for plants with a small genome (*Arabidopsis thaliana*: 10^8 bp) and those with a large genome (*Nicotiana tabacum*: 5×10^9 bp), which suggests that T-DNA integrates preferentially in potentially transcriptionally active areas. By the analysis of the gene expression pattern of the reporter construct in the tagged transgenic plants genes may be identified that have a tissue- or organ-specific expression pattern. Besides these promoter/enhancer trap constructs more recently also other novel types of T-DNA vectors were constructed, such as promoter/enhancer-out constructs which have a strong promoter/enhancer near one of the border repeats [174, 181]. It is hoped that with these genes can be identified involved in the regulation of growth and development. Tumour formation in mammalian systems is often due to the unregulated (over) expression of genes involved in the control of growth, and it can therefore be imagined that activation of similar genes in plants by an outward directed promoter/enhancer in the T-DNA may lead to tumour formation or an otherwise aberrant development. Another novel type of T-DNA vector contains a promoterless toxic gene such as that for diphtheria toxin, which is toxic to plant cells [40], near the border repeat [174]. It can be imagined that integration into a tissue- or developmentally-specific gene will lead to ablation of the tissue or a halt in development at a specific stage. With this latter type of construct also cell ablation experiments can be done by fusion of the toxin gene to well characterized promoters in the same way as is done in mice [24]. Another novel type of *Agrobacterium*/T-DNA vector system was recently described by Ludwig *et al.* [90]. This concerned the construction of a special *Agrobacterium* strain that expressed the *E. coli lamB* gene and since it expressed the LamB protein became sensitive for *E. coli* bacteriophage λ. Thus cosmids can easily be introduced into this strain. A special cosmid vector was constructed containing between T-DNA borders plant selectable markers and a *cos* site. A cosmid bank containing genomic fragments from *Arabidopsis thaliana* was established in this *Agrobacterium* strain, which will no doubt become important in complementation experiments in the near future.

Looking back on developments in the field of plant molecular biology in the last decade we like to conclude with saying that the development of plant vectors on the basis of what was known about the *Agrobacterium* T-DNA transfer system in the early stages of this decade was one of the important factors that made a vast increase in knowledge in the field of plant molecular biology possible during the past ten years. We sincerely believe that further detailed knowledge of the molecular mechanism of T-DNA transfer will contribute to the further development of the field of plant molecular biology by making the genetic modification of plants more precise and sophisticated.

References

1. Abel PP, Nelson RS, De B, Hoffmann N, Rogers SG, Fraley RT, Beachy RN: Delay of disease development in transgenic plants that express the tobacco mosaic virus coat protein gene. Science 232: 738–743 (1986).
2. Aiba H, Mizuno T: Phosphorylation of a bacterial activator protein, OmpR, by a protein kinase, EnvZ, stimulates the transcription of the *ompF* and *ompC* genes in *Escherichia coli*. FEBS Lett 261: 19–22 (1990).
3. Albano M, Breitling R, Dubnau DA: Nucleotide sequence and genetic organization of the *Bacillus subtilis comG* operon. J Bact 171: 5386–5404 (1989).
4. Alt-Moerbe J, Neddermann P, Von Lintig J, Weiler EW, Schröder J: Temperature-sensitive step in Ti plasmid vir region induction and correlation with cytokinin secretion by *Agrobacteria*. Mol Gen Genet 213: 1–8 (1988).
5. Amasino RM, Powell ALT, Gordon MP: Changes in T-DNA methylation and expression are associated with phenotypic variation and plant regeneration in a crown gall tumor line. Mol Gen Genet 197: 437–446 (1984).
6. Ambros PF, Matzke AJM, Matzke MA: Localization of *Agrobacterium rhizogenes* T-DNA in plant chromosomes by *in situ* hybridization. EMBO J 5: 2073–2077 (1986).
7. Ankenbauer RG, Nester EW: Sugar-mediated induction

of *Agrobacterium tumefaciens* virulence genes: structural specificity and activities of monosaccharides. J Bact 172: 6442–6446 (1990).

8. Anzai H, Yoneyama K, Yamaguchi I: Transgenic tobacco resistant to a bacterial disease by the detoxification of a pathogenic toxin. Mol Gen Genet 219: 492–494 (1989).
9. Bakkeren G, Koukolikova-Nicola Z, Grimsley N, Hohn B: Recovery of *Agrobacterium tumefaciens* T-DNA molecules from whole plants early after transfer. Cell 57: 847–857 (1989).
10. Barker RF, Idler KB, Thompson DV, Kemp JD: Nucleotide sequence of the T-DNA region from *Agrobacterium tumefaciens* octopine Ti plasmid pTi15955. Plant Mol Biol 2: 335–350 (1983).
11. Barry GF, Rogers SG, Fraley RT, Brand L: Identification of a cloned cytokinin biosynthetic gene. Proc. Natl. Acad. Sci. USA 81: 4776–4780 (1984).
12. Beaty JS, Powell GK, Lica L, Regier DA, MacDonald EMS, Hommes NG, Morris RO: *Tzs*, a nopaline Ti plasmid gene from *Agrobacterium tumefaciens* associated with transzeatin biosynthesis. Mol Gen Genet 203: 274–280 (1986).
13. Beijersbergen A, den Dulk-Ras A, Schilperoort RA, Hooykaas PJJ: Conjugative transfer by the virulence system of *Agrobacterium tumefaciens*. Science (in press).
14. Beijersbergen A, Hooykaas PJJ: Topology of the VirB complex analysed via PhoA fusions (in preparation).
15. Beijersbergen A, Idler KB, Melchers LS, Thompson DV, Hooykaas PJJ: The complete nucleotide sequence of the octopine Ti plasmid virulence region (in preparation).
16. Bevan M: Binary *Agrobacterium* vectors for plant transformation. Nucl Acids Res 12: 8711–8721 (1984).
17. Bevan MW, Chilton M-D: T-DNA of the *Agrobacterium* Ti and Ri plasmids. Annu Rev Genet 16: 357–384 (1982).
18. Bevan MW, Flavell RB, Chilton M-D: A chimaeric antibiotic resistance gene as a selectable marker for plant cell transformation. Nature 304: 184–187 (1983).
19. Bomhoff G, Klapwijk PM, Kester HCM, Schilperoort RA, Hernalsteens JP, Schell J: Octopine and nopaline synthesis and breakdown genetically controlled by a plasmid of *Agrobacterium tumefaciens*. Mol Gen Genet 145: 177–181 (1976).
20. Boulton MI, Buchholz WG, Marks MS, Parkham PG, Davies JW: Specificity of *Agrobacterium* mediated delivery of maize streak virus DNA to members of the Gramineae. Plant Mol Biol 12: 31–40 (1989).
21. Bourret RB, Hess JF, Simon MI: Conserved aspartate residues and phosphorylation in signal transduction by the chemotaxis protein CheY. Proc Natl Acad Sci USA 87: 41–45 (1990).
22. Braun AC: Thermal studies on the factors responsible for tumor initiation in crown gall. Am J Bot 34: 234–240 (1947).
23. Braun AC: A physiological basis for autonomous growth of crown gall tumor cell. Proc Natl Acad Sci USA 44: 344–349 (1958).
24. Breitman ML, Rombola H, Maxwell IH, Klintworth GD, Bernstein A: Genetic ablation in transgenic mice with an attenuated diphtheria toxin A gene. Mol Cell Biol 10: 474–479 (1990).
25. Buchanan-Wollaston V, Passiatore JE, Cannon F: The *mob* and *oriT* mobilization functions of a bacterial plasmid promote its transfer to plants. Nature 328: 172–175 (1987).
26. Buchholz WG, Thomashow MF: Comparison of T-DNA oncogene complements of *Agrobacterium tumefaciens* tumor-inducing plasmids with limited and wide host ranges. J Bact 160: 319–326 (1984).
27. Bytebier B, Deboeck F, De Greve H, Van Montagu M, Hernalsteens J-P: T-DNA organization in tumor cultures and transgenic plants of the monocotyledon *Asparagus officinalis*. Proc Natl Acad Sci USA 84: 5345–5349 (1987).
28. Cangelosi GA, Ankenbauer RG, Nester EW: Sugars induce the *Agrobacterium tumefaciens* virulence genes via a periplasmic binding protein and the VirA protein. Proc Natl Acad Sci USA 87: 6708–6712 (1990).
29. Cangelosi GA, Martinetti G, Leigh JA, Lee CC, Theines C, Nester EW: Role of *Agrobacterium tumefaciens* ChvA protein in export of β-1,2 glucan. J Bact 171: 1609–1615 (1989).
30. Capecchi MR: Altering the genome by homologous recombination. Science 244: 1288–1292 (1989).
31. Charest PJ, Holbrook LA, Gabard J, Iyer VN, Miki BL: *Agrobacterium* mediated transformation of thin cell layer explants from *Brassica napus* L. Theor Appl Genet 75: 438–445 (1988).
32. Chilton M-D, Drummond MH, Merlo DJ, Sciaky D, Montoya AL, Gordon MP, Nester EW: Stable incorporation of plasmid DNA into higher plant cells: the molecular basis of crown gall tumorigenesis. Cell 11: 263–271 (1977).
33. Chilton M-D, Montoya AL, Merlo DJ, Drummond MH, Nutter R, Gordon MP, Nester EW: Restriction endonuclease mapping of a plasmid that confers oncogenicity upon *Agrobacterium tumefaciens* strain B6–806. Plasmid 1: 254–269 (1978).
34. Chilton M-D, Saiki RK, Yadav N, Gordon MP, Quetier F: T- DNA from *Agrobacterium* Ti plasmid is in the nuclear DNA fraction of crown gall tumor cells. Proc Natl Acad Sci USA 77: 4060–4064 (1980).
35. Christie PJ, Ward JE, Gordon MP, Nester EW: A gene required for transfer of T-DNA to plants encodes an ATPase with autophosphorylating activity. Proc Natl Acad Sci USA 86: 9677–9681 (1989).
36. Christou P, Ford TL, Kofron M: Production of transgenic rice *(Oryza sativa* L.) from agronomically important indica and japonica varieties via electric discharge particle acceleration of exogenous DNA into immature zygotic embryos. Bio/technology 9: 957–962 (1991).

37. Christou P, Platt SG, Ackerman MC: Opine synthesis in wild-type plant tissue. Plant Physiol 82: 218–221 (1986).
38. Chyi Y-S, Jorgensen RA, Goldstein D, Tanksley SD, Loaiza-Figueroa F: Locations and stability of *Agrobacterium* mediated T-DNA insertions in the *Lycopersicon* genome. Mol Gen Genet 204: 64–69 (1986).
39. Citovsky V, Wong ML, Zambryski P: Cooperative interaction of *Agrobacterium* VirE2 protein with single-stranded DNA: implications for the T-DNA transfer process. Proc Natl Acad Sci USA 86: 1193–1197 (1989).
40. Czako M, An G: Expression of DNA coding for diphtheria toxin chain A is toxic to plant cells. Plant Physiol 95: 687–692 (1991).
41. Dale EC, Ow DW: Intra- and intermolecular site-specific recombination in plant cells mediated by bacteriophage P1 recombinase. Gene 91: 79–85 (1990).
42. De Block M, Botterman J, Vandewiele M, Dockx J, Thoen C, Gosselé V, Movva NR, Thompson C, Van Montagu M, Leemans J: Engineering herbicide resistance in plants by expression of a detoxifying enzyme. EMBO J 6: 2513–2518 (1987).
43. De Beuckeleer M, Lemmers M, De Vos G, Willmitzer L, Van Montagu M, Schell J: Further insight on the transferred- DNA of octopine crown gall. Mol Gen Genet 183: 283–288 (1981).
44. De Cleene M, De Ley J: The host range of crown gall. Bot Rev 42: 389–466 (1976).
45. De Framond AJ, Barton KA, Chilton M-D: Mini-Ti: a new vector strategy for plant genetic engineering. Bio/technology 1: 262–269 (1983).
46. Douglas CJ, Staneloni RJ, Rubin RA, Nester EW: Identification and genetic analysis of an *Agrobacterium tumefaciens* chromosomal virulence region. J Bact 161: 850–860 (1985).
47. Dürrenberger F, Crameri A, Hohn B, Koukolikova-Nicola Z: Covalently bound VirD2 protein of *Agrobacterium tumefaciens* protects the T-DNA from exonucleolytic degradation. Proc Natl Acad Sci USA 86: 9154–9158 (1989).
48. Eichholtz DA, Rogers SG, Horsch RB, Klee HJ, Hayford M, Hoffmann NL, Bradford SB, Fink C, Flick J, O'Connell KM, Fraley RT: Expression of mouse dihydrofolate reductase gene confers methotrexate resistance in transgenic petunia plants. Som Cell Mol Genet 13: 67–76 (1987).
49. Feldmann KA: T-DNA insertion mutagenesis in *Arabidopsis*: mutational spectrum. Plant J 1: 71– 82 (1991).
50. Forst SA, Delgado J, Inouye M: DNA-binding properties of the transcripton activator (OmpR) for the upstream sequences of *ompF* in *Escherichia coli* are altered by *envZ* mutations and medium osmolarity. J Bact 171: 2949–2955 (1989).
51. Fraser CM: Site-directed mutagenesis of β-adrenergic receptors. Identification of conserved cysteine residues that independently affect ligand binding and receptor activation. J Biol Chem 264: 9266–9270 (1989).
52. Garfinkel DJ, Simpson RB, Ream LW, White FF, Gordon MP, Nester EW: Genetic analysis of crown gall: fine structure map of the T-DNA by site-directed mutagenesis. Cell 27: 143–153 (1981).
53. Gelvin SB, Thomashow MF, McPherson JC, Gordon MP, Nester EW: Sizes and map positions of several plasmid DNA-encoded transcripts in octopine-type crown gall tumors. Proc Natl Acad Sci USA 79: 76–80 (1982).
54. Gheysen G, Villarroel R, Van Montagu M: Illegitimate recombination in plants: a model for T-DNA integration. Genes Devel 5: 287–297 (1991).
55. Graves ACF, Goldman SL: The transformation of *Zea mays* seedlings with *Agrobacterium tumefaciens*. Plant Mol Biol 7: 43–50 (1986).
56. Grimsley N, Hohn B, Hohn T, Walden R: 'Agroinfection' an alternative route for viral infection of plants by using the Ti plasmid. Proc Natl Acad Sci USA 83: 3282– 3286 (1986).
57. Grimsley N, Hohn T, Davies JW, Hohn B: *Agrobacterium* mediated delivery of infectious maize streak virus into maize plants. Nature 325: 177–179 (1987).
58. Hall G, Allen GC, Loer DS, Thompson WF, Spiker S: Nuclear scaffolds and scaffold-attachment regions in higher plants. Proc Natl Acad Sci USA 88: 9320–9324 (1991).
59. Heinemann JA, Sprague GF: Bacterial conjugative plasmids mobilize DNA transfer between bacteria and yeast. Nature 340: 205–209 (1989).
60. Hemstad PR, Reisch BI: *In vitro* production of galls induced by *Agrobacterium tumefaciens* and *Agrobacterium rhizogenes* on *Vitis* and *Rubus*. J Plant Physiol 120: 9–17 (1985).
61. Hepburn AG, Clarke LE, Pearson L, White J: The role of cytosine methylation in the control of nopaline synthase gene expression in a plant tumor. J Mol Appl Genet 2: 315–329 (1983).
62. Hernalsteens J-P, Thia-Toong L, Schell J, Van Montagu M: An *Agrobacterium* transformed cell culture from the monocot *Asparagus officinalis*. EMBO J 3: 3039–3041 (1984).
63. Herrera-Estrella A, Chen Z, Van Montagu M, Wang K: VirD proteins of *Agrobacterium tumefaciens* are required for the formation of a covalent DNA-protein complex at the 5′ terminus of T strand molecules. EMBO J 7: 4055–4062 (1988).
64. Herrera-Estrella A, Van Montagu M, Wang K: A bacterial peptide acting as a plant nuclear targeting signal: the amino-terminal portion of *Agrobacterium* VirD2 protein directs a β-galactosidase fusion protein into tobacco nuclei. Proc Natl Acad Sci USA 87: 9534–9537 (1990).
65. Hightower R, Baden C, Penzes E, Lund P, Dunsmuir P: Expression of antifreeze proteins in transgenic plants. Plant Mol Biol 17: 1013–1021 (1991).

66. Hille J, Van Kan J, Klasen I, Schilperoort R: Site- directed mutagenesis in *Escherichia coli* of a stable R772::Ti cointegrate plasmid from *Agrobacterium tumefaciens*. J Bact 154: 693–701 (1983).
67. Hille J, Wullems G, Schilperoort RA: Non-oncogenic T-region mutants of *Agrobacterium tumefaciens* do transfer T-DNA into plant cells. Plant Mol Biol 2: 155–163 (1983).
68. Hoekema A, de Pater BS, Fellinger AJ, Hooykaas PJJ, Schilperoort RA: The limited host range of an *Agrobacterium tumefaciens* strain extended by a cytokinin gene from a wide host range T region. EMBO J 3: 3043–3047 (1984).
69. Hoekema A, Hirsch PR, Hooykaas PJJ, Schilperoort RA: A binary plant vector strategy based on separation of *vir* and T-region of the *Agrobacterium tumefaciens* Ti-plasmid. Nature 303: 179–180 (1983).
70. Hooykaas PJJ, den Dulk-Ras H, Schilperoort RA: The *Agrobacterium tumefaciens* T-DNA gene 6b is an *onc* gene. Plant Mol Biol 11: 791–794 (1988).
71. Hooykaas PJJ, Klapwijk PM, Nuti MP, Schilperoort RA, Rörsch A: Transfer of the *Agrobacterium tumefaciens* Ti plasmid to avirulent agrobacteria and to *Rhizobium ex planta*. J Gen Microbiol 98: 477–484 (1977).
72. Hooykaas-Van Slogteren GMS, Hooykaas PJJ, Schilperoort RA: Expression of Ti plasmid genes in monocotyledonous plants infected with *Agrobacterium tumefaciens*. Nature 311: 763–764 (1984).
73. Horsch RB, Fry JE, Hoffmann NL, Eichholtz D, Rogers SG, Fraley RT: A simple and general method for transferring genes into plants. Science 227: 1229–1231 (1985).
74. Huang Y, Morel P, Powell B, Kado CI: VirA, a coregulator of Ti-specified virulence genes, is phosphorylated in vitro. J Bact 172: 1142–1144 (1990).
75. Janssen BJ, Gardner RC: Localized transient expression of GUS in leaf discs following cocultivation with *Agrobacterium*. Plant Mol Biol 14: 61–72 (1989).
76. Jefferson RA, Kavanagh TA, Bevan MW: GUS fusions: *β*- glucuronidase as a sensitive and versatile gene fusion marker in higher plants. EMBO J 6: 3901–3907 (1987).
77. Jin S, Prusti RK, Roitsch T, Ankenbauer RG, Nester EW: The VirG protein of *Agrobacterium tumefaciens* is phosphorylated by the autophosphorylated VirA protein and this is essential for its biological activity. J Bact 172: 4945–4950 (1990).
78. Jin S, Roitsch T, Ankenbauer RG, Gordon MP, Nester EW: The VirA protein of *Agrobacterium tumefaciens* is autophosphorylated and is essential for *vir* gene regulation. J Bact 172: 525–530 (1990).
79. Kanemoto RH, Powell AT, Akiyoshi DE, Regier DA, Kerstetter RA, Nester EW, Hawes MC, Gordon MP: Nucleotide sequence and analysis of the plant-inducible locus *pinF* from *Agrobacterium tumefaciens*. J Bact 171: 2506–2512 (1989).
80. Keener J, Kustu S: Protein kinase and phosphoprotein phosphatase activities of nitrogen regulatory proteins NTRB and NTRC of enteric bacteria: roles of the conserved amino-terminal domain of NTRC. Proc Natl Acad Sci USA 85: 4976–4980 (1988).
81. Kerr A: Transfer of virulence between isolates of *Agrobacterium*. Nature 223: 1175–1176 (1969).
82. Klapwijk PM, Scheulderman T, Schilperoort RA: Coordinated regulation of octopine degradation and conjugative transfer of Ti plasmids in *Agrobacterium tumefaciens*: evidence for a common regulatory gene and separate operons. J Bact 136: 775–785 (1978).
83. Klein TM, Wolf ED, Wu R, Sanford JC: High-velocity microprojectiles for delivery of nucleic acids into living cells. Nature 327: 70–73 (1987).
84. Koekman BP, Ooms G, Klapwijk PM, Schilperoort RA: Genetic map of an octopine Ti-plasmid. Plasmid 2: 347–357 (1979).
85. Komro CT, DiRita VJ, Gelvin SB, Kemp JD: Site-specific mutagenesis in the TR-DNA region of octopine-type Ti plasmids. Plant Mol Biol 4: 253–263 (1985).
86. Koncz C, Martini N, Mayerhofer R, Koncz-Kalman Z, Körber H, Redei GP, Schell J: High-frequency T-DNA-mediated gene tagging in plants. Proc Natl Acad Sci USA 86: 8467–8471 (1989).
87. Krens FA, Mans RMW, van Slogteren TMS, Hoge JHC, Wullems GJ, Schilperoort RA: Structure and expression of DNA transferred to tobacco via transformation of protoplasts with Ti-plasmid DNA: co-transfer of T-DNA and non-T-DNA sequences. Plant Mol Biol 5: 223–234 (1985).
88. Kuldau GA, De Vos G, Owen J, McCaffrey G, Zambryski P: The *virB* operon of *Agrobacterium tumefaciens* pTiC58 encodes 11 open reading frames. Mol Gen Genet 221: 256–266 (1990).
89. Körber H, Strizhov N, Staiger D, Feldwisch J, Olsson O, Sandberg G, Palme K, Schell J, Koncz C: T-DNA gene 5 of *Agrobacterium* modulates auxin response by autoregulated synthesis of a growth hormone antagonist in plants. EMBO J 10: 3983–3991 (1991).
90. Lazo GR, Stein PA, Ludwig RA: A DNA transformation- competent *Arabidopsis* genomic library in *Agrobacterium*. Bio/technology 9: 963–967 (1991).
91. Ledeboer AM: Large plasmids in Rhizobiaceae. Studies on the transcription of the tumour inducing plasmid from *Agrobacterium tumefaciens* in sterile crown gall tumour cells. Thesis, University of Leiden, Netherlands, p 180 (1978).
92. Lee KY, Lund P, Lowe K, Dunsmuir P: Homologous recombination in plant cells after *Agrobacterium* mediated transformation. Plant Cell 2: 415–425 (1990).
93. Leemans J, Deblaere R, Willmitzer L, de Greve H, Hernalsteens JP, Van Montagu M, Schell J: Genetic identification of functions of TL-DNA transcripts in octopine crown galls. EMBO J 1: 147–152 (1982).
94. Leroux B, Yanofsky MF, Winans SC, Ward JE, Ziegler SF, Nester EW: Characterization of the *virA*

locus of *Agrobacterium tumefaciens*: a transcriptional regulator and host range determinant. EMBO J 6: 849–856 (1987).
95. Lippincott BB, Lippincott JA: Bacterial attachment to a specific wound site as an essential stage in tumor initiation by *Agrobacterium tumefaciens*. J Bact 97: 620–628 (1969).
96. Manoil C, Beckwith J: Tn *phoA*: a transposon probe for protein export signals. Proc Natl Acad Sci USA 82: 8129– 8133 (1985).
97. Mariani C, De Beuckeleer M, Truettner J, Leemans J, Goldberg RB: Induction of male sterility in plants by a chimaeric ribonuclease gene. Nature 347: 737–741 (1990).
98. Matsumoto S, Ito Y, Hosoi T, Takahashi Y, Machida Y: Integration of *Agrobacterium* T-DNA into a tobacco chromosome: possible involvement of DNA homology between T-DNA and plant DNA. Mol Gen Genet 224: 309–316 (1990).
99. Matzke MA, Primig M, Trnovsky J, Matzke AJM: Reversible methylation and inactivation of marker genes in sequentially transformed tobacco plants. EMBO J 8: 643– 649 (1989).
100. Mayerhofer R, Koncz-Kalman Z, Nawrath C, Bakkeren G, Crameri A, Angelis K, Redei GP, Schell J, Hohn B, Koncz C: T-DNA integration: a mode of illegitimate recombination in plants. EMBO J 10: 697–704 (1991).
101. McCabe DE, Swain MF, Martinell BJ, Christou P: Stable transformation of soybean *(Glycine max)* by particle acceleration. Bio/technology 6: 923–926 (1988).
102. Melchers LS, Maroney MJ, den Dulk-Ras A, Thompson DV, van Vuuren HAJ, Schilperoort RA, Hooykaas PJJ: Octopine and nopaline strains of *Agrobacterium tumefaciens* differ in virulence; molecular characterization of the *virF*-locus. Plant Mol Biol 14: 249–259 (1990).
103. Melchers LS, Regensburg-Tuïnk AJG, Schilperoort RA, Hooykaas PJJ: Specificity of signal molecules in the activation of *Agrobacterium* virulence gene expression. Mol Microbiol 3: 969–977 (1989).
104. Melchers LS, Regensburg-Tuïnk TJG, Bourret RB, Sedee NJA, Schilperoort RA, Hooykaas PJJ: Membrane topology and functional analysis of the sensory protein VirA of *Agrobacterium tumefaciens*. EMBO J 8: 1919–1925 (1989).
105. Melchers LS, Thompson DV, Idler KB, Neuteboom STC, De Maagd RA, Schilperoort RA, Hooykaas PJJ: Molecular characterization of the virulence gene *virA* of the *Agrobacterium tumefaciens* octopine Ti plasmid. Plant Mol Biol 9: 635–645 (1987).
106. Melchers LS, Thompson DV, Idler KB, Schilperoort RA, Hooykaas PJJ: Nucleotide sequence of the virulence gene *vir*G of the *Agrobacterium tumefaciens* octopine Ti plasmid: significant homology between *vir*G and the regulatory genes *omp*R, *pho*B and dye of *E. coli*. Nucl Acids Res 14: 9933–9942 (1986).
107. Messens E, Dekeyser R, Stachel SE: A nontransformable *Triticum monococcum* monocotyledonous culture produces the potent *Agrobacterium vir*-inducing compound ethyl ferulate. Proc Natl Acad Sci USA 87: 4368–4372 (1990).
108. Nixon BT, Ronson CW, Ausubel FM: Two-component regulatory systems responsive to enviromental stimuli share strongly conserved domains with the nitrogen assimilation regulatory genes *ntr*B and *ntr*C. Proc Natl Acad Sci USA 83: 7850–7854 (1986).
109. Oakes JV, Shewmaker CK, Stalker DM: Production of cyclodextrins, a novel carbohydrate, in the tubers of transgenic potato plants. Bio/technology 9: 982–986 (1991).
110. Offringa R, de Groot MJA, Haagsman HJ, Does MP, van den Elzen PJ, Hooykaas PJJ: Extrachromosomal homologous recombination and gene targeting in plant cells after *Agrobacterium* mediated transformation. EMBO J 9: 3077– 3084 (1990).
111. Ooms G, Hooykaas PJJ, Moolenaar G, Schilperoort RA: Crown gall plant tumors of abnormal morphology, induced by *Agrobacterium tumefaciens* carrying mutated octopine Ti plasmids: analysis of T-DNA functions. Gene 14: 33–50 (1981).
112. Ooms G, Hooykaas PJJ, van Veen RJM, van Beelen P, Regensburg-Tuïnk AJG, Schilperoort RA: Octopine Ti-plasmid deletion mutants of *Agrobacterium tumefaciens* with emphasis on the right side of the T region. Plasmid 7: 15–29 (1982).
113. Otten LABM, Schilperoort RA: A rapid micro scale method for the detection of lysopine and nopaline dehydrogenase activities. Biochim Biophys Acta 527: 497–500 (1978).
114. Ow DW, Wood KV, De Luca M, De Wet JR, Helinski DR, Howell SH: Transient and stable expression of the firefly luciferase gene in plant cells and transgenic plants. Science 234: 856–859 (1986).
115. Panagopoulos CG, Psallidas PG: Characteristics of Greek isolates of *Agrobacterium tumefaciens* (Smith and Townsend) Conn J Appl Bact 36: 233–240 (1973).
116. Pansegrau W, Lanka E: Common sequence motifs in the DNA relaxases and nick regions from a variety of DNA transfer systems. Nucl Acids Res 19: 3455 (1991).
117. Paszkowski J, Baur B, Bogucki A, Potrykus I: Gene targeting in plants. EMBO J 7: 4021–4026 (1988).
118. Pazour GJ, Das A: *VirG*, an *Agrobacterium tumefaciens* transcriptional activator, initiates translation at a UUG codon and is a sequence-specific DNA-binding protein. J Bact 172: 1241–1249 (1990).
119. Peach C, Velten J: Transgene expression variability (position effect) of CAT and GUS reporter genes driven by linked divergent T-DNA promoters. Plant Mol Biol 17: 49–60 (1991).
120. Peerbolte R, Leenhouts K, Hooykaas-van Slogteren GMS, Hoge JHC, Wullems GJ, Schilperoort RA: Clones from a shooty tobacco crown gall tumor I: deletions, rearrangements and amplifications resulting in

irregular T-DNA structures and organizations. Plant Mol Biol 7: 265–284 (1986).

121. Peralta EG, Hellmiss R, Ream W: *Overdrive*, a T-DNA transmission enhancer on the *A. tumefaciens* tumour-inducing plasmid. EMBO J 5: 1137–1142 (1986).

122. Peralta EG, Ream LW: T-DNA border sequences required for crown gall tumorigenesis. Proc Natl Acad Sci USA 82: 5112–5116 (1985).

123. Petit A, Delhaye S, Tempé J, Morel G: Recherches sur les guanidines des tissus de crown-gall. Mise en evidence d'une relation biochimique spécifique entre les souches d'*Agrobacterium tumefaciens* et les tumeurs qu'elles induisent. Physiol Végét 8: 205–213 (1970).

124. Regensburg-Tuïnk AJG, Mozo T, Hooykaas PJJ: The virulence protein VirF of *Agrobacterium tumefaciens* is transferred to and active in plant cells (in preparation).

125. Rodenburg K, Vriend G, Schilperoort RA, Hooykaas PJJ: A model for VirG based on the 3 dimensional structure of the *E. coli* CheY protein (in preparation).

126. Rodenburg KW, de Groot MJA, Schilperoort RA, Hooykaas PJJ: Single stranded DNA used as an efficient new vehicle for plant protoplast transformation. Plant Mol Biol 13: 711–719 (1989).

127. Rogowsky PM, Close TJ, Chimera JA, Shaw JJ, Kado CI: Regulation of the *vir* genes of *Agrobacterium tumefaciens* plasmid pTiC58. J Bact 169: 5101–5112 (1987).

128. Rogowsky PM, Powell BS, Shirasu K, Lin T-S, Morel P, Zyprian EM, Steck TR, Kado CI: Molecular characterization of the *vir* regulon of *Agrobacterium tumefaciens*: complete nucleotide sequence and gene organization of the 28.63–kbp regulon cloned as a single unit. Plasmid 23: 85–106 (1990).

129. Rueb S, Hensgens LAM, Schilperoort RA: Transgenic rice plants obtained after particle gun transformation (in preparation).

130. Sahi SV, Chilton M-D, Chilton WS: Corn metabolites affect growth and virulence of *Agrobacterium tumefaciens*. Proc Natl Acad Sci USA 87: 3879–3883 (1990).

131. Schilperoort RA: Investigations on plant tumors – Crown gall. On the biochemistry of tumor induction by *Agrobacterium tumefaciens*. Thesis, Leiden University, Netherlands (1969).

132. Schröder G, Waffenschmidt S, Weiler EW, Schröder J: The T-region of Ti-plasmids codes for an enzyme synthesizing idole-3–acetic acid. Eur J Biochem 138: 387–391 (1984).

133. Schäfer W, Görz A, Kahl G: T-DNA integration and expression in a monocot crop plant after induction of *Agrobacterium*. Nature 327: 529–532 (1987).

134. Shaw CH, Watson MD, Carter GH, Shaw CH: The right hand copy of the nopaline Ti-plasmid 25 bp repeat is required for tumour formation. Nucl Acids Res 12: 6031–6041 (1984).

135. Sheikholeslam SN, Weeks DP: Acetosyringone promotes high efficiency transformation of *Arabidopsis thaliana* explants by *Agrobacterium tumefaciens*. Plant Mol Biol 8: 291–298 (1987).

136. Shimoda N, Toyoda-Yamamoto A, Nagamine J, Usami S, Katayama M, Sakagami Y, Machida Y: Control of expression of Agrobacterium vir genes by synergistic actions of phenolic signal molecules and monosaccharides. Proc Natl Acad Sci USA 87: 6684–6688 (1990).

137. Sikorski RS, Michaud W, Levin HL, Boeke JD, Hieter P: Trans-kingdom promiscuity. Nature 345: 581–582 (1990).

138. Skoog F, Miller CO: Chemical regulation of growth and organ formation in plant tissues cultured in vitro. Symp Soc Exp Biol 11: 118–131 (1957).

139. Smit G, Logman TJJ, Boerrigter MET, Kijne JW, Lugtenberg BJJ: Purification and partial characterization of the *Rhizobium leguminosarum* biovar *viciae* Ca^{2+} -dependent adhesin, which mediates the first step in attachment of cells of the family Rhizobiaceae to plant root hair tips. J Bact 171: 4054–4062 (1989).

140. Smith CJS, Watson CF, Ray J, Bird CR, Morris PC, Schuch W, Grierson D: Antisense RNA inhibition of polygalacturonase gene expression in transgenic tomatoes. Nature 334: 724–726 (1988).

141. Smith EF, Townsend CO: A plant tumor of bacterial origin. Science 25: 671–673 (1907).

142. Song Y-N, Shibuya M, Ebizuka Y, Sankawa U: Synergistic action of phenolic signal compounds and carbohydrates in the induction of virulence gene expression in *Agrobacterium tumefaciens*. Chem Pharm Bull 39: 2613– 2616 (1991).

143. Song Y-N, Shibuya M, Ebrizuka Y, Sankawa U: Identification of plant factors inducing virulence gene expression in *Agrobacterium tumefaciens*. Chem Pharm Bull 39: 2347–2350 (1991).

144. Southern E: Detection of specific sequences among DNA fragments separated by gel electrophoresis. J Mol Biol 98: 503–518 (1975).

145. Spencer PA, Towers GHN: Specificity of signal compounds detected by *Agrobacterium tumefaciens*. Phytochemistry 27: 2781–2785 (1988).

146. Spencer PA, Towers GHN: Restricted occurrence of acetophenone signal compounds. Phytochemistry 30: 2933–2937 (1991).

147. Spielmann A, Simpson RB: T-DNA structure in transgenic tobacco plants with multiple independent integration sites. Mol Gen Genet 205: 34–41 (1986).

148. Stachel SE, Messens E, Van Montagu M, Zambryski P: Identification of the signal molecules produced by wounded plant cells that activate T-DNA transfer in *Agrobacterium tumefaciens*. Nature 318: 624–629 (1985).

149. Stachel SE, Timmerman B, Zambryski P: Generation of single-stranded T-DNA molecules during the initial stages of T-DNA transfer from *Agrobacterium tumefaciens* to plant cells. Nature 322: 706–712 (1986).

150. Stachel SE, Zambryski PC: *virA* and *virG* control the plant-induced activation of the T-DNA transfer process

of *Agrobacterium tumefaciens*. Cell 46: 325–333 (1986).

151. Steck TR, Close TJ, Kado CI: High levels of double-stranded transferred DNA (T-DNA) processing from an intact nopaline Ti plasmid. Proc Natl Acad Sci USA 86: 2133–2137 (1989).
152. Stief A, Winter DM, Strätling WH, Sippel AE: A nuclear DNA attachment element mediates elevated and position-independent gene activity. Nature 341: 343–345 (1989).
153. Stiekema WJ, Heidekamp F, Louwerse JD, Verhoeven HA, Dijkhuis P: Introduction of foreign genes into potato cultivars Bintje and Desirée using an *Agrobacterium tumefaciens* binary vector. Plant Cell Rep 7: 47–50 (1988).
154. Strader CD, Candelore MR, Hill WS, Sigal IS, Dixon RAF: Identification of two serine residues involved in agonist activation of the β-adrenergic receptor. J Biol Chem 264: 13572–13578 (1989).
155. Tamamoto S, Aoyama T, Takanami M, Oka A: Binding of the regulatory protein VirG to the phased signal sequences upstream from virulence genes on the hairy-root-inducing plasmid. J Mol Biol 215: 537–547 (1990).
156. Teeri TH, Herrera-Estrella L, Depicker A, Van Montagu M, Palva ET: Identification of plant promoters *in situ* by T-DNA mediated transcriptional fusions to the nptII gene. EMBO J 5: 1755–1760 (1986).
157. Teeri TH, Lehväslaiho H, Franck M, Uotila J, Heino P, Palva ET, Van Montagu M, Herrera-Estrella L: Gene fusions to *lacZ* reveal new expression patterns of chimeric genes in transgenic plants. EMBO J 8: 343–350 (1989).
158. Tempé J, Goldmann A: Occurrence and biosynthesis of opines. In: Kahl G, Schell J (eds) Molecular Biology of Plant Tumors, pp. 427–449. Academic Press, New York (1982).
159. Thomashow MF, Hugly S, Buchholz WG, Thomashow LS: Molecular basis for the auxin-independent phenotype of crown gall tumor tissues. Science 231: 616–618 (1986).
160. Thomashow MF, Karlinsey JE, Marks JR, Hurlbert RE: Identification of a new virulence locus in *Agrobacterium tumefaciens* that affects polysaccharide composition and plant cell attachment. J Bact 169: 3209–3216 (1987).
161. Thomashow MF, Nutter R, Montoya AL, Gordon MP, Nester EW: Integration and organization of Ti plasmid sequences in crown gall tumors. Cell 19: 729–739 (1980).
162. Thompson DV, Melchers LS, Idler KB, Schilperoort RA, Hooykaas PJJ: Analysis of the complete nucleotide sequence of the *Agrobacterium tumefaciens virB* operon. Nucl Acids Res 16: 4621–4636 (1988).
163. Timmerman B, Van Montagu M, Zambryski P: *vir* induced recombination in *Agrobacterium*. Physical characterization of precise and imprecise T-circle formation. J Mol Biol 203: 373–384 (1988).
164. Tinland B, Huss B, Paulus F, Bonnard G, Otten L: *Agrobacterium tumefaciens* 6b genes are strain-specific and affect the activity of auxin as well as cytokinin genes. Mol Gen Genet 219: 217–224 (1989).
165. Toro N, Datta A, Carmi OA, Young C, Prusti RK, Nester EW: The *Agrobacterium tumefaciens virC1* gene product binds to overdrive, a T-DNA transfer enhancer. J Bact 171: 6845–6849 (1989).
166. Turk S, Hooykaas PJJ: Effect of C-terminal deletions on the VirA function (in preparation).
167. Turk SCHJ, Melchers LS, den Dulk-Ras H, Regensburg-Tuïnk AJG, Hooykaas PJJ: Environmental conditions differentially affect *vir* gene induction in different *Agrobacterium* strains. Role of the *Vir*A sensor protein. Plant Mol Biol 16: 1051–1059 (1991).
168. van Haaren MJJ, Pronk JT, Schilperoort RA, Hooykaas PJJ: Functional analysis of the *Agrobacterium tumefaciens* octopine Ti-plasmid left and right T-region border fragments. Plant Mol Biol 8: 95–104 (1987).
169. van Haaren MJJ, Sedee NJA, de Boer HA, Schilperoort RA, Hooykaas PJJ: Mutational analysis of the conserved domains of a T-region border repeat of *Agrobacterium tumefaciens*. Plant Mol Biol 13: 523–531 (1989).
170. van Haaren MJJ, Sedee NJA, Schilperoort RA, Hooykaas PJJ: Overdrive is a T-region transfer enhancer which stimulates T-strand production in *Agrobacterium tumefaciens*. Nucl Acids Res 15: 8983–8997 (1987).
171. van Larebeke N, Genetello C, Schell J, Schilperoort RA, Hermans AK, Hernalsteens JP, Van Montagu M: Acquisition of tumour-inducing ability by non-oncogenic agrobacteria as a result of plasmid transfer. Nature 255: 742–743 (1975).
172. van Veen RJM, den Dulk-Ras H, Bisseling T, Schilperoort RA, Hooykaas PJJ: Grown gall tumor and root nodule formation by the bacterium *Phyllobacterium myrsinacearum* after the introduction of an *Agrobacterium* Ti plasmid or a *Rhizobium* Sym plasmid. Mol Plant-Microbe Interact 1: 231–234 (1988).
173. van den Elzen PJM, Townsend J, Lee KY, Bedbrook JR: A chimaeric hygromycin resistance gene as a selectable marker in plant cells. Plant Mol Biol 5: 299–302 (1985).
174. van der Graaff E, Hooykaas PJJ: Construction of special-purpose T-DNA vectors (in preparation).
175. van der Krol AR, Lenting PE, Veenstra J, van der Meer IM, Koes RE, Gerats AGM, Mol JNM, Stuitje AR: An anti-sense chalcone synthase gene in transgenic plants inhibits flower pigmentation. Nature 333: 866–869 (1988).
176. van Slogteren GMS, Hooykaas PJJ, Schilperoort RA. Silent T-DNA genes in plant lines transformed by *Agrobacterium tumefaciens* are activated by grafting and by 5–azacytidine treatment. Plant Mol Biol 3: 333–336 (1984).
177. van Slogteren GMS, Hoge JHC, Hooykaas PJJ, Schilperoort RA: Clonal analysis of heterogeneous crown gall

tumor tissues induced by wildtype and shooter mutant strains of *Agrobacterium tumefaciens* expression of T-DNA genes. Plant Mol Biol 2: 321–333 (1983).

178. Vancanneyt G, Schmidt R, O'Conner-Sanchez A, Willmitzer L, Rocha-Sosa M: Construction of an intron-containing marker gene: splicing of the intron in transgenic plants and its use in monitoring early events in *Agrobacterium* mediated plant transformation. Mol Gen Genet 220: 245–250 (1990).
179. Visser RGF, Somhorst I, Kuipers GJ, Ruys NJ, Feenstra WJ, Jacobsen E: Inhibition of the expression of the gene for granule-bound starch synthase in potato by antisense constructs. Mol Gen Genet 225: 289–296 (1991).
180. Von Schaewen A, Stitt M, Schmidt R, Sonnewald U, Willmitzer L: Expression of a yeast-derived invertase in the cell wall of tobacco and *Arabidopsis* plants leads to accumulation of carbohydrate and inhibition of photosynthesis and strongly influences growth and phenotype of transgenic tobacco plants. EMBO J 9: 3033–3044 (1990).
181. Walden R, Hayashi H, Schell J: T-DNA as a gene tag. Plant J 1: 281–288 (1991).
182. Wan-yin D, Xiao-ying L, Qi-quan S: *Agrobacterium tumefaciens* can transform *Triticum aestivum* and *Hordeum vulgare* of Gramineae. Science China Ser B 33: 27–34 (1990).
183. Ward ER, Barnes WM: VirD2 protein of *Agrobacterium tumefaciens* very tightly linked to the 5′ end of T-strand DNA. Science 242: 927–930 (1988).
184. Ward JE, Dale EM, Nester EW, Binns AN: Identification of a VirB10 protein aggregate in the inner membrane of *Agrobacterium tumefaciens*. J Bact 172: 5200–5210 (1990).
184a. Ward JE, Akiyoshi DE, Reiger D, Dalta A, Gordon MP, Nester EW: Characterization of the virB operon from an *Agrobacterium tumefaciens* Ti plasmid. J Biol Chem 263: 5804–5814 (1988).
185. Willetts N, Wilkins B: Processing of plasmid DNA during bacterial conjugation. Microbiol Rev 48: 24–41 (1984).
186. Willmitzer L, Simons G, Schell J: The TL-DNA in octopine crown-gall tumours codes for seven well-defined polyadenylated transcripts. EMBO J 1: 139–146 (1982).
187. Winans SC: An *Agrobacterium* two-component regulatory system for the detection of chemicals released from plant wounds. Mol Microbiol 5: 2345–2350 (1991).
188. Winans SC, Ebert PR, Stachel SE, Gordon MP, Nester EW: A gene essential for *Agrobacterium* virulence is homologous to a family of positive regulatory loci. Proc Natl Acad Sci USA 83: 8278–8282 (1986).
189. Winans SC, Kerstetter RA, Ward JE, Nester EW: A protein required for transcriptional regulation of *Agrobacterium* virulence genes spans the cytoplasmic membrane. J Bact 171: 1616–1622 (1989).
190. Yadav NS, Vanderleyden J, Bennett DR, Barnes WM, Chilton M-D: Short direct repeats flank the T-DNA on a nopaline Ti plasmid. Proc Natl Acad Sci USA 79: 6322–6326 (1982).
191. Yamada T, Palm CJ, Brooks B, Kosuge T: Nucleotide sequences of the *Pseudomonas savastanoi* indole acetic acid gene show homology with *Agrobacterium tumefaciens* T-DNA. Proc Natl Acad Sci USA 82: 6522–6526 (1985).
192. Yamashita T, Tida A, Morikawa H: Evidence that more than 90% of β-glucuronidase-expressing cells after particle bombardment directly receive the foreign gene in their nucleus. Plant Physiol. 97: 829–831 (1991).
193. Yanofsky MF, Ma H, Bowman JL, Drews GN, Feldmann KA, Meyerowitz EM: The protein encoded by the *Arabidopsis* homeotic gene *agamous* resembles transcription factors. Nature 346: 35–39 (1990).
194. Yanofsky MF, Porter SG, Young C, Albright LM, Gordon MP, Nester EW: The *vir*D operon of *Agrobacterium tumefaciens* encodes a site-specific andonuclease. Cell 47: 471–477 (1986).
195. Zaenen I, Van Larebeke N, Teuchy H, Van Montagu M, Schell J: Supercoiled circular DNA in crown-gall inducing *Agrobacterium* strains. J Mol Biol 86: 109–127 (1974).
196. Zambryski P, Joos H, Genetello C, Leemans J, Van Montagu M, Schell J: Ti plasmid vector for the introduction of DNA into plant cells without alteration of their normal regeneration capacity. EMBO J 2: 2143–2150 (1983).
197. Zerback R, Dressler K, Hess D: Flavonoid compounds from pollen and stigma of *Petunia hybrida*: inducers of the *vir* region of the *Agrobacterium tumefaciens* Ti plasmid. Plant Sci 62: 83–91 (1989).
198. Zhan X, Jones DA, Kerr A: The pTiC58 *tzs* gene promotes high-efficiency root induction by agropine strain 1855 of *Agrobacterium rhizogenes*. Plant Mol Biol 14: 785–792 (1990).
199. Ziegelin G, Pansegrau W, Strack B, Balzer D, Kroger M, Kruft V, Lanka E: Nucleotide sequence and organization of genes flanking the transfer origin of promiscuous plasmid RP4. DNA Sequence 1: 303–327 (1991).
200. Zorreguieta A, Ugalde RA: Formation in *Rhizobium* and *Agrobacterium* spp. of a 235–Kilodalton protein intermediate in β-*d*(1–2)glucan synthesis. J Bact 167: 947–951 (1986).

Plant Molecular Biology **19**: 39–49, 1992.
© 1992 *Kluwer Academic Publishers. Printed in Belgium.*

Plant-transposable elements and gene tagging

Alfons Gierl and Heinz Saedler*
Max-Planck-Institut für Züchtungsforschung, Carl-von-Linné-Weg 10, D-5000 Köln 30, Germany
(author for correspondence)*

Introduction

While most genes are lined up on chromosomes like pearls on a string, one group of genes, so-called transposons or transposable elements, are highly mobile. Transposons frequently alter their chromosomal location and consequently often induce mutations, by integrating into other genes and destroying their structural integrity. An additional feature of transposon-induced mutations is their somatic instability. It is by this property that the presence of transposable elements is recognized. While the organism develops, the transposable element leaves its site of integration within the gene and hence the function of that gene is restored in this cell as well as in its progeny. These excision events result in a mosaic tissue of mutant and revertant (wild-type) cells. Such mosaic or variegation patterns strike the eye, provided genes involved in coloration were affected by the transposable element.

One of the first reports on colour variegation in a plant was given in 1588 by Jacob Theodor from Bergzabern, a village south of Strasbourg. He described herbal colour variegation in kernels of *Zea mays*. Since then patterns of variegation have been reported for many plant species, though not all of these patterns reflect the action of transposable elements. Cases of mineral deficiencies, viral infections or non-Mendelian transmissions of variegation patterns, and especially the influence of DNA methylation during somatic development [39], have to be excluded. Therefore, Mendelian inheritance is an important criterion to identify a variegation pattern resulting from transposable elements, though other cases have been reported [26, 34, 56].

Based on the inheritance of instability, Barbara McClintock [47, 48] developed the concept of transposable elements as mobile genetic entities. She used the term 'controlling elements', since they seemed capable of influencing the expression of a given locus when integrated into or nearby. By the criteria of inheritance instability, transposable elements have been described in at least 35 mono- and dicotyledonous plant species. A list of these is given in Nevers *et al.* [56].

Given the concept of transposable elements, it was not until similar elements were detected in bacteria [reviewed in 75] that molecular analysis of such elements became feasible, first in bacteria [32] and more than a decade later in plants by a few laboratories (Burr, Fedoroff, Peacock, Starlinger, Saedler). In the past 10 years numerous transposable element systems, mostly from *Zea mays* and *Antirrhinum majus*, have been studied molecularly. Two distinct classes of transposable elements have emerged that transpose by different mechanisms. While in the case of retro-transposons a copy of the element seems to move via an RNA intermediate, classical transposable elements transpose via excision and re-integration directly from DNA to DNA.

Since not many active retro-transposons have been described in plants, they are excluded from this review (the reader is referred to Grandbastian [22] for details), whereas we wish to concentrate on classical transposable elements, even though numerous reviews on this subject have appeared [7, 13, 17, 21, 56, 61]. Molecular results for two transposable element systems have, however, indicated common features for their mechanism of transposition. Transposition is the result of an interaction of proteins (*trans*-acting factors) with the termini (*cis* determinants) of transposable elements. In the first part, therefore, the *cis*/*trans* requirements for transposition are summarized, using the Ac and En/Spm elements of *Zea mays* as examples. Not only are these elements relatively well characterized, but they also represent prototypes of transposon families that have been found throughout in the plant kingdom. In addition, due to its different structure, the Mu element of *Zea mays* is briefly considered. In the second part the different strategies using transposable elements as tools for the isolation of genes are described.

Structural and functional features of Ac, En/Spm and Mu

The Ac and En/Spm element families each include autonomous and non-autonomous elements. The autonomous elements encode functions required for their own transposition. The non-autonomous elements (receptor or defective elements) are often internal deletion derivatives of the autonomous elements. Defective elements that have retained the *cis* determinants are still mobile, if the functions expressed by the autonomous element of the same family are provided in *trans*. This combination of autonomous and non-autonomous elements is referred to as a two-component system. As described below, the reconstitution of these two-component systems in transgenic tobacco provided the possibility to identify the *trans*-acting factors encoded by the autonomous elements and to define the *cis* determinants for excision. In addition, the functions of the proteins encoded by Ac and En/Spm were analysed by *in vitro* DNA binding studies.

The Ac transposable element family

The Activator (Ac) element [47] of *Zea mays* represents a relatively simple transposable element system (Fig. 1). The autonomous Ac element is 4565 bp long [54, 62] and expresses one mRNA of 3.5 kb [36]. This transcript is produced at a low level [17] and it is assumed that the 92 kDa protein is the only product encoded by Ac. This product, termed Ac transposase, is sufficient for transposition of Ac in a growing number of transgenic heterologous plant species [for review 1, 25]. Interestingly, 101 amino acids can be removed from the N terminus of the protein without impairing excision [38]. The *cis* determinants for transposition of Ac are represented by about 200 bp of each end [10]. Mutational alterations within these regions, reduce or abolish excision of Ac elements. DNA-binding tests showed that the Ac transposase binds to the hexamer

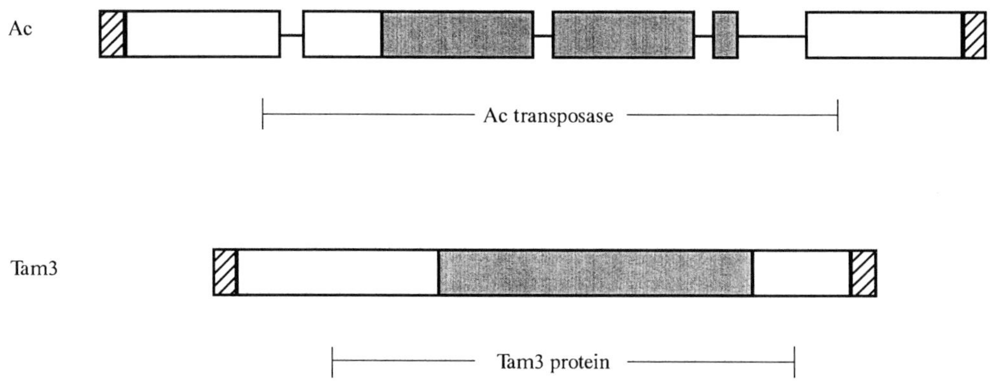

Fig. 1. Ac of *Zea mays* and Tam3 of *Antirrhinum majus* belong to one family of transposable elements that have similar TIRs and generate a 8 bp target site duplication upon insertion. About 200 bp (hatched boxes) of each terminus represent the *cis* determinants required for excision. Ac transposase is encoded by five exons. The protein encoded by Tam3 shares significant similarity with Ac transposase (shaded areas), its coding region, however, is not interrupted by introns.

motif AAACGG that occurs in clusters within these regions [35]. The other *cis* determinant that is clearly defined and absolutely required for excision [10, 29] is represented by the 11 bp terminal inverted repeats (TIRs) which seem not to be recognized by Ac transposase. The asymmetry of the two Ac ends is important with respect to excision. Ac elements with identical ends (two 5′ or two 3′ ends) are defective in excision [10].

The defective elements of the Ac family, termed Dissociation (Ds) [46], form a rather heterogeneous group. Common to all Ds elements is the 11 bp TIR. Some Ds elements are internal deletion derivatives of Ac. Other elements, such as Ds2 [52], share only rudimentary homology with Ac or have little more than the 11 bp TIR and one or two binding motifs for transposase in common with Ac, as in the case of Ds1 [76]. Other elements are more complex derivatives of Ac sequences and the so-called 'double Ds' element [14] seems to induce Ac-dependent chromosome breakage, by which the Ac-Ds family was identified [46].

Ac seems to be the prototype of a family of elements whose members are widely distributed in plants (Table 1). These members have similar TIRs and usually generate an eight nucleotide target site duplication upon integration. However, mainly defective elements have been isolated and characterized from various plant species, except for the autonomous elements Bg from *Zea mays* [27] and Tam3 from *Antirrhinum majus* [28]. A stretch of 520 amino acids of the putative protein encoded by Tam3 is about 30% identical to Ac transposase, and patches of much higher homology are included in this region (Fig. 1) [28]. This probably indicates that both proteins share a common function. In addition, substantial amino acid similarity of these two proteins with a protein encoded by a hobo element of *Drosophila* was reported [3] and it was suggested that this conservation represents an example of horizontal transmission of genetic information between plants and animals [3].

The En/Spm transposable element family

In contrast to Ac, the 8287 bp long autonomous Enhancer [59, 60] or Suppressor-mutator element [48] (En/Spm) encodes at least two functional products (Fig. 2), which are derived by alternative splicing from a precursor transcript [43]. The two proteins have been termed TNPA and TNPD [20, 43]. TNPA is 67 kDa and, although expressed at a similar low level as Ac transposase, is about 100 times more abundant than the 131 kDa TNPD protein [20]. Both proteins are absolutely required for transposition of En/Spm [18, 44]. About 200 bp of the 5′ end and 300 bp of the 3′ end of En/Spm represent the *cis* determinants for excision [for review, 21]. Well contained within these regions are several reiterations of a 12 bp sequence motif that is recognized by

Table 1. The Ac transposable element family.

Element[a]	TIR[b]	TSD[c]	Size[d]	Species	Ref.
Ac	$\frac{C}{T}$AGGGATGAAA	8	4563	*Z. mays*	54, 62
Bg	CAGGG	8	4869	*Z. mays*	27
Tam3	TAAAGATGTGAA	8	3629	*A. majus*	28
Tpc1	TAGGGTGTAAA	8	927	*P. crispum*	30
Ips-r	TAGGGGTGGCAA	8	0.8 kb	*P. sativum*	2
Tst1	CAGGGGCGTAT	8	736	*S. tuberosum*	33

[a] Autonomous elements are in bold type.
[2] Terminal inverted repeat.
[3] Target site duplication upon insertion in bp.
[4] Sizes in kb refer to elements that have not been completely sequenced.

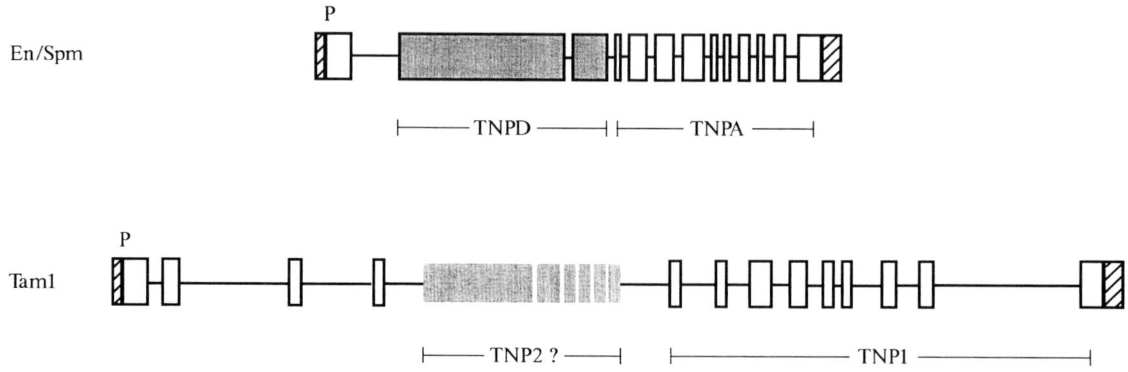

Fig. 2. En/Spm of *Zea mays* and Tam1 of *Antirrhinum majus* are members of the so-called CACTA family of transposable elements. These elements have very similar TIRs and generate a 3 bp target site duplication upon insertion. About 200 bp of the 5′ end 300 bp of the 3′ end (hatched boxes) represent the *cis* determinants required for excision. A single pre-mRNA is initiated at the promotor (P) of En/Spm and is differentially spliced into *tnpA* and *tnpD* mRNA. The exons encoding TNPA (open boxes) and TNPD (shaded boxes) protein products are indicated. Tam1 probably also encodes two proteins and has a similar structure like En/Spm. Tam1 contains in the center region open reading frames (shaded boxes) that encode the putative protein TNP2. TNP2 and TNPD share significant similarity at the amino acid level. In contrast, TNPA and TNP1 are not homologous at the sequence level, but may be functionally equivalent.

the TNPA protein [18]. Six motifs are present at the 5′ end and eight at the 3′ end. The two 13 bp TIRs of En/Spm are the other *cis*-acting sequences that are absolutely required for excision. Similar to Ac, the asymmetry at the two ends is also required for En/Spm. An element with two 5′ ends does not excise (Reinecke, unpublished). A possible autoregulatory role for TNPA has been suggested, because the innermost TNPA-binding motif at the 5′ end overlaps with the 'TATA' box of the single En/Spm promotor [20]. It has been speculated, from a comparison with En/Spm-related elements (see below) from other plant species, that TNPD interacts with 13 bp TIRs and in fact may accomplish endonucleolytic cleavage at the element's ends [18, 55].

En/Spm belongs to the so-called 'CACTA' family [5] of elements (Table 2). These elements produce a 3 bp target site duplication upon insertion and share nearly identical 13 bp TIRs. In addition, when En/Spm is compared to the autonomous element Tam1 from *A. majus*, the

Table 2. The CACTA transposable element family.

Element[a]	TIR[b]	TSD[c]	Size[d]	Species	Ref.
En/Spm	CACTACAAGAAAA	3	8287	*Z. mays*	59
MpI1	CACTACCGGAATT	3	9 kb	*Z. mays*	80
Tam1	CACTACAACAAAA	3	15 164	*A. majus*	55
Tam2	CACTACAACAAAA	3	5187	*A. majus*	34
Tam4	CACTACAAAAAAA	3	4329	*A. majus*	40
Tam7	CACTACAACAAAA	3	7 kb	*A. majus*	e
Tam8	CACTACAACAAAA	3	3 kb	*A. majus*	e
Tam9	CACTACAACAAAA	3	5.5 kb	*A. majus*	e
Tgm	CACTATTAGAAAA	3	1.6–12 kb	*G. max*	64
Pis1	CACTACGCCAAA	3	2.5 kb	*P. sativum*	73

[a] to [d], see Table 1.
[e] Zs. Schwarz-Sommer and H. Sommer, personal communication.

structural organization of the two elements appears to be strikingly similar (Fig. 2). Tam1 encodes also two products, termed TNP1 and TNP2 [55]. The putative TNP2 protein is about 45% identical to TNPD at the amino acid level. Another element belonging to the CACTA family is Tgm1 from *Glycine max*. However, no autonomous Tgm element has yet been reported. Nevertheless, an open reading frame was found in some Tgm elements that shares a similar amount of homology (at the amino acid level) to TNPD [64] and TNP2 [55]. Since the 13 bp TIRs of En/Spm, Tam1 and Tgm are very similar, it was hypothesized that the conserved protein interacts with the conserved TIRs [18, 55].

In contrast, TNPA of En/Spm and TNP1 of Tam1 share no homology. Nonetheless, these proteins might have a similar function. TNPA recognizes a 12 bp motif clustered at the subtermini of En/Spm. Tam1 has a similar structure at its ends [55]. There is preliminary evidence that TNP1 binds to the sequence motif that is repeated in the subterminal regions of Tam1 (Trentmann, personal communication).

The Mu transposable element family

While Ac and En/Spm are characterized by short TIR, the Mutator (Mu) [65] elements from *Zea mays* have long TIRs of approximately 200 to 500 bp. Nine different Mu elements have been isolated [for review 77] that share the first 200 bp of the TIRs, but are heterogeneous in size and contain more or less unrelated internal sequences. These Mu elements seem to be defective elements that encode no functional product. Upon insertion Mu elements typically generate a 9 bp target site duplication. The inheritance of Mutator activity seems to be non-Mendelian [65]. Therefore, the isolation of an autonomous Mu element has been hampered, and no information about the *cis*/*trans* requirements is yet available. By repeated cycles of outcrossing to a Mu inactive line, however, it was recently possible to isolate Mu activity at a single locus [66, 70]. Therefore, cloning of the autonomous Mu element will probably be accomplished soon. Some encouraging candidates have been published [63].

Mechanism of transposition

Transposition of Ac, En/Spm and Mu probably occurs via excision and integration. One essential requirement for such a mechanism is healing of the chromosome breaks that are generated during excision, thereby avoiding chromosomal loss. Healing of chromosomal breaks is most easily envisaged, if the distant ends of the transposable element come into close physical proximity. This would allow excision of the element and rejoining of the broken chromosome. Two functions have therefore been postulated [16]: one that is required for association of the two ends of the transposable element, and one that cuts close to the ends of the element.

With respect to excision, the transposable elements Ac and En/Spm have three features in common:

1) an element encoded protein that binds to subterminal *cis* determinants, which
2) have to be located asymmetrically within the ends, and
3) TIRs that are probably the substrates for endonucleolytic cleavage.

Based on these features a model for excision of elements like En/Spm and Ac has been proposed [16]. According to this hypothesis the element-encoded function that binds to the subterminal binding motifs (TNPA or Ac transposase) acts like a 'glue' in complex formation of the two ends. The asymmetric nature of the two ends leads to correct positioning of the TIRs by a 'zipper'-like mechanism. The TIRs in this complex are then recognized by an endonucleolytic activity that cuts close to the element's ends resulting in the release of the element. The endonuclease could either be encoded by the element itself, as is the case with the En/Spm-encoded TNPD, or be a cellular protein, which is more likely for the Ac element.

Due to the long TIRs of Mu, the ends of this element are symmetrical. Hence, the mechanism that leads to a putative association of the Mu ends should be different from that of Ac and En/Spm.

Soon after the first excision products (empty donor sites) of plant-transposable elements were analysed [5, 62, 67, 79], it became apparent that the wild-type sequence is usually not precisely re-established after excision. Therefore gene function may be restored to varying degrees, ranging from zero to full activity. In a systematic analysis of En/Spm excisions, Schwarz-Sommer *et al.* [71] found that only one in nine excision events was precise. The analysis of these altered sequences, so-called footprints, led to the formulation of excision models [6, 7, 68].

According to the model of Saedler and Nevers [68], staggered cuts are introduced at the ends of the target site duplication by an element-specific endonuclease. Footprints then result from the action of cellular DNA repair enzymes (exonuclease, DNA polymerase, ligase), which act on the protruding single-stranded fringes at the excision site. A similar staggered cut is formed during integration of the element by the same endonuclease causing the target site duplication, whose length is specific for different families of elements (Tables 1, 2).

In the model of Coen *et al.* [4, 5] the endonucleolytic activity for excision and integration is different. A staggered cut is only formed during integration, while for excision cutting occurs more precisely at the ends of the element. Excision generates hairpin structures at the broken chromosomal ends. Footprints are generated by the resolution of these hairpins and ligation of the resulting products.

It is probably necessary to develop an *in vitro* assay system to allow biochemical dissection of the cutting reaction in order to prove or disprove the models described above.

Transposons as tools for the isolation of genes

Once a transposable element has been isolated molecularly, it can be used as a probe to clone genes that are mutated by insertion of this element. First, the mutation caused by the transposon has to be identified in a genetic screening procedure. Subsequently the gene can be isolated molecularly by cloning the DNA sequences flanking the transposable element insertion. This technique, called transposon tagging, has been widely used to isolate genes from *Zea mays* and *Antirrhinum majus* (Table 3). It requires no knowledge about the nature of the product of the tagged gene, it rather depends only upon the expression of a mutant phenotype. Therefore, genes involved in development, pathogen resistance and certain physiological and biochemical processes have been isolated by transposon tagging or are the subject of current isolation programmes. Two approaches are used: targeted and non-targeted transposon tagging.

Targeted transposon tagging

If only one particular gene is desired, then targeted tagging seems to be the procedure of choice.

Table 3. Genes cloned or identified with transposable elements.

Gene	Function	Element	Ref.
Zea mays			
A1	NADPH-dep. reductase	En, Mu	57
A2	athocyanin pathway	En, rcy	51
Bz1	uDP-glycosyltransferase	Ac	16
Bz2	anthocyanin pathway	Ds2, Mu	50, 78
C1	regulatory gene	Spm, En	9, 58
C2	chalcone synthase	Spm	81
hcf-106	chloroplast development	Mu	41
Kn1	regulatory gene	Ds2	24
02	regulatory gene	Ac, Spm	53, 69
P	regulatory gene	Ac	37
R	regulatory gene	Ac	11
Vp1	regulatory gene	Mu	45
A. majus			
deficiens	regulatory gene	Tam7	74
delila	regulatory gene	Tam2	40
floricaula	regulatory gene	Tam3	8
globosa	regulatory gene	Tam7, Tam9	72
incolorata	anthocyanin pathway	Tam1	40
olive	chloroplast development	Tam3	40
pallida	NADPH-dep. reductase	Tam3	42

Wild-type plants containing active transposable elements are crossed with plants that are homozygous recessive for the gene in question. The majority of the F1 progeny will exhibit wild-type phenotype, since these plants are heterozygous for this gene. In exceptional cases, however, a transposable element in the wild-type parent will have been inserted into the gene and be transmitted in the gametes. These rare events are then directly uncovered in the F1 generation by individuals which exhibit the mutant phenotype. The frequency of mutations caused by transposable elements may vary. It depends mainly on the nature of the active element system used and on the position of the element(s) in the chromosome. Common belief is that short intra-chromosomal transpositions occur at higher rates than inter-chromosomal transpositions [12, 23, 31]. In spite of these peculiarities the insertional frequencies for most genes are in the range of 10^{-4} to 10^{-5}. Therefore, although targeted tagging leads to a relatively direct identification of a particular gene, it requires a very large number of crosses, and a large population of plants must be screened in order to isolate a single gene.

Non-targeted transposon tagging

If instead of a particular gene, a group of genes that belong for example to a particular physiological pathway or to a developmental programme are to be isolated, non-targeted tagging (Fig. 3) is an alternative. This strategy requires an extra generation in order to identify putative transposon-induced mutations. In the first step a plant population is generated that contains active transposable elements. Many individuals of this population will contain a transposable element at a new chromosomal location [4, 40]. In addition, it can be tried to enrich for such transposition events by selecting germinal revertants of a trans-

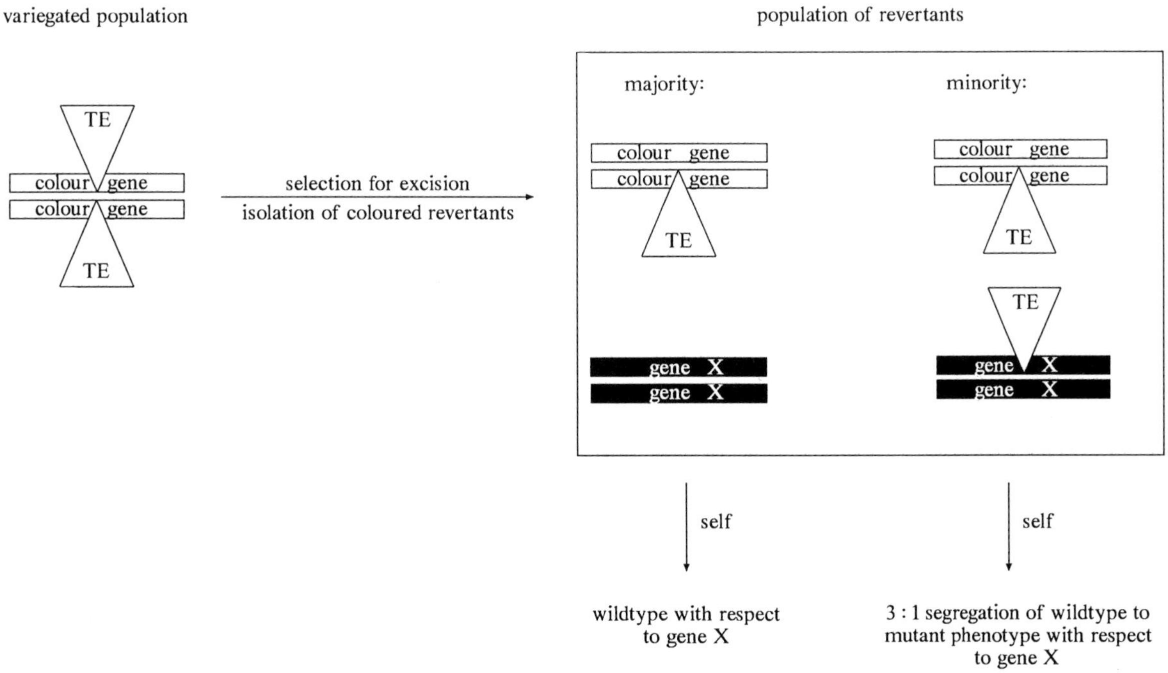

Fig. 3. Non-targeted transposon tagging. The first step in this procedure is the isolation of germinal transposon excision events in order to enrich for transposon integrations at new chromosomal locations. With a certain probability any given gene *X* is a target for the transposable element. Germinal revertants are very easily recognized if the element resides for example in a gene required for flower pigmentation. Insertion of the transposable element into gene *X* is uncovered upon self-fertilization of the population and segregation of the mutant phenotype in the progeny.

posable element integrated at a known locus are isolated. The rationale is that such liberated transposons are now available for reintegration into any new locus. In a diploid organism the resulting mutant phenotype of a recessive mutation, however, would only become apparent in the M_2 generation upon selfing of the individuals of the population. Hence, new phenotypes will segregate in such selfed populations. The frequency with which new phenotypes in a preselection programme are observed can be rather high (up to 1%, Lönning, personal communication) and depends on the number of genes involved in the pathway as well as on the transposons and their locations used.

Independent of the tagging strategy used, the mutants isolated from a tagging programme have to be characterized. The question to be answered is which transposable element insert has caused the mutation? This can be difficult, since autonomous as well as defective elements can transpose. In addition, species like *Z. mays* and *A. majus* contain several different active transposable element systems and these elements occur in more than one copy per genome. In any case, cosegregation of a transposon copy with the mutant phenotype has to be demonstrated. This is done by analysing DNA from the progenitor, mutant and revertant plants in genomic Southern experiments. The success of gene tagging therefore increases with the extent to which the different, potentially mobile transposable elements of a given organism are molecularly characterized. Aside from using transposable elements in the isolation of genes, unstable mutations have been very useful in the process of defining and identifying genes that were isolated by conventional means.

The maize elements Ac and En/Spm are also being used to extend transposon tagging to plant species in which no suitable elements have been identified or exist [for review, see 1, 23]. This is possible because Ac and En/Spm have been shown to transpose in a growing number of plant species [23]. Attempts are now in progress to engineer these systems to achieve the best conditions for tagging. One advantage of gene tagging in heterologous species is that the transposon introduced will be present in a low copy number (1 to 5 copies). This should make the identification of mutants much easier. The progress made in this area, especially for Ac, will probably soon lead to the first isolation of a gene using a transposable element in a heterologous plant system.

Acknowledgements

We would like to thank George Coupland and Robert Masterson for critical reading of the manuscript.

References

1. Balcells L, Swinburne J, Coupland G: Transposons as tools for the isolation of genes. Trends Biotechnol 9: 31–37 (1991).
2. Bhattacharyya MK, Smith AM, Ellis THN, Hedley C, Martin C: The wrinkled-seed character of pea described by Mendel is caused by a transposon-like insert in a gene encoding starch-branching enzyme. Cell 60: 115–120 (1990).
3. Calvi BR, Hong TJ, Findley SD, Gelbart WM: Evidence for a common evolutionary origin of inverted repeat transposons in *Drosophila* and plants: hobo, Activator, and Tam3. Cell 66: 465–471 (1991).
4. Carpenter R, Coen ES: Floral homeotic mutations produced by transposon-mutagenesis in *Antirrhinum majus*. Genes Devel 4: 1483–1493 (1990).
5. Bonas U, Sommer H, Saedler H: The 17 kb Tam1 element of *Antirrhinum majus* induces a 3 bp duplication upon integration into chalcon synthase gene. EMBO J 3: 1015–1019 (1984).
6. Coen ES, Carpenter R, Martin C: Transposable elements generate novel spatial patterns of gene expression in *Antirrhinum majus*. Cell 47: 285–296 (1986).
7. Coen ES, Robbins TP, Almeida J, Hudson A, Carpenter R: Consequences and mechanisms of transposition in *Antirrhinum majus*. In: Berg DE, Howe MM (eds) Mobile DNA, pp. 413–437. American Society for Microbiology (1989).
8. Coen ES, Romero JM, Doyle S, Elliott R, Murphy G, Carpenter R: Floricaula: a homeotic gene required for flower development in *Antirrhinum majus*. Cell 63: 1311–1322 (1990).
9. Cone KC, Burr F, Burr B: Molecular analysis of the maize regulatory locus C1. Proc Natl Acad Sci USA 83: 9631–9635 (1986).
10. Coupland G, Plum C, Chatterjee S, Post A, Starlinger P:

Sequences near the termini are required for transposition of the maize transposon Ac in transgenic tobacco plants. Proc Natl Acad Sci USA 86: 9385–9388 (1989).
11. Dellaporta SL, Greenblatt I, Kermicle J, Hicks JB, Wessler S: Molecular cloning and structural characterization of the maize R-nj allele. Stadler Symp 18: 262–282 (1987).
12. Dooner HK, Belachew A: Transposition pattern of the maize element Ac from the bz-m2 (Ac) allele. Genetics 113: 1021–1036 (1989).
13. Döring H-P, Starlinger P: Molecular genetics of transposable elements in plants. Annu Rev Genet 20: 175–200 (1986).
14. Döring H-P, Nelsen-Salz B, Garber R, Tillmann E: Double Ds elements are involved in specific chromosome breakage. Mol Gen Genet 219: 299–305 (1989).
15. Fincham JRS, Sastry GRK: Controlling elements in maize. Annu Rev Genet 8: 15–50 (1974).
16. Fedoroff N, Furtek D, Nelson OE: Cloning of the bronze locus in maize by a simple and generizable procedure using the transposable controlling element Ac. Proc Nat Acad Sci USA 81: 3825–3829 (1984).
17. Fedoroff N: Maize transposable elements. In: Berg DE, Howe MM (eds) Mobile DNA, pp. 375–411. American Society for Microbiology (1989).
18. Frey M, Reinicke J, Grant S, Saedler H, Gierl A: Excision of En/Spm transposable element of *Zea mays* requires two element-encoded proteins. EMBO J 9: 4037–4044 (1990).
19. Fusswinkel H, Schein S, Courage U, Starlinger P, Kunze R: Detection and abundance of mRNA and protein encoded by transposable element activator (Ac) in maize. Mol Gen Genet 225: 186–192 (1991).
20. Gierl A, Lütticke S, Saedler H: TnpA product encoded by the transposable element En-1 of *Zea mays* is a DNA binding protein. EMBO J 7: 4045–4053 (1988).
21. Gierl A, Saedler H, Peterson PA: Maize transposable elements. Annu Rev Genet 23: 71–85 (1989).
22. Grandbastian M-A: Retroelements in higher plants. Trends Genet, in press (1991).
23. Greenblatt I: A chromosome replication pattern deduced from pericarp phenotypes resulting from movement of the transposable element Modulator in maize. Genetics 108: 471–485 (1984).
24. Hake S, Vollbrecht EV, Freeling M: Cloning of knotted, the dominant morphological mutant in maize using Ds2 as a transposon tag. EMBO J 8: 15–22 (1989).
25. Haring MA, Caius MT, Rommens H, Nijkamp HJJ, Hille J: The use of transgenic plants to understand transposition mechanisms and to develop transposon tagging strategies. Plant Mol Biol 16: 449–461 (1991).
26. Harrison BJ, Carpenter R: A comparison of the instabilities at the nivea and pallida loci in *Antirrhinum majus*. Heredity 31: 309–323 (1973).
27. Hartings H, Spilmont C, Lazzaroni N, Rossi V, Salamini F, Thompson RD, Motto M: Molecular analysis of the Bg-rbg transposable element system of *Zea mays* L. Mol Gen Genet 227: 91–96 (1991).
28. Hehl R, Nacken WKF, Krause A, Saedler H, Sommer H: Structural analysis of Tam3, a transposable element from *Antirrhinum majus*, reveals homologies to the Ac element from maize. Plant Mol Biol 16: 369–371 (1991).
29. Hehl R, Baker B: Induced transposition of Ds by a stable Ac in crosses of transgenic tobacco plants. Mol Gen Genet 217: 53–59 (1989).
30. Herrmann A, Schulz W, Hahlbrock K: Two alleles of the single copy chalcon synthase gene in parsley differ by a transposon-like element. Mol Gen Genet 212: 93–98 (1988).
31. Jones JDG, Cerland F, Lim E, Ralston E, Dooner HK: Preferential transposition of the maize element Activator to linked chromosomal locations in tobacco. Plant Cell 2: 701–707 (1990).
32. Jordan E, Saedler H, Starlinger P: O^0 and strong polar mutations in the gal operon are insertions. Mol Gen Genet 102: 353–363 (1968).
33. Köster-Töpfer M, Frommer WB, Rocha-Sosa M, Willmitzer L: Presence of a transposon-like element in the promoter region of an inactive patatin gene in *Solanum tuberosum* L. Plant Mol Biol 14: 239–247 (1990).
34. Krebbers E, Hehl R, Pioterowiak R, Lönnig E-E, Sommer H, Saedler H: Molecular analysis of paramutant plants of *Antirrhinum majus* and the involvement of transposable elements. Mol Gen Genet 209: 499–507 (1987).
35. Kunze R, Starlinger P: The putative transposase of transposable element Ac from *Zea mays* interacts with subterminal sequences of Ac. Embo J 8: 3177–3185 (1989).
36. Kunze R, Stochaj U, Lauf J, Starlinger P: Transcription of transposable element activator (Ac) of *Zea mays* L. EMBO J 6: 1555–1563 (1987).
37. Lechelt C, Peterson T, Laird A, Chen LC, Dellaporta SL, Dennis E, Peacock WJ, Starlinger P: Isolation and molecular analysis of the maize P-locus. Mol Gen Genet 219: 225–234 (1989).
38. Li M-G, Starlinger P: Mutational analysis of the N terminus of the protein of maize transposable element Ac. Proc Natl Acad Sci USA 87: 6044–6048 (1990).
39. Linn F, Heidmann I, Saedler H, Meyer P: Epigenetic changes in the expression of the maize A1 gene in *Petunia hybrida*: Role of numbers of integrated gene copies and state of methylation. Mol Gen Genet 222: 329–336 (1990).
40. Luo D, Coen ES, Doyle S, Carpenter R: Pigmentation mutants produced by transposon mutagenesis in *Antirrhinum majus*. Plant J 1: 59–69 (1991).
41. Martienssen RA, Barkan A, Freeling M, Taylor WC: Molecular cloning of a maize gene involved in photosynthetic membrane organization that is regulated by Robertson's Mutator. EMBO J 8: 1633–1639 (1989).
42. Martin CR, Carpenter R, Sommer H, Saedler H, Coen ES: Molecular analysis of instability in flower pigmenta-

tion of *Antirrhinum majus*, following isolation of the pallida locus by transposon tagging. EMBO J 4: 1625–1630 (1985).
43. Masson P, Rutherford G, Banks JA, Fedoroff N: Essential large transcripts of the maize Spm transposable element are generated by alternative splicing. Cell 58: 755–765 (1989).
44. Masson P, Strem M, Fedoroff N: The tnpA and tnpD gene products of the Spm element are required for transposition in tobacco. Plant Cell 33: 73-85 (1991).
45. McCarty DR, Carson CB, Stinard PS, Robertson DS: Molecular analysis of viviparous-1: an abscisic acid-insensitive mutant of maize. Plant Cell 1: 523–532 (1989).
46. McClintock B: Cytogenetic studies of maize and neurospora. Carnegie Inst Wash Yearb 46: 146–152 (1947).
47. McClintock B: Mutable loci in Maize. Carnegie Inst Wash Yearb 47: 155–169 (1948).
48. McClintock B: Chromosome organization and genic expression. Cold Spring Harbor Symp Quant Biol 16: 13–47 (1951).
49. McClintock B: Mutations in maize and chromosomal aberrations in neurospora. Carnegie Inst Wash Yearb 53: 254–260 (1954).
50. McLaughlin M, Walbot V: Cloning of a mutable bz2 allele of maize by transposon tagging and differential hybridization. Genetics 117: 771–776 (1987).
51. Menssen A, Höhmann S, Martin W, Schnable PS, Peterson PA, Saedler H, Gierl A: The En/Spm transposable element of *Zea mays* contains splice sites at the termini generating a novel intron from a dSpm element in the A2 gene. EMBO J 9: 3051–3057 (1990).
52. Merckelbach A, Döring H-P, Starlinger P: The aberrant Ds element in the Adh1-2F11: Ds allele. Maydica 31: 109–122 (1986).
53. Motto M, Maddaloni M, Ponziani G, Brembilla M, Marrota R, DiFonzo N, Soave C, Thompson R, Salamini F: Molecular analysis of the Bg-rbg transposable element system of *Zea mays* L. Mol Gen Genet 212: 488–494 (1988).
54. Müller-Neumann M, Yoder JI, Starlinger P: The DNA sequence of the transposable element of Ac of *Zea mays* L. Mol Gen Genet 198: 19–24 (1984).
55. Nacken WKF, Piotrowiak R, Saedler H, Sommer H: The transposable element Tam1 from *Antirrhinum majus* shows structural homology to the maize transposon En/Spm and has no sequence specificity of insertion. Mol Gen Genet 228: 201–208 (1991).
56. Nevers P, Shepherd NA, Saedler H: Plant transposable elements. Adv Bot Res 12: 102–203 (1986).
57. O'Reilly C, Shepherd NS, Pereira A, Schwarz-Sommer Zs, Bertram I, Robertson DS, Peterson PA, Saedler H: Molecular cloning of the a1 locus of *Zea mays* using the transposable elements En and Mu1. EMBO J 4: 877–882 (1985).
58. Paz-Arez J, Wienand U, Peterson PA, Saedler H: Molecular cloning of the c locus of *Zea mays*: a locus regulating the anthocyanin pathway. EMBO J 5: 829–833 (1986).
59. Pereira A, Cuypers H, Gierl A, Schwarz-Sommer Zs, Saedler H: Molecular analysis of the En/Spm transposable element system of *Zea mays*. EMBO J 5: 835–841 (1986).
60. Peterson PA: A mutable pale green locus in maize. Genetics 38: 682–683 (1953).
61. Peterson PA: Mobile elements in plants. CRC Crit Rev Plant Sci 6: 105–208 (1987).
62. Pohlman R, Fedoroff N, Messing J: The nucleotide sequence of the maize controlling element Activator. Cell 37: 635–642 (1984).
63. Qin M, Ellingboe AH: A transcript associated with Mutator activity. Mol Gen Genet 224: 357–363 (1990).
64. Rhodes PR, Vodkin L: Organization of the Tgm family of transposable elements in soybean. Genetics 120: 597–604 (1988).
65. Robertson DS: Characterization of a mutator system in maize. Mutat Res 51: 21–28 (1978).
66. Robertson DS, Stinard P: Genetic analyses of putative two-element systems regulating somatic mutability in Mutator-induced aleurone mutants of maize. Dev Genet 10: 482–506 (1989).
67. Sachs MM, Peacock WJ, Dennis ES, Gerlach WL: Maize Ac/Ds controlling elements – a molecular viewpoint. Maydica 28: 289–303 (1983).
68. Saedler H, Nevers P: Transposition in plants: a molecular model. EMBO J 4: 585–590 (1985).
69. Schmidt RJ, Burr FA, Burr B: Transposon tagging and molecular analysis of the maize regulatory locus opaque-2. Science 238: 930–936 (1987).
70. Schnable PS, Peterson PA: The Mutator-related Cy transposable element of *Zea mays* L. behaves like a near-Mendelian factor. Genetics 120: 587–596 (1988).
71. Schwarz-Sommer Zs, Gierl A, Cuypers H, Peterson PA, Saedler H: Plant transposable elements generate the DNA sequence diversity needed in evolution. EMBO J 4: 591–597 (1985).
72. Schwarz-Sommer Zs, Huijser P, Nacken W, Saedler H, Sommer H: Genetic control of flower development by homeotic genes in *Antirrhinum majus*. Science 250: 931–936 (1990).
73. Shirsat AH: A transposon-like structure in the 5′ flanking sequence of a legumin gene from *Pisum sativum*. Mol Gen Genet 212: 129–133 (1988).
74. Sommer H, Beltran J-P, Huijser P, Pape H, Lönning W-E, Saedler H, Schwarz-Sommer Zs: Deficiens, a homeotic gene involved in the control of flower morphogenesis in *Antirrhinum majus*: the protein shows homology to transcription factors. EMBO J 9: 605–613 (1990).
75. Starlinger P, Saedler H: Insertion mutations in microorganisms. Biochemie 54: 177–185 (1972).
76. Sutton WD, Gerlach WL, Schwartz D, Peacock WJ: Molecular analysis of Ds controlling element mutations at the Adh1 locus of maize. Science 223: 1265–1268 (1984).

77. Talbert LE, Patterson GI, Chandler VL: Mu transposable elements are structurally diverse and distributed throughout the genus *Zea*. J Mol Evol 29: 28–39 (1989).
78. Theres K, Scheele T, Starlinger P: Cloning of the Bz2 locus of *Zea mays* using the transposable element Ds as a gene tag. Mol Gen Genet 209: 193–197 (1987).
79. Weck E, Courage U, Döring H-P, Fedoroff N, Starlinger P: Analysis of sh-m6233, a mutation induced by the transposable element Ds in the sucrose synthase gene of *Zea mays*. EMBO J 3: 1713–1716 (1984).
80. Weydemann U, Wienand U, Niesbach-Klösgen U, Peterson PA, Saedler H: Cloning of the transposable element Mpi1 from c2-m3. Maize Genet Coop Newslett 62: 48 (1987).
81. Wienand U, Weydemann U, Niesbach-Klösgen U, Peterson PA, Saedler H: Molecular cloning of the c2 locus of *Zea mays*, the gene coding for chalcon synthase. Mol Gen Genet 203: 202–207 (1986).

Plant Molecular Biology **19**: 51–68, 1992.
© 1992 *Kluwer Academic Publishers. Printed in Belgium.*

Plant and organ development

R.F. Lyndon [1] and D. Francis [2]
[1] *Institute of Cell and Molecular Biology, Division of Biological Sciences, University of Edinburgh, Mayfield Road, Edinburgh, EH9 3JH, UK;* [2] *School of Pure and Applied Biology, University of Wales, P.O. Box 915, Cardiff, CF1 3TL, UK*

Key words: cell division, flowering, homeotic genes, meristems, plant development

Introduction

The work to be discussed here will focus on three main areas. The first concerns the selective expression of specific genes, regulated according to what type of cells they are in (tissue- and cell-specific expression) and according to which temporal stage of development has been reached. The second considers the expression of genes that control developmental switching and the determination of cell, tissue and organ fate, these being the preconditions for tissue- and cell-specific gene expression. The third briefly reviews the question of the control and regulation of gene activity by hormonal factors.

Gene expression during development

Cell-, tissue- and organ-specific gene expression

Particular proteins, such as seed storage proteins, chloroplast proteins or enzymes concerned with the formation of pigments and storage products in fruits, tend to be restricted to particular structures or organs within the plant and their synthesis to particular periods during development. Using this knowledge as a starting point, the techniques of plant molecular biology, especially the use of cDNAs in detecting specific mRNAs, have led to a much better understanding of the control of gene expression which is specific to certain cells, tissues and organs [23, 30]. Typically, expression of genes is specific to seeds, embryos, leaves (where expression may also be regulated by light), roots, and floral organs. In leaves of C4 plants, genes for enzymes specific to the C3 photosynthetic enzymes are expressed in the bundle sheath, and genes for the C4 pathways only in the mesophyll, which accords with the leaf anatomy and physiology [49]. Gene expression specific to the stamen is confined to the anther, genes specific for the pistil may be confined to the ovule, or the integuments, or the stigma and style. Within each organ, specific gene expression may be restricted to particular tissues, such as the tapetum of the anther, or the pollen, or to vascular tissues or the epidermis (petal pigment), and may be limited to a few cell types or only one. In tomato anthers a region of the anther wall can be distinguished only by the hybridization to it of five different pollen cDNA clones [88]. Similarly, in the tomato ovule there is a gene expressed predominantly in only the innermost layers of the integument, cells that otherwise appear to be similar to the other cells of the integument [30]. Presumably there are other genes, not yet investigated, which will be found to be expressed only in specific cell layers or regions of tissues where the cells are biochemically differentiated but otherwise appear identical.

Constructs made from reporter genes and the upstream DNA sequences from developmentally expressed genes have shown that the expression of the latter is governed by promoter/enhancer regions probably common to large groups of plants, if not universal. Further work will undoubtedly focus on the nature and expression of these regulatory sequences and the transcription factors that bind to them, and also their specificity for cells, organs, organisms, and stages of development.

Seed proteins are known to be characteristic of the seeds and have obvious agronomic importance because of the promise of being able to engineer the protein composition of cereals. In particular, it is highly desirable to improve the quality of seeds which are poor in nutritionally important amino acids. The use of site-directed mutagenesis has enabled modifications to be made to the zein seed storage genes so that at particular sites within the genes lysine is encoded rather than leucine [16]. Whilst these modified genes have been expressed in transgenic animals [90] they have yet to be routinely expressed in cereals because of the lack of a reliable transformation system. In a developmental context, site-directed mutagenesis could be used to alter the coding capacity of genes which are important for specific changes in morphogenesis. The use of appropriate synthetic oligonucleotides containing altered base composition at strategic sites of a gene, coupled with the polymerase chain reaction (PCR), may facilitate the construction of altered genes. If these could be successfully integrated into the plant genome in a manner which allows their expression without altering the expression of other, native, genes then it should be possible to examine their expression in relation to specific developmental pathways. This approach would both verify the fidelity of putative master developmental genes and give new insight into developmental patterns. However, little is known about the identity of such master genes.

Temporal control of gene expression

Cell-specific gene expression is also probably always temporally regulated. In the broadest sense it is limited to the period of existence of the organ. But it may be more time-limited than this, to a specific time window during organ development, as it is during embryogenesis and seed formation [33]. Gene expression obviously changes during growth when particular enzymes are formed and are active for a short time during the plant's life. This is seen especially during germination when storage products are mobilised for growth of the young seedling. In the germination of seeds in which lipids are major storage substances, as in the cotyledons of the sunflower or cucumber seed, enzymes of the glyoxylate cycle, involved in the conversion of lipid to sugar, increase in activity for a few days during germination and then decrease. A key enzyme is malate synthase (MS), which reaches peak activity on the 4th day of germination in cucumber seeds. The gene for MS has been isolated, sequenced and cloned. The 5′ upstream regulator sequence has been excised and transferred to the GUS reporter gene. In transgenic plants this becomes activated during germination when MS activation would be expected and in the same tissues. MS activity is therefore regulated at the transcriptional level [36]. The nature of the regulatory process can now be investigated. This gene is also expressed, at lower levels, in petals, and also in senescing tissues [20], and may be activated by the products of lipid breakdown.

With this, and other temporally regulated genes, we need to know what substances can induce gene expression and what the nature of the transduction chain is. This is the same problem raised by the activation or inhibition of gene expression by plant growth substances and other effectors or inhibitors. The growth substances gibberellic acid and abscisic acid, respectively, promote and inhibit transcription of the MS gene [75]. However, this regulation could be very indirect, through their general effects of promoting or inhibiting germination. If the time scale of the developmental event is altered by the growth substance, then the timing of transcription of developmentally regulated genes, of which MS is one, will necessarily also be regulated. What we need to know is the nature of the primary regulatory action on the gene regulator sequences. The action of plant growth regulators, which is not very gene-specific, may be some considerable way back through the transduction chain.

Are there indeed any genes that are not expressed developmentally? Even constitutive enzymes and proteins will be developmentally regulated as they are probably synthesized at different rates at different times during cell development. The control of the expression of genes for constitutive components has been especially studied in the formation and development of plastids and mitochondria [64, 66].

Three levels of regulation of gene expression can therefore be identified: organ or tissue; cell; and time during development. Organ-, tissue- and cell-specific gene expression depends on the existence, and hence on the formation and development, of the specific organ, tissue and cell types. This will be controlled by other, regulatory genes such as the homeotic genes (see below).

Combinatorial control of transcription could be adequate to account for gene programming in plants

The possibility of the combinatorial regulation of genes in plants by the simultaneous activation of more than one promoter sequence, perhaps in conjunction with other, enhancer, sequences, is of the order required for specification of cell and tissue determination and for the temporal control of gene activity [6].

In a study of the efficacy of the cauliflower mosaic virus 35S promoter, Benfey and Chua [6] have shown that it contains seven interacting domains in a 351 bp stretch of DNA i.e. ca. 50 bp per domain. If there are synergistic effects of approximately 50 bp lengths of DNA and each gene has on average something like 1 kb of upstream regulatory DNA, then there could be upwards of 20 interacting domains. Assuming interactions were only between 2 domains at a time this would still give 2^{20} possibilities for different effects. Even with only 2^{10} this is > 1000 choices, more than enough to specify where and when each gene should act plus redundancy of information and the possibility of gene families with different activation requirements. (This is ignoring the further possibilities for quantitative up- and down-regulation of genes.) During differentiation any plant cell may have to make about 10 binary choices (Fig. 1). Evidence that genes are actually controlled in a combinatorial fashion comes from alterations in the tissue specificity of the *Pal* gene in *Antirrhinum* and the way it responds to a second *trans*-acting gene [1].

Gene expression where development originates: in meristems

The meristems are those regions of dividing cells at the extremities of the plant that give rise to all the rest of the plant and where morphogenetic programs and patterning are determined. The expression of a root meristem-specific gene has been shown to be regulated by a *cis* element. The 5′ sequence of −636 bp upstream from the gene restricted the expression of the GUS reporter gene to the root meristem and some of the immediate cell derivatives in the young vascular system [92]. Study of this and similar regulatory sequences should begin to show exactly how it is that root and shoot meristems differ at the molecular level and why each is so stable that they very rarely transform one into the other.

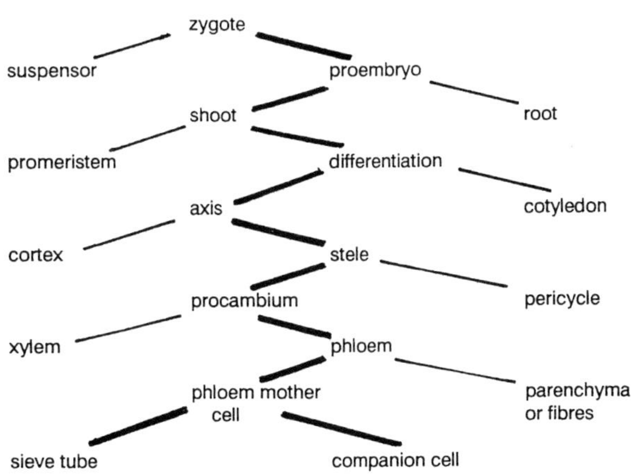

Fig. 1. Possible pathway of developmental choices open to a plant cell during embryogenesis and development. Only one pathway is shown here; from each branch, except the final one, further pathways could be constructed.

Because of its minute size, the shoot meristem does not lend itself easily to molecular and biochemical techniques. However, using cauliflower heads, which are very highly branched shoots, and are a collection of thousands of shoot meristems, and the powerful PCR technique, some cDNAs were isolated which were expressed in the *Arabidopsis* shoot meristem but not in leaves [58]. As shown by the fusion of its promoter region to the GUS reporter gene, one of these clones, meri-5, was most strongly expressed at the branching points in the stem, but was also strongly expressed in the shoot meristem and young primordia in both vegetative and floral apices. However, it was also expressed at lower levels in the root meristem, the vascular tissues (especially the internal phloem), and the receptacle and pedicel of the flower. Another gene (meri-1) coded for a histone, H3. Its promoter region directed GUS expression in transgenic tobacco particularly on the flanks of the shoot apex, where primordia are formed, but not in the central zone at the summit of the apex where the initial cells are found. The lack of expression in the summit of the apex is consistent with the low rates of cell division often observed in this region of shoot meristems (e.g. [18]). It also was expressed strongly in the procambium. This distribution parallels that found often for general nucleic acid and protein stains and is probably indicative of the regions of small, dense and dividing cells, rather than anything more specific. We will need to compare different experimental systems so that the gene expression intrinsic to a particular developmental event can be distinguished from gene expression necessary for a supporting process such as cell division, which will be coincident in time and space with cell and organ formation. In this respect, it will be important to use *in situ* hybridization techniques on all of the meristems of the plant.

The cyclic formation of new organs during normal iterative growth in the plant may not involve the genes directly. The process of leaf initiation at the shoot apex has been modelled on the assumption that young primordia are sources of an inhibitor which prevents formation of another primordium in the immediate vicinity. It is supposed that only when the inhibitory effects of a young primordium are dissipated by radial expansion of the apex and metabolism of the inhibitor, can a new primordium form at a minimum in the inhibitor field [61, 86, 89]. These models assume the existence of otherwise hypothetical activators and inhibitors of the production and destruction of substances necessary for primordia to form at a given site. What these models do not address is the nature of the activators or inhibitors. If these prove to be substances which directly regulate the activity of proteins, then there is no need to involve the genes directly in this regulation: the genes will have set up a self-regulating system. Only if metabolic regulation is assumed to involve transcriptional or translational events does the problem pass from the realm of cellular biochemistry to molecular biology.

We can envisage the change of form involved in the initiation of leaf and organ being the operation of systems built by gene action but then operating automatically. Changes in gene expression may be involved only in developmental switching. This would be when new types of organs are first formed during embryogenesis and during flower development, and when cells and tissues differentiate during organ development. Perhaps the iterative development so characteristic of plants is that part of growth and development that is regulated by metabolic feedback controls and for the moment does not depend on changes in the expression of master genes specifically concerned with regulating development. If so, then plants have phases of development that have no counterparts in animals and so may not necessarily be controlled in the same way. Indeed it may be because plants have evolved a way of producing units repetitively and without recourse to immediate gene control, like an automatic production line, that they have been able to adopt the life style they have. An important step forward in understanding the control of plant development would be to find out whether all developmental events depend on changes in gene expression (at the transcriptional and translational levels), or whether some events do not, but are self-regulating at the metabolic, rather than the genic, level.

Gene expression during flower differentiation

There are two occasions during the development of the plant when whole new sets of genes are expressed. The first is during embryogenesis, the second is during flowering. Gene expression in the shoot meristem changes spectacularly when it transforms into the inflorescence or into a flower meristem, with the formation of a completely new set of organs which make up the flower.

Each type of organ within the flower probably expresses about 25000 genes. The anther and ovary contain about 10000 genes not expressed elsewhere in the plant. The petal, on the other hand, has about 7000 genes in common with the leaf, but with no other plant organs, but also has distinctive genes of its own [22]. Generally, some of the RNA nuclear transcripts are not detectable in the cytoplasm implying at least some degree of post-transcriptional control. In the ovaries, however, all nuclear transcripts were also found in the cytoplasm, implying overwhelmingly transcriptional control of gene expression. The stem tissues showed the opposite picture. Stem-specific mRNAs were not detectable in the nuclear transcripts, implying post-transcriptional control of stem gene expression [32].

Differences in protein composition of different floral organs were first shown by PAGE in tulips [3]. The floral organs had different protein profiles, the clearest differences being between the stamens and ovaries. Floral organs also differed in their isozyme patterns for malate dehydrogenase, and to a lesser extent, for esterase. In tomato, there were greater differences in protein complements between the parts of the pistil (stigma, style and ovary) than there were between ovaries and stamens [77]. In a comparison of the protein composition of the floral organs of several solanaceous species, one-dimensional SDS-PAGE was inadequate to show clear and consistent differences between organ types, and no organ-specific polypeptides were detected [10]. In a more detailed analysis of the protein composition of the floral organs of *Pharbitis* with two-dimensional PAGE, the number of spots varied from 209 for the corolla extracts to 479 for the pistil [4]. Of about 800 identifiable spots, only about 4% or about 30 in each floral organ (corolla, stamen, pistil) were unique to each organ (Fig. 2). Sepals shared a greater proportion of spots with leaves than did corollas, stamens or pistils, and these latter 3 seemed to form a distinct subset. Presumably the relatively low proportion of organ-specific polypeptides explains why they were not found in the one-dimensional gels of Solanaceae flower parts.

Many fewer proteins are detected immunologically. In *Sinapis alba* only 3 antigens specific to the flowers were detected [44]. Using protein extracts from whole tobacco flowers, eleven antibodies were raised which could be shown to react to floral proteins, but only a few of these were totally organ-specific [25]. Those with the greatest differential reactivity were those targeted against carbohydrate moieties, suggesting a possible general role for glycosylation in plant differentiation, as well as in self-incompatibility.

The differences in protein composition of different floral organs are evidence for changes in

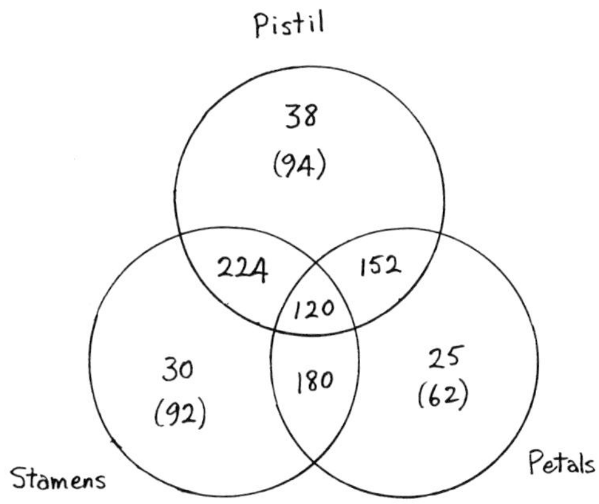

Fig. 2. Numbers of polypeptides distinguished in two-dimensional PAGE gels of extracts of *Pharbitis* floral organs. Polypeptides different in corolla, stamens, and pistil, but also occurring in stems, leaves, or sepals, shown in paretheses. (After Bassett *et al.* [4].)

gene expression but do not tell us at what level between the gene and the protein the control is being exercised. Direct evidence for changes in gene expression at or near the level of transcription come from experiments with cDNAs [30]. Tissue-specific cDNA has been isolated from tomato pistils. Nine cDNA clones were characterised from immature (premeiotic to early meiosis) and mature (at anthesis) pistils. Four of these were also expressed in anthers. *In situ* hybridisation [78] showed that the expression of these genes was regulated very tightly both temporally and spatially. One gene was expressed only in the transmitting tissue of the styles. A second gene was expressed only in 2–3 cell layers of the ovules for a period of less than 8 days [31].

Similarly, genes specific to stamens have also been identified. Five anther-specific cDNA transcripts were detected in developing tomato anthers. Transcripts increased during gametogenesis and reached maximal levels in the mature pollen grains. Since transcripts also persisted in growing pollen tubes, towards their growing tips, and were also in the anther wall, this showed that the genes were expressed in both gametophytic and sporophytic derived tissues [88]. A clone, FA3, which increased in concentration during the transition to flowering in tobacco apical meristems [45] was in parenchyma cells of the anther and carpel but was not in the developing pollen sac or the placenta and so seems to be concerned with the development of the reproductive organs but not with gametogenesis.

Gene expression is also tissue- and time-specific in stamens. In *Sinapis* a group of 7 cDNA transcripts, confined to the tapetum of the anthers, appeared 10 days after the beginning of floral induction and declined after 15 days [62]. Other anther-specific cDNA clones, in sunflower, are expressed only in the anther epidermis [26a]. Indeed, not unsurprisingly, there appear to be several sets of anther-specific genes that are each expressed in a different tissue of the anther and at different times during anther development. One of the genes involved, TA29, is regulated primarily at the transcriptional level and its expression regulated by a 122 bp 5′ region upstream of the gene [47]. Genes specific to stamens all seem to be confined to the anther, which is the part of the stamen where the male gametes are formed. The tissues of the filament may not be different from similar tissues in other organs. It would be useful to know when the organ-specific genes in the floral organs start to be expressed; is it when the organs are first initiated or only after they have begun development? Or only after they have become determined (committed) as that particular type of organ?

These experiments provide direct evidence at the molecular level that genes are expressed specifically in floral organs, but also tell us that the control of expression of these genes is probably at the level of transcription. They are therefore the floral analogues of the genes for seed proteins or leaf proteins which are expressed tissue-specifically and at given developmental time windows.

Necessarily, the genes are recognized as being specific to the cells, tissues or organs only when these are clearly distinguishable and are therefore developing or mature. Because site-specific genes can be shown to be expressed only when the particular cells themselves are present, it seems probable that the expression of these genes depends on the presence or activation of transcription factors, or combinations of transcription factors, specific to each cell type [30]. This raises the question of how genes are involved in specifying cell, tissue and organ types in plants, in which cell types and organ types can grade into each other or where a single cell or organ can consist partly of one type and partly of another. For instance, cells at the edge of sclerenchyma can sometimes have a fibrous wall on the sclerenchyma side and a parenchymatous wall on the parenchyma side. Organs intermediate between types can be found in some homeotic mutant plants and in some florally reverted plants [5]. The involvement of genes in patterning is presumably in specifying the structure of the self-regulating system which may then operate independently of further changes at the transcriptional level, perhaps through the action of molecules such as plant growth substances.

Developmental switching. Phase change and determination at the apex

The process of differentiation of cells, tissues and organs is not instantaneous but, in plants, occurs usually over several days, during which the cells become increasingly committed to the differentiated state, and therefore become determined [53, 57]. Cells are regarded as determined when they can no longer be experimentally diverted into an alternative developmental pathway or prevented from following the new one. Determination is thus the process by which cells become locked into expressing a new set or combination of genes. The biochemical and molecular processes occurring during determination, and when cells alter their competence to become determined, are completely unknown. These processes of competence and determination are common to all developing and differentiating cells but have not yet achieved due recognition as key processes in locking gene expression into a particular mode. Plants may be better than animals to investigate these processes because developmental pathways in plants are manipulable in a way that is not possible in animals.

Floral determination has been investigated in day-neutral tobacco by excising buds from different positions on the plant at various times during the transition to flowering [57]. The state of determination of the meristems is shown by the subsequent mode of growth of the buds, once isolated. If they immediately form flowers they are determined for flowering. If they produce many leaves then they are still determined in the vegetative state. The physiological studies on this, and similar, systems could now be complemented by the use of PCR to show which genes are activated or switched off, and when, in relation to changes in competence and determination.

Changes in gene expression at the switch from vegetative to flowering growth

Proteins

Changes in gene expression in the shoot apex on transition to flowering should be detectable as changes in protein complement on stained gels after two-dimensional PAGE. In order to focus on the cells directly involved in the vegetative to floral transition, experiments with *Silene* were carefully restricted to the shoot apex itself (comprising the apical dome plus the youngest pair of leaf primordia). Even so, no changes in total protein composition were detected during the first 5 days of floral induction and evocation (commitment of the apex to flowering). Only after 8 days, when evocation is complete (7 long days give 100% flowering) were there sufficient changes to produce a change in the overall protein composition of the apex [28, 85]. 107 new proteins were found in the evoked apices at the end of evocation, on day 8, and 182 identical to those found in the uninduced plants (in short days); 17 unique to the uninduced apices disappeared [28]. Similar results were obtained in confirmatory experiments [85].

In *Chrysanthemum* also, 10 new polypeptides appeared in the prefloral meristem and 2 (unique to the shoot apex) disappeared [74]. In reproductive meristems 4 new spots appeared and 2 disappeared (1 in both vegetative and prefloral meristems, the other specific to prefloral meristems). Of the >500 spots on the gels these changes involved <2% of the proteins. The main changes occurred at the transition from vegetative to prefloral (i.e. bract initiation) growth at the apex.

Staining therefore shows relatively few differences, and then only in the apical meristem itself. When whole apical buds are used these differences are completely masked [17]. Labelling would be expected to show differences earlier and more clearly. In *Sinapis*, commitment of the shoot apex to flower becomes irreversible 44 h after the beginning of induction [46]. A comparison of the fluorographs of two-dimensional (2-D) PAGE gels of extracts of apical meristems of vegetative plants (kept in non-inductive SD) and evoked plants supplied with [^{35}S]methionine for 2 h, at 50 h after the beginning of induction and about 6 h after the completion of evocation, showed that 6 proteins (out of 400–500 detected) were synthesised less in the evoked meristems and 14 proteins were synthesised relatively more than the rest in the evoked meristems and 16 new proteins were synthesised [55]. These are relatively small changes in the proteins synthesised, presumably in a few crucial proteins, and only in those cells of the apical dome that are about to make new primordia or produce the floral or inflorescence meristem.

In similar experiments with *Pharbitis*, about 1000 polypeptides were detected on 2-D gels either by silver staining, or by autoradiography after labelling of the apices for 2 h with [^{35}S]methionine [2]. But only 5 polypeptides were different between buds of plants evoked to flower by being given 1 short day (SD), and the buds of vegetative plants prevented from flowering by being given a short burst of light in the middle of the night (NB; night break treatment) after the SD. Two of these proteins increased in SD but remained low after NB, a third increased in SD then decreased (but showed no change in the labelling experiment), and the 4th and 5th decreased after NB. These differences were retained or were magnified with time from day 1 to day 16. Also in *Pharbitis*, the translation products of mRNA of induced and non-induced apices were examined, but with very similar results [67]. Although 400 or so polypeptides could be distinguished on the gels, there were only 3 that were present in the induced apices and not in the non-induced, even though the inductive treatment was greater (given on three successive cycles, i.e. 3SD v. 3NB). There were minimal differences between leaves, cotyledons, petioles, hypocotyls and roots (all of course vegetative organs), although there were at least 6 organ-specific polypeptides. The main message from these experiments is that remarkably few changes can be detected in the protein complement or in the proteins synthesised as a first result of evocation at the shoot apex.

The relatively few changes in polypeptides revealed by PAGE must, in part, reflect the limited resolution of the technique. One only has to compare the number of genes active in an individual cell with the maximum of about 1000 polypeptides detected on 2-D gels to realise that where differences in polypeptide profiles emerge they are probably under-estimates. Where no differences are resolved they may still mask subtle differences at the mRNA level. As mentioned earlier, mRNA extractions coupled with PCR will enable closer scrutiny of temporal changes in gene expression that occur during evocation. However, care must be taken to distinguish qualitative changes that commit the apex to a floral mode of growth from those changes that are linked to growth. Model photoperiodic systems where treatments exist which are transient photoperiodic perturbations of otherwise inductive conditions (e.g. the NB in *Pharbitis*, the dark interruption of LDs in *Silene*) will be critical in this regard. Moreover, *in situ* hybridization in these systems will resolve tissue-specific changes at the shoot meristem from those that are linked to cell division *per se*.

Genes

Changes in gene expression have also been shown by cDNA cloning. In *Sinapis alba* a group of 6 cDNA transcripts were identified which were expressed only at low concentration in vegetative apices but accumulated to a maximum 2–10 days after the beginning of induction [62]. These transcripts did not seem to be involved in cell division as such: although they were more prevalent on the flanks of the apex they were in very low concentration at the summit of the floral apex, in which cell division is also relatively rapid in *Sinapis* [12].

In tobacco (cv. Samsun; a day-neutral plant) transcripts of a cDNA clone (FA2), were present in only very small amounts in the mature vegetative apex, but increased in the shoot apical meristem during the transition to flowering and were highly expressed in developing petals, stamens and pistils [45].

Changes in gene expression were followed in cultures of thin cell layer explants (TCLE) of cv. Samsun tobacco [59]. Six different gene families (FB-1 to FB 7-6) were identified. FB7 transcripts were poorly expressed in TCLE on the vegetative program (in which flowers formed only late in culture) but had increased by day 7 in TCLE given kinetin, which induced early flower formation. This transcript then decreased but increased again by days 23–25, when the flowers were forming. Note, however, that none of the transcripts were floral meristem-specific. This is perhaps not too surprising since the whole process of TCLE is designed to elicit only certain programs from TCLE that are competent for all [87]. This in itself might be telling us that the changes in gene expression in different organ types are quantitative and that qualitative changes are only associated with changes in competence of cells during developmental switching.

Any changes in gene expression that are found at the transition to flowering will need to be checked to see that they are specific for this process and not concerned with changes in the cell cycle or any other associated but non-specific processes that also occur at this time. Little attention has been paid yet to differences in gene expression between leaves and sepals; although the differences may be minimal those that do occur should be particularly linked to the transition to flowering. It is far from clear that major changes in gene expression should in any case be expected at the commitment to flowering. Perhaps only a few genes could be involved, for instance concerned with hormone synthesis or synthesis of hormone receptors. If there are master genes which initiate a cascade of biochemical events leading to phase change, then they may be only transiently expressed and may escape all but the most rigorous of PCR protocols. The earliest changes in the apex on flowering are a change in the size and arrangement of primordia and the non-initiation of internodes [5, 51]. It is not known whether there are genes specific to internodes, which would be expected to be no longer expressed in the formation of individual flowers.

Homeotic mutants and flower development

The expression of specific genes in the floral organs parallels specific gene expression in vegetative and seed organs, tissues and cells. These genes, which are expressed during flower differentiation, must be distinguished from those causing the formation of the organs. Only in flowering do we so far have available genes that direct the formation and development of new organs and so are concerned with developmental switching. These are the homeotic genes, which therefore seem to be regulatory genes.

The *apetala* mutants in *Arabidopsis* cause the formation of homeotic flowers in which petals are suppressed or modified to other organs. The *apetala-1* gene causes failure of petal initiation. Sepals become bracts with apetalous flowers in their axils and the sepals on these flowers repeat this structure, so that flowers within flowers are formed [47]. The wild-type *AP2* gene determines the identity of the perianth organs in *Arabidopsis* [48]. This is deduced from the effects of three different mutant alleles (*Ap2-5*, *Ap2-6* and *Ap2-7*), all of which cause sepals to be transformed into carpels, and *Ap2-5* petals to stamens and *Ap2-6* petals to carpels. A schema of action is proposed as a branching diagram (Fig. 3).

A most interesting development is the finding that the homeotic gene *agamous* in *Arabidopsis* [93] and the *deficiens* gene in *Antirrhinum* [79] show DNA sequence homology with DNA transcription factors isolated from yeast and humans. The *agamous* sequence in *Arabidopsis* was iso-

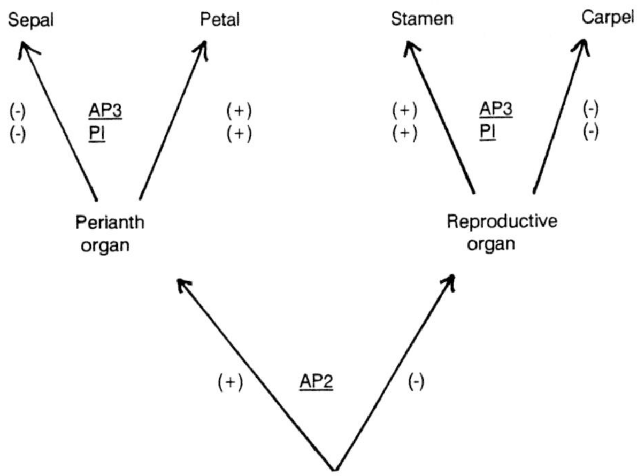

Fig. 3. Action of homeotic genes in *Arabidopsis*. Determination of organ identity could be hierarchical. The APETALA2 gene is required for the formation of perianth organs (sepals and petals); in the *apetala2* mutant reproductive organs are formed. This interpretation, which implies a temporal sequence of gene actions, may be compared with that shown in Fig. 4, which emphasises spatial action. (After Kunst *et al.* [48].)

lated by the use of T-DNA insertions which caused the mutant phenotype in wild-type plants [93], and the *def* gene by differential cDNA screening [79].

Homeotic genes often seem to act not on a single whorl but on two successive floral whorls, as though the flower consists of three overlapping regions – (a) sepals + petals, (b) petals + stamens, and (c) stamens + carpels – each influenced by a set of homeotic genes [15]. Homeotic genes may also interact with each other [13, 22], but also almost certainly interact with other 'target' genes concerned with specification of organ form. However, the nature of these 'target' genes or their interactions is for the moment conjectural. Although homeotic genes are assumed to determine positional effects (Fig. 4), they could just as well be determining temporal expression of different genes [73]. Because the formation of the floral whorls represents a succession not only in space on the meristem but also in time, any map of homeotic gene action could also be redrawn as a time scale, with three main periods of action, each overlapping the production of more than one set of organs (Fig. 4). Homeotic genes may therefore not be positional genes but timing genes, regulating development by controlling the timing of transcription of other genes.

Although the expression of homeotic genes must be crucial to pattern formation, those cells in which such genes are expressed have not so far been identified during organ initiation and the earliest stages of development. The phase change which meristems undergo in order to form flowers must depend upon the activation of founder cells for each floral whorl. Their location is in those parts of the apex in which surface microstructure changes before the organs are formed, but what genes are involved in these processes, and how they are influenced by floral stimuli, is so far a complete mystery. As noted by Steeves and Sussex [83], more refined cell biological techniques are necessary in order to resolve the identity of cells in the apex which are crucial for changes in development. Are there 'special' cells in which the homeotic genes function?

In animal development clones of specific embryonic cell types can be traced to specific founder cells. Davidson [19] noted that in relation to sea urchin development 'specific functional characters of these cell types requires that particular genes be expressed in them'. Moreover, the founder cells of these lineages occupy 'spatial domains' within which identified genes will be active. Clearly, vegetative plant meristems are much more developmentally plastic than animal embryos. Apical initials are not permanent [83] and, for example, the number of layers of cells which constitute the tunica and corpus can fluctuate during the course of leaf initiation [50]. Moreover, the initial floral parts, the sepals, are more similar in polypeptide composition (see earlier) and presumably in underlying levels of gene expression, to leaves than to petals, stamens or carpels. However, once a meristem is committed to become floral, its growth is determinate and the

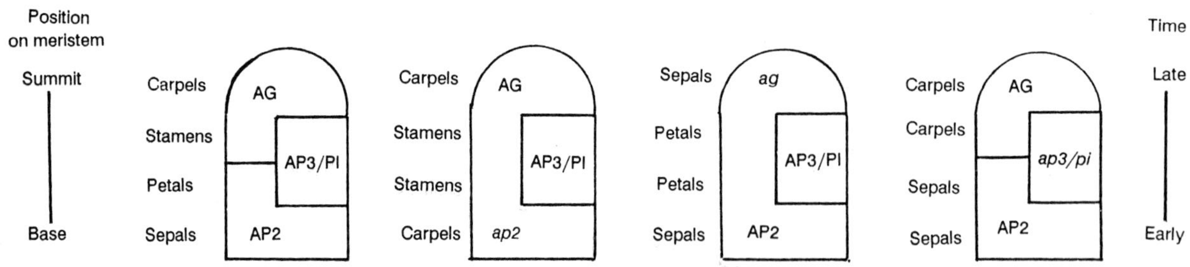

Fig. 4. Interactions of homeotic genes in *Arabidopsis*. The APETALA2, APETALA3/PISTILLATA, and AGAMOUS genes appear to act on three overlapping spatial domains in the shoot apex; at the base, the midregion, and the summit, respectively. AGAMOUS and APETALA2 are assumed to be mutually exclusive in their actions on each other's territory. If one of them is mutant, then the other can exert its action on all domains. The spatial sequence also represents a time sequence, which allows a different interpretation (cf. Fig. 3).

notion that spatial domains, which initiate early floral whorls, exist in these meristems would not be dissimilar to the situation described for the sea urchin embryo. In that system entire groups of genes and gene products are recognized as a pattern of lineage-specific gene expression [7]. A pattern emerges of:

(a) genes activated very soon after the founder cell populations form,

(b) genes that are suppressed in the lineage derived from a particular founder cell population,

(c) genes that are apparently expressed everywhere initially but where expression gradually diminishes in a particular founder cell lineage with time,

(d) genes that are expressed late in the development of a particular lineage.

Presumably we would assign the expression of homeotic genes for flower development into class (a), leaf-specific genes to class (b), house-keeping genes, linked perhaps to photosynthesis for example, to class (c), and genes necessary for colouration, or sex expression, to class (d), Genes in (c) and (d) would be switched on by (a) genes whilst (b) genes could, conceivably, be repressed by (a) gene products. Floral evocation may sensitize discrete domains within the apex to facilitate the expression of the homeotic genes. Evocation, in effect, would establish a co-ordinated cellular network capable of initiating the floral parts in a rhythmic manner. This notion argues for a temporal activation of founder cells for each floral whorl. When floral induction is perturbed, meristems remain vegetative or, in some cases, partially floral meristems can revert to vegetative structures. According to the above model, class (a) genes in founder cells would be repressed or down-regulated by factors that induce floral reversion. The extent of the down-regulation would be reflected in whether the meristem switches immediately from initiating floral to vegetative primordia (acute down-regulation/suppression) or whether the meristem switches to initiating aberrant floral parts such as petalloid stamens (mild down-regulation/suppression). Clearly, it should be possible to test such models with the aid of genes, such as the *def* gene in *Antirrhinum* [14], and gene products that are responsible for specific organ formation.

In some plants homeotic-type changes can be induced by the environment. In those plants showing flower reversion, a flower starts to be formed but during flower formation the meristem reverts to a vegetative mode of functioning so that leaves or other vegetative organs (e.g. leafy bracts or similar organs) are formed in the middle of the flower, or growth is continued from the middle of the flower as a leafy shoot [5]. Flower reversion can be brought about in a controlled manner in *Impatiens* and *Anagallis* by photoperiod. When the plants are induced to flower by the appropriate photoperiod, flower development continues normally as long as the plants are kept in inductive photoperiods. If they are transferred to non-inductive photoperiods while flower development is going on, then the apex stops forming floral organs and reverts to making leaves. Since the pre-existing leaves are the site of perception of the photoperiod this implies that a continued supply of substances from the leaves is required for floral development to continue. Reverted plants resemble certain homeotic mutants and suggest that the control of organ type at the apex can depend on substances produced by either the leaves (in revertable plants) or in the apex itself in most plants, in which the apex becomes autonomous for flower formation once the plant has been induced to flower. This poses the question of whether homeotic mutants may operate via the action of plant growth substances.

Possible modes of action of genes in development – interaction with biophysical parameters

Genetic and molecular techniques allow the identification of the site and phenotypic nature of gene action and the sequence structure of the gene products. Where these resemble transcription factors we may even know the function of these gene products. However, regulatory genes themselves must be regulated and no genes can act in isolation from cellular structure. Gene action also has to be translated into three-dimensional biological form. In part this will be through the direction of gene products to specific locations within the cell by the action of chaperone molecules [24]. It will also depend to a considerable degree on biophysical features which may be involved in determining the position and shape of organs [37]. We know nothing about how the biophysical structure of the cells is specified by the genes. Presumably this could be through genic specification of the types of molecules synthesized in the cell wall or in the cytoskeleton that may be concerned with directing and orienting cell wall structure [82]. As noted by Wick [91], more work is required to examine the cytoskeletal or 'motor protein inhibitors' on the component phases of division site regulation. Moreover, nothing is known about cellular/transcellular signals which are responsible for spatial regulation of cytokinesis.

Cell cycle and developmental switching

Changes in the cell cycle are characteristic of developmental switching. At the transition to flowering the cell cycle shortens for one or more cycles [34, 63] and in the formation of the successive whorls of floral organs in *Silene* the cell cycle seems to alternate between shorter and longer cycles [54]. The shorter cell cycle at the end of floral evocation in *Silene* may be associated not specifically with evocation (the change in commitment of the meristem from vegetative to flowering) but with flower morphogenesis, the change in functioning of each individual meristem to form a flower [52]. In *Sinapis*, the shortened cell cycle occupies much of the period of evocation, including the 36 h or so before the start of floral morphogenesis itself. During this period, secondary initiation sites of DNA replication are switched on [43], S-phase shortens [34] and the cell cycle becomes synchronous [8]. In *Sinapis*, the activation of secondary origins is a response to a component of the floral stimulus which may be a cytokinin [40]. What might be the function of such an activation is unknown, although the reordering of initiation sites may well facilitate changes in gene expression which are necessary for flower initiation to begin.

In *Silene* there may also be a necessity for a changed cell cycle during the very early stages of developmental switching. *Silene* plants require 4–7 LD for complete floral induction. Although the first 3LD are in themselves not sufficient to cause flowering, they cause changes in the cell cycle. A very early change is seen at the very beginning of induction, which involves keeping the plants in continuous light instead of on a light/dark cycle. The first 5 minutes of extended light is sufficient to trigger a shortening of the cell cycle from 19 to 13 h (Ormrod and Francis, 1985). If the plants are kept in darkness for 20 min at the beginning of each extended light period, then flowering is all but eliminated [69] and the change in cell cycle does not happen [71]. This shortening of the cell cycle can be triggered by both R or FR light, but only FR light alters the cycle in the same way as inductive photoperiods, by shortening G1 and S (DNA synthesis). The reduction of S is by a doubling of the rate of fork movement during DNA replication, and not by any effect on replicon number, which remains constant. This effect of light can be detected experimentally in 30 min and probably begins within a few minutes of exposure to the extended light period [70].

The problem in assessing the significance of such changes is to know whether they are causal to the floral process or derivative effects depending on some underlying change (such as a change in hormone or calcium concentrations) which have so far remained undetected. The possibility of transferring cloned *cdc* genes from yeast into higher plants will provide the opportunity to alter cell cycles in specific ways and to test whether specific changes in the cell cycle can cause developmental switching in reproductive or vegetative growth, or whether the cell cycle changes are the consequences of developmental switching by other (unknown) mechanisms.

Synchronisation of the cell cycle

Synchronisation of the cell cycle occurs at four points in plant development: (1) in the young pro-embryo, when all cells are recent derivatives of the same mother cell and so have similar cell cycles which after 5 or 6 divisions begin to lose their synchrony; (2) in the free-nuclear endosperm where the tissue is essentially a coenocyte, divisions are synchronous, and the cell cycle consists solely of M-S-M transitions [29]; (3) in the pre-meiotic cells of the anther, in which cells are held in pre-meiotic interphase until all have reached it and can then progress through meiosis synchronously; and (4) on the transition to flowering when one or more synchronised cell cycles are characteristic of evocation and flower initiation [8, 27, 38, 52].

Studies of the molecular control of the meiotic process in plants have shown that in lily meiocytes the pre-meiotic S-phase is characterised by an under-replication of about 0.2% of the genome. During subsequent meiosis the residual DNA is replicated at zygotene (Z-DNA) and is followed by DNA repair at pachytene (P-DNA). At the same time the synthesis of meiosis-specific poly-A RNA was detected. Treating such meiocytes with cyclohexamide impairs both Z- and P-DNA and the chromatids fail to form chiasmata. The molecular controls that block replication of this tiny fraction of the lily genome before meiosis are unknown, but it is clear that Z- and P-DNA synthesis are essential for successful meiosis [39, 84]. The fact that Z-DNA replication is scattered throughout the genome suggests that it may be essential for recombination events during prophase I of meiosis.

In flowering, the synchronised cell cycle seems to be an essential part of the flowering transition in *Sinapis* since it has not been possible to prevent or delay floral induction without also preventing or delaying the synchronous cell cycle [9]. In *Silene*, however, synchrony seems to be an adjunct of flowering rather than essential or causal, because it can be prevented by keeping the plants in darkness for 48 h after induction is complete [38].

In *Silene* the possibility was raised that the observed changes in the protein complement in the shoot apex on the transition to flowering may have been related to the synchronisation of cell division which occurs on days 7–8 rather than to flowering *per se* [28]. When the synchronous cell divisions in the apex were suppressed, and flowering was delayed but not prevented, by inducing plants with 7LD, then placing them in darkness for 48 h, the same changes in protein complement occurred as found on day 8. This showed that the changes in protein complement were intrinsic to flowering and not concerned with division synchronisation [85].

Coincident with the synchronous divisions is a reduction of the molecular exclusion limit within the shoot apex (from 600 to about 500 Da) [35]. This is also prevented by 48 h darkness [76] and so is linked to the non-essential synchronous divisions rather than specifically to the transition to flowering. It is not known how either the synchronous divisions or the reduction of the molecular exclusion limit are controlled. They are presumably indications of some other, as yet undetected, changes, perhaps in hormone status or in the activity of regulatory genes. Again, this highlights the care that has to be taken to ensure by experiment that cellular events or changes in gene expression occurring at the same time as developmental switches are indeed causal rather than consequential.

Growth substances and the control of organ development

Organ formation occurs in the pericycle, where roots are initiated, and at the shoot apex, where leaves and floral organs are produced. So far these processes have not been investigated at the molecular level, although the initiation of lateral roots can be experimentally controlled by auxin [11]. How, or even whether, gene expression changes during root or leaf initiation is simply not known.

Important information about the way initiation and development of meristems is controlled and how gene expression changes, should be obtainable from the use of thin cell layer explants (TCLE) [87]. These are slivers of stem outer tissue which can be made to grow and develop in different ways by altering the nutritional (including hormonal) factors in the culture medium. According to what is supplied, the TCLE can be made to follow the root, shoot, callus, or flowering programmes at will. However, since the explants are competent for all developmental programmes, the channelling into only one of the programmes is essentially the suppression, or inhibition, of unwanted programmes rather than the stimulation of a new programme *de novo*. This may provide valuable insights into the control of development by nutrients and growth regulators in a flowering plant, but may not tell us about the mechanisms of developmental switching required to bring new programmes into play.

Homeotic mutants have been produced by screening for lesions in polyamine synthesis [56], suggesting that polyamines may be involved in the regulation of expression of homeotic genes in flowering. Different mutants showed anthers partially converted to petals, ovules transformed into stamens (stamenoid ovules), stigmoid anthers, and the nested doll phenotype where flower formation starts again within the pistil of the developing flower and this structure is repeated. Polyamines have been implicated as possible natural growth regulators [26]. Clearly the relationships between polyamine (and other growth regulator) changes in apical meristems and the action of genes known to control flowering would be well worth knowing, but the methods for the detection of growth substance changes in the meristem itself are not yet really available.

If habituation for growth substance synthesis is a normal feature of developing plant systems [42] then small and temporary changes in growth substance production in the cells, or in cells in some other part of the plant, could be effective in altering developmental patterns in the meristems. The action of regulator genes could, at least partly, be on genes for enzymes involved in the synthesis or destruction of growth substances. Action of a single gene could thus have many pleiotropic effects. Regulator genes could also perhaps act to change the specific type of hormone molecule being synthesised. If morphogenesis at the meristems depends, even only in part, on the existence of diffusion fields of morphogens, then the substitution of one morphogen by another having identical physiological properties but having a different diffusion constant could theoretically result in a change in the patterning and positioning of organs [86, 89].

In order to know how growth substances (hormones) act in initiating organs we really need to know about growth substance distribution and activity (depending on tissue sensitivity) within the meristems themselves. One method which shows promise is immunocytochemistry to show the distribution of soluble compounds, such as growth substances, at the cellular level. The difficulties of the technique are in immobilising such soluble molecules within the cells and tissues and at the same time preserving their antigenicity. This has been done for abscisic acid and cytokinins [80, 81]. However, the technique has so far not been developed to the point at which the detailed images at the cellular level are sufficiently clear to allow variations in distribution and concentration within the shoot apical meristem to be described with any certainty [81]. The technique would also depend on a prior knowledge of the growth substances involved with initiation of a particular organ although this may not be necessary so long as applied substances (such as NPA and PCIB [60]) can be used to alter the positions and timing of primordium initiation.

The action of plant growth substances in many ways resembles that of trans acting factors. Do plant growth substances perhaps modify production of trans acting factors, or activators of the sort that may be involved in transcription [72]? It would be useful to be able to investigate the action of genes required for plant growth substance activity and response, and a start is being made in this direction [65].

Conclusions

1. The spectacular increase in the number of papers about plant development at the molecular level reflects the massive training in techniques in the early and mid 1980s and now the application of these techniques to development. Of these

techniques, PCR is a very powerful one for investigating gene expression not only qualitatively, but also quantitatively, so that the precise time at which genes begin to be activated could be pinpointed much more precisely. No doubt it will also be refined to provide quantitative data on the progression of gene activation as genes are up- and down- regulated. However, in the increasing application of DNA cloning and PCR techniques to meristems, careful interpretation will be required on the tissue and developmental specificity of cDNA clones. These trends promise an exciting future for plant development research.

2. We can expect an increasing understanding of the nature of combinatorial control of gene activity by the simultaneous activation of more than one promotory sequence. We can also expect better characterisation of the modular action of promoters which regulate the responses to environmental factors such as light, wounding, sucrose, and plant growth substances. The interaction of different classes of promoters will undoubtedly reveal a complexity of possible controls reflecting the versatility of a plant's response to the environment, which may not be paralleled in animals.

3. The discovery of homeotic genes and their resemblance to DNA transcription factors in yeast and humans has led to models of concentric domains of gene activity regulating floral morphogenesis. There is now a need to relate spatial gene activity to the cellular domains in which they function.

4. Cell cycle changes may not be simply reflecting shifts in growth programs, but may be integral to them, for differentiation starts in meristems. We will need to know how spatial control of morphogens is linked with cell cycle changes. In this regard, progress on the molecular controls of the cell cycle has been rapid, and studies on the natural substrates for key cell division cycle gene products, such as the nuclear lamins or microtubules, may offer a way forward.

5. The role of plant growth substances still needs clarification. They are unlikely to be primary elicitors of gene expression, but they can clearly modify developmental programs and can perhaps trigger them. Some transduction mechanisms seem likely to depend on ionic (Ca^{2+}) fluxes. The relationships between transduction mechanisms in the cell, primary and secondary messengers and the control of gene expression will undoubtedly be one of the most complex areas but one of the most rewarding in the near future.

Acknowledgements

We would like to thank Andrew Hudson and Steve Smith for their helpful comments and constructive criticisms.

References

1. Almeida J, Carpenter R, Robbins TP, Martin C, Coen ES: Genetic interactions underlying flower patterns in *Antirrhinum majus*. Genes Devel 3: 1758–1567 (1989).
2. Araki T, Komeda Y: Electrophoretic analysis of florally-evoked meristems of *Pharbitis nil* Choisy cv. Violet. Plant Cell Physiol 31: 137–144 (1990).
3. Barber JT, Steward FC: The proteins of *Tulipa* and their relation to morphogenesis. Devel Biol 17: 326–349 (1968).
4. Bassett CL, Mothershed CP, Galau GA: Floral-specific polypeptides in the Japanese morning glory. Planta 175: 221–228 (1988).
5. Battey NH, Lyndon RF: Reversion of flowering. Bot Rev 56: 162–89 (1990).
6. Benfey PN, Chua N-H: The cauliflower mosaic virus 35S promoter: combinatorial regulation of transcription in plants. Science 250: 959–966 (1990).
7. Benson S, Sucov H, Stephens L, Davidson E, Wilt F: A lineage-specific gene encoding a major matrix protein of the sea urchin embryo spicule. Devel Biol 120: 499–506 (1987).
8. Bernier G, Kinet J-M, Bronchart R: Cellular events at the meristem during floral induction in *Sinapis alba* L. Physiol Végét 5: 311–324 (1967).
9. Bernier G, Kinet J-M, Bodson M, Rouma Y, Jacqmard A: Experimental studies on the mitotic activity of the shoot apical meristem and its relation to floral evocation and morphogenesis in *Sinapis alba*. Bot Gaz 135: 345–352 (1974).
10. Bhadula SK, Sawhney VK: Protein analysis of floral organs of some members of Solanaceae. Bot Mag Tokyo 102: 85–91 (1989).
11. Blakely LM, Blakely RM, Colowit PM, Elliott DS: Experimental studies on lateral root formation in radish seedling roots. II. Analysis of the dose-response to endogenous auxin. Plant Physiol 87: 414–419 (1988).
12. Bodson M: Variation in the rate of cell division in the

apical meristem of *Sinapis alba* during transition to flowering. Ann Bot 39: 547–554 (1975).

13. Bowman JL, Smyth DR, Meyerowitz EM: Genetic interactions among floral homeotic genes of *Arabidopsis*. Development 112: 1–20 (1991).
14. Carpenter R, Coen ES: Floral homeotic mutations produced by transposon mutagenesis in *Antirrhinum majus*. Genes Devel 4: 1483–1493 (1990).
15. Coen ES: The role of homeotic genes in flower development and evolution. Annu Rev Plant Physiol Plant Mol Biol 42: 241–279 (1991).
16. Creamer LK, Jimines-Flores R, Richardson T: Genetic modifications of food proteins. Trends Biotechnol 6: 163–169 (1988).
17. Cremer F, Van de Walle C, Bernier G: Two-dimensional gel electrophoresis of polypeptides from vegetative and reproductive buds of *Sinapis alba*. Arch Int Physiol Biochim 94: 9–10 (1985).
18. Davies EL, Rennie P, Steeves TA: Further analytical and experimental studies on the shoot apex of *Helianthus annuus*: variable activity in the central zone. Canad J Bot 57: 971–980 (1979).
19. Davidson EH: Lineage-specific gene expression and the regulative capacities of the sea urchin embryo: a proposed mechanism. Development 105: 421–445 (1989).
20. De Bellis L, Nishimura M: Development of enzymes of the glyoxylate cycle during senescence of pumpkin cotyledons. Plant Cell Physiol 32: 555–561 (1991).
21. Drews GN, Bowman JL, Meyerowitz EM: Negative regulation of the *Arabidopsis* homeotic gene AGAMOUS by the APETALA2 product. Cell 65: 991–1002 (1991).
22. Drews GN, Goldberg RB: Genetic controls of flower development. Trends Genet 5: 256–261 (1989).
23. Edwards JW, Coruzzi GM: Cell-specific gene expression in plants. Annu Rev Genet 24: 275–30 (1990).
24. Ellis RJ: Molecular chaperones. Annu Rev Biochem 60: 321–347 (1991).
25. Evans PT, Holaway BL, Malmberg RL: Biochemical differentiation in the tobacco flower probed with monoclonal antibodies. Planta 175: 259–269 (1988).
26. Evans PT, Malmberg RL: Do polyamines have roles in plant development? Annu Rev Plant Physiol Plant Mol Biol 40: 235–269 (1989).

26a. Evrard JL, Jako C, Saint-Guily A, Weil JH, Kuntz M: Anther-specific, developmentally regulated expression of genes encoding a new class of prolinerich proteins in sunflower. Plant Mol Biol 16: 271–281 (1991).

27. Francis D, Lyndon RF: Synchronisation of cell division in the shoot apex of *Silene* in relation to flower initiation. Planta 145: 151–7 (1979).
28. Francis D, Rembur J, Nougarède A: Changements dans la composition polypeptidique du méristème de *Silene coeli-rosa* (L.) au cours de l'induction florale. Comptes Rendues Acad Sci Paris, Ser III 307: 763–770 (1988).
29. Gao X-P, Francis D, Ormrod JC, Bennett MD: Unpublished data (1991).
30. Gasser CS: Molecular studies on the differentiation of floral organs. Annu Rev Plant Physiol Plant Mol Biol 42: 621–649 (1991).
31. Gasser CS, Budelier KA, Smith AG, Shah DM, Fraley RT: Isolation of tissue-specific cDNAs from tomato pistils. Plant Cell 1: 15–24 (1989).
32. Goldberg RB: Regulation of plant gene expression. Phil Trans Royal Soc London B 314: 343–353 (1986).
33. Goldberg RB, Barker SJ, Perez-Grau L: Regulation of gene expression during plant embryogenesis. Cell 56: 149–160 (1989).
34. Gonthier R, Jacqmard A, Bernier G: Changes in cell-cycle duration and growth fraction in the shoot meristem of *Sinapis* during floral transition. Planta 170: 55–59 (1987).
35. Goodwin PB, Lyndon RF: Synchronisation of cell division during transition to flowering in *Silene* apices not due to increased symplast permeability. Protoplasma 116: 219–222 (1983).
36. Graham IA, Smith LM, Leaver CJ, Smith SM: Developmental regulation of expression of the malate synthase gene in transgenic plants. Plant Mol Biol 15: 539–549 (1990).
37. Green PB: Plasticity in shoot development: a biophysical view. Symp Soc Exp Biol 40: 211–232 (1986).
38. Grose S, Lyndon RF: Inhibition of growth and synchronised cell division in the shoot apex in relation to flowering in *Silene*. Planta 161: 289–294 (1984).
39. Hotta Y, de la Pena A, Stern H: Control of enzyme accessibility to specific DNA sequences during meiotic prophase by alterations in chromatin structure. Cytologia 50: 611–620 (1985).
40. Houssa C, Jacqmard A, Bernier G: Activation of replicon origins as a possible target for cytokinins in shoot meristems of *Sinapis*. Planta 181: 324–326 (1990).
41. Irish VF, Sussex IM: Function of the *apetala-1* gene during *Arabidopsis* floral development. Plant Cell 2: 741–753 (1990).
42. Jackson JA, Lyndon RF: Habituation: cultural curiosity or developmental determinant? Physiol Plant 79: 579–583 (1990).
43. Jacqmard A, Houssa C: DNA fiber replication during a morphogenetic switch in the shoot meristematic cells of a higher plant. Exp Cell Res 179: 454–461 (1988).
44. Jacqmard A, Lyndon RF, Salmon J: Appearance of specific antigenic proteins in the maturing sexual organs of *Sinapis* flowers. J Cell Sci 68: 195–209 (1984).
45. Kelly AJ, Zagotta MT, White RA, Chang C, Weeks-Wagner DR: Identification of genes expressed in the tobacco shoot apex during the floral transition. Plant Cell 2: 963–972 (1990).
46. Kinet J-M, Bodson M, Alvinia AM, Bernier G: The inhibition of flowering in *Sinapis alba* after the arrival of the floral stimulus at the meristem. Z Pflanzenphysiol 66: 49–63 (1971).
47. Koltunow AM, Truettner J, Cox KN, Wallroth M, Gold-

berg RB: Different temporal and spatial gene expression patterns occur during anther development. Plant Cell 2: 1201–1224 (1990).
48. Kunst L, Klenz JE, Martinez-Zapater J, Haughn GW: *Ap 2* gene determines the identity of perianth organs in flowers of *Arabidopsis thaliana*. Plant Cell 1: 1195–1208 (1989).
49. Langdale JA, Rothermel BA, Nelson T: Cellular pattern of photosynthetic gene expression in developing maize leaves. Genes Devel 2: 106–115 (1988).
50. Lyndon RF: The shoot apex. In: Yeoman MM (ed) Cell Division in Higher Plants, pp. 285–314. Academic Press, New York (1976).
51. Lyndon RF: Initiation and growth of internodes and stem and flower frusta in *Silene coeli-rosa*. In: Atherton J (ed) The Manipulation of Flowering, pp. 301–314. Butterworth, London (1987).
52. Lyndon RF: Synchronization of cell division during flower initiation in third-order buds of *Silene*. Ann Bot 59: 67–72 (1987).
53. Lyndon RF: Plant Development: The Cellular Basis, 320 pp. Unwin Hyman, London (1990).
54. Lyndon RF, Cunninghame ME: Control of shoot apical development via cell division. Symp Soc Exp Biol 40: 233–255 (1986).
55. Lyndon RF, Jacqmard A, Bernier G: Changes in protein composition of the shoot meristem during floral evocation in *Sinapis alba*. Physiol Plant 59: 476–480 (1983).
56. Malmberg RL, Mclndoo J, Hiatt J, Lowe BA: Genetics of polyamine synthesis in tobacco: genetic switches in the flower. Cold Spring Harb Symp Quant Biol 50: 475–482 (1985).
57. McDaniel CN: Competence, determination, and induction in plant development. In: Malacinski GM (ed) Pattern Formation: A Primer in Developmental Biology, pp. 393–412. Macmillan, New York (1984).
58. Medford JI, Elmer JS, Klee HJ: Molecular cloning and characterization of genes expressed in shoot apical meristems. Plant Cell 3: 359–370 (1991).
59. Meeks-Wagner DR, Dennis ES, Tran Thanh Van K, Peacock WJ: Tobacco genes expressed during *in vitro* floral initiation and their expression during normal plant development. Plant Cell 1: 25–35 (1989).
60. Meicenheimer RD: Changes in *Epliobium* phyllotaxy induced by N-1-naphthylphthalamic acid and A-4-chlorophenoxyisobutyric acid. Am J Bot 68: 1139–1154 (1981).
61. Meinhardt H: Models of Biological Pattern Formation. Academic Press, London (1982).
62. Melzer S, Majewski DM, Apel K: Early changes in gene expression during the transition from vegetative to generative growth in the long-day plant *Sinapis alba*. Plant Cell 2: 953–961 (1990).
63. Miller MB, Lyndon RF: The cell cycle in vegetative and floral shoot meristems measured by a double labelling technique. Planta 126: 37–43 (1975).
64. Mullet JE: Chloroplast development and gene expression. Annu Rev Plant Physiol Plant Mol Biol 39: 475–502 (1988).
65. Napier RM, Venis MA: Receptors for plant growth regulators: recent advances. J Plant Growth Regul 9: 113–126 (1990).
66. Newton KJ: Plant mitochondrial genomes: organization, expression and variation. Annu Rev Plant Physiol Plant Mol Biol 39: 503–532 (1988).
67. Ono M, Okazaki M, Harada H, Uchimiya H: In vitro translated polypeptides of different organs of *Pharbitis nil* Chois, strain Violet under flower-inductive and non-inductive conditions. Plant Sci 58: 1–7 (1988).
68. Ormrod JC, Francis D: Effects of light on the cell cycle in the shoot apex of *Silene coeli-rosa* L. on the first day of floral induction. Protoplasma 124: 96–105 (1985).
69. Ormrod JC, Francis D: Cell cycle responses to red or far-red light, or darkness, in the shoot apex of *Silene coeli-rosa* L. during floral induction. Ann Bot 57: 91–100 (1986).
70. Ormrod JC, Francis D: Mean rate of DNA replication and replicon size in the shoot apex of *Silene coeli-rosa* L. during the initial 120 minutes of the first day of floral induction. Protoplasma 130: 206–210 (1986).
71. Ormrod JC, Francis D: Effects of interpolated dark periods during the first long day of floral induction on the cell cycle in the shoot apex of *Silene coeli-rosa*. Physiol Plant 71: 372–378 (1987).
72. Ptashne M, Gann AAF: Activators and targets. Nature 346: 329–331 (1990).
73. Poethig RS: Phase change and the regulation of shoot morphogenesis in plants. Science 250: 923–930 (1990).
74. Rembur J, Nougarède A: Changes in the polypeptide composition during the ontogenic development of the shoot apex of *Chrysanthemum segetum* L. analyzed by two-dimensional mini-gel electrophoresis. Plant Cell Physiol 30: 359–363 (1989).
75. Rodriguez D, Dommes J, Northcote DH: Effect of abscisic and gibberellic acids on malate synthase transcripts in germinating castor bean seeds. Plant Mol Biol 9: 227–235 (1987).
76. Santiago JF, Goodwin PB: Restricted cell/cell communication in the shoot apex of *Silene coeli-rosa* during the transition to flowering is associated with a high mitotic index rather than with evocation. Protoplasma 146: 52–60 (1988).
77. Sawhney VK, Chen K, Sussex IM: Soluble proteins of the mature floral organs of tomato (*Lycopersicon esculentum* Mill.). J Plant Physiol 121: 265–271 (1985).
78. Smith AG, Hinchee M, Horsch R: Cell and tissue specific expression localized by *in situ* RNA hybridization in floral tissues. Plant Mol Biol Rep 5: 237–241 (1987).
79. Sommer H, Beltran J-P, Huitser P, Pape H, Lönnig W-E, Saedler H, Schwarz-Sommer Z: *Deficiens*, a homeotic gene involved in the control of flower morphogenesis in

Antirrhinum majus: The protein shows homology to transcription factors. EMBO J 9: 605–613 (1990).
80. Sossountzov L, Maldiney R, Sotta B, Sabbagh I, Habricot Y, Bonnet, Miginiac E: Immunocytochemical localization of cytokinins in Craigella tomato and a sideshootless mutant. Planta 175: 291–304 (1988).
81. Sotta B, Sossountzov L, Maldiney R, Sabbagh I, Tachon P, Miginiac E: Abscisic acid localization by light microscopic immunohistochemistry in *Chenopodium polyspermum* L. Histochem Cytochem 33: 201–208 (1985).
82. Staiger CJ, Lloyd CW: The plant cytoskeleton. Curr Opinion Cell Biol 3: 33–42 (1991).
83. Steeves TA, Sussex I: Patterns in plant development. Cambridge University Press, Cambridge (1989).
84. Stern H, Hotta Y: Chromosome organization in the regulation of meiotic prophase. In: Dickinson HG (ed) Controlling Events in Meiosis, 38th Symposium of the Society for Experimental Biology, pp. 161–175. Cambridge University Press, Cambridge (1984).
85. Taylor M, Francis D, Rembur J, Nougarède A: Changes to proteins in the shoot meristem of *Silene coeli-rosa* during the transition to flowering. Plant Cell Physiol 31: 1169–1176 (1990).
86. Thornley JHM: Phyllotaxis. I. A mechanistic model. Ann Bot 39: 491–507 (1975).
87. Tran Thanh Van KM: Control of morphogenesis in *in-vitro* cultures. Annu Rev Plant Physiol 32: 291–311 (1981).
88. Ursin VM, Yamaguchi J, McCormick S: Gametophytic and sporophytic expression of anther-specific genes in developing tomato anthers. Plant Cell 1: 727–736 (1989).
89. Veen AH, Lindenmayer A: Diffusion mechanism for phyllotaxis. Theoretical physico-chemical and computer study. Plant Physiol 60: 127–139 (1977).
90. Wallace JC, Gallili G, Kawata EE, Cuellar RE, Shotwell MA, Larkins BA: Aggregation of lysine-containing zeins into protein bodies in *Xenopus* oocytes. Science 240: 662–664 (1988).
91. Wick SM: Spatial aspects of cytokinesis in plant cells. Curr Opinion Cell Biol 3: 253–260 (1991).
92. Yamamoto YT, Taylor CG, Acedo GN, Chen C-L, Conkling MA: Characterization of *cis*-acting sequences regulating root-specific gene expression in tobacco. Plant Cell 3: 371–382 (1991).
93. Yanofsky MF, Ma H, Bowmann JL, Drews GN, Feldmann KA, Meyerowitz EM: The protein encoded by the *Arabidopsis* homeotic gene *agamous* resembles transcription factors. Nature 346: 35–39 (1990).

Plant Molecular Biology **19**: 69–87, 1992.
© 1992 *Kluwer Academic Publishers. Printed in Belgium.*

Molecular biology of fruit ripening and its manipulation with antisense genes

Julie Gray [1], Steve Picton [1], Junaid Shabbeer [1], Wolfgang Schuch [2] and Don Grierson [1]*
[1] *AFRC Research Group in Plant Gene Regulation, Department of Physiology and Environmental Science, Nottingham University School of Agriculture, Sutton Bonington, Loughborough, LE12 5RD, UK (*author for correspondence);* [2] *ICI Plant Biotechnology Group, ICI Seeds, Jealott's Hill Research Station, Bracknell, Berkshire, RG12 6EY, UK*

Key words: antisense RNA, carotenoids, ethylene, polygalacturonase, tomato, transgenic, ethylene forming enzyme, ripening genes

Abstract

Considerable progress in tomato molecular biology has been made over the past five years. At least 19 different mRNAs which increase in amount during tomato fruit ripening have been cloned and genes for enzymes involved in cell wall degradation (polygalacturonase and pectinesterase) and ethylene synthesis (ACC synthase) have been identified by conventional procedures. Transgenic plants have been used to identify regions of DNA flanking fruit-specific, ripening-related and ethylene-regulated genes and *trans*-acting factors which bind to these promoters have also been identified.

Antisense genes expressed in transgenic plants have proved to be highly effective for inhibiting the specific expression of ripening-related genes. These experiments have changed our understanding of how softening occurs in tomato fruit. Antisense techniques have also been used to identify genes encoding enzymes for carotenoid biosynthesis (phytoene synthase) and ethylene biosynthesis (the ethylene-forming enzyme). The altered characteristics of fruit transformed with specific antisense genes, such as retarded ripening and resistance to splitting, may prove to be of value to fruit growers, processors and ultimately the consumer.

1. Introduction

Many flowering plants invest large amounts of energy in the production of fleshy fruits in order to promote dispersal of their seed. During the final stage in their development, fruit undergo a complex series of physiological and biochemical events involving changes in colour, flavour, aroma and texture that make them both attractive and tasty to eat. These ripening processes are due to alterations in the activity of several biochemical pathways, and involve all cell compartments. The ripening programme requires differential gene expression and is modulated by environmental conditions and phytohormones. Knowledge of the function and regulation of gene expression during ripening is important for understanding fruit production, quality and storage, and may add to our overall understanding of how developmental processes in plants are controlled. The molecular basis of fruit ripening has been most widely studied in tomato (*Lycopersicon esculentum*). This is due to the availability of extensive chromosome maps [7, 63, 17], a variety of ripening mutants [17, 39] and a good transformation system [49, 59, 10, 29], combined with the commercial importance of the crop.

Here we review recent progress in the molecular biology of tomato ripening and highlight experiments where the expression of specific ripening genes has been dramatically reduced by introduction of the corresponding antisense genes. This has enabled the function of a number of ripening genes to be tested, has led to the identification of others, and has allowed aspects of fruit ripening in transgenic tomato plants to be altered [84, 85, 86, 81, 41, 11, 78, 65].

2. Biochemical changes during tomato fruit ripening

2.1. *Alterations in colour, flavour and texture*

Changes in fruit colour during tomato ripening are primarily brought about by the transition of chloroplasts into chromoplasts [44, 5]. At an early stage in the ripening process chloroplast thylakoid membranes, starch grains and chlorophyll pigments are broken down, and new carotenoid pigments such as β-carotene and lycopene, which are responsible for the orange and red colours of tomatoes, accumulate in the plastid [44, 32]. A number of tomato ripening mutants have been identified which have altered fruit colour. Some of these mutations, such as *yellow flesh*, *greenflesh* and *tangerine*, only affect carotenoid biosynthesis or chlorophyll degradation, whereas others, such as *ripening inhibitor* (*rin*), *Neverripe* (*Nr*), and *nonripening* (*nor*), also affect other aspects of ripening such as flavour and softening [39].

High concentrations of sugars (principally fructose and glucose) and high concentrations of acids (especially citric acid but also malic acid) make an important contribution to tomato flavour [38]. During ripening the ratio of starch polymers to sugar molecules [31] decreases, due to stored starch being metabolised into sugars within the fruit, or sugars being translocated from other parts of the plant. In addition to this increase in sweetness, the levels of other flavour components such as organic acids and many aromatic compounds increase during ripening to produce the final unique balance of sensory components of the ripe fruit.

Breakdown of stored starch and, later, loss of turgor contribute to fruit softening during ripening but the most important factor responsible for the softer texture of ripe fruit is believed to be changes to the structure of the celluloses, hemicelluloses and pectins which are the major constituents of the fruit cell wall [12]. The enzymes primarily thought to be involved in fruit cell wall metabolism during ripening are cellulase and polygalacturonase (PG) as activities of both of these enzymes increase during ripening. A number of other enzymes such as pectinesterase and β-galactosidase may also be involved. The relative importance of these various cell wall hydrolases may vary among species; for example, cellulase activity correlates well with softening of avocado fruit [15] whereas PG correlates best with softening of tomato fruits [45]. In addition, it is likely that no one enzyme is exclusively responsible for softening, as discussed in section 6.1.

2.2. *Role of ethylene in ripening*

In climacteric fruit (eg tomato, apple, banana, and avocado) the onset of ripening is accompanied by an increase in respiration called the respiratory climacteric [8]. This burst of respiration may be necessary to fuel the changes which occur during ripening of climacteric fruit. The onset of the respiratory increase is stimulated by the phytohormone ethylene, [58] and a peak of ethylene synthesis by the fruit accompanies the respiratory climacteric. Furthermore, addition of exogenous ethylene can stimulate mature fruit to ripen earlier than normal [58] and inhibition of ethylene synthesis [100] or removal of ethylene from the environment of picked green fruit delays the onset of ripening [50]. Ethylene promotes the appearance of new mRNA transcripts when applied to unripe fruit [34, 62, 55] and this is prevented by chemicals that interfere with ethylene perception or action [55, 92, 18]. These observations are consistent with the suggestion that ethylene regulates ripening by stimulating changes in gene expression. Ethylene is known to be involved in the regulation of many other plant developmental processes such as germination, flowering, abcission, and senescence, and also is associated with the response to environmental stresses such as mechanical wounding, infection and waterlogging [1, 99]. On the other hand, non-climacteric fruit such as citrus and strawberry ripen without an accompanying peak of respiration or ethylene synthesis [8]. At least some new enzymes are synthesised during ripening of non-climacteric fruit [30], indicating that changes in gene expression are involved in this process also, although it is not known what factor substitutes for ethylene during ripening in these species.

2.3. mRNA and protein synthesis

Studies on protein synthesis *in vivo* and *in vitro* have shown that the levels of a number of proteins [34] and mRNAs alter during ripening [69, 88, 35, 9, 14, 82, 60, 55], with the amounts of specific mRNAs and proteins either increasing, decreasing or remaining the same. A few proteins synthesised during ripening have been identified, such as polygalacturonase [91], but the majority remain unknown. mRNAs which would be expected to increase in amount during ripening include those encoding enzymes involved in pigment biosynthesis, ethylene biosynthesis and cell wall metabolism. mRNAs encoding a number of chloroplast enzymes would be expected to decrease during ripening and those encoding 'housekeeping' enzymes necessary for normal cellular metabolism would be expected to be maintained at similar levels. These observed changes in mRNA levels confirmed the view that the control of fruit ripening is at the level of gene expression and paved the way for cDNA cloning experiments.

3. Ripening-related cDNA libraries from tomato

Several groups have constructed cDNA libraries from mRNA extracted from ripening fruit. Screening of these libraries and subsequent analysis has resulted in the identification and characterisation of a number of ripening-related cDNAs (Tables 1 and 2).

Nineteen ripening-related clones (the pTOM series) were isolated by Slater *et al.* [82] from a cDNA library prepared from mRNA from ripe fruit. A further clone (pPE1), encoding a mRNA present in green and ripe fruit, was subsequently isolated from the same library [71].

Four clones homologous to mRNAs present at enhanced levels in ripe fruit (22.2, 24.3, 28.8, 9.24) were isolated by Mansson *et al.* [60]. A further cDNA clone for an mRNA present in green and ripe fruit, known as 2A11, has also been isolated from this library [67]. Lincoln *et al.* [55] constructed a cDNA library from mRNA extracted from tomato fruit at an early stage in ripening and identified four cDNA clones which were homologous to ethylene-inducible mRNAs (E4, E8, E17, J49) and one clone for a mRNA which was not ethylene-inducible (E41). Schaffer and Fischer [76], whilst studying chilling-induced genes, identified three clones (C14, C17, C19) representing mRNAs which are only expressed at low levels during ripening. Other tomato fruit cDNA libraries have been used to isolate specific clones [23, 80, 94].

3.1. Identification and characterisation of ripening-related cDNAs

A number of ripening-related cDNAs have been studied in detail, leading to the estimation of the sizes of their homologous mRNAs and the polypeptides they encode, gene copy number, and chromosome location (Table 1).

Screening of expression libraries, sequencing, and other approaches has led to the identification of some of the enzymes encoded by these cDNAs. pTOM 6 has been shown, by homology to an N-terminal amino acid sequence, to encode the

Table 1. cDNA clones from tomato fruit[1]

Clone	Approximate cDNA size (bp)	Transcript size (kb)	Polypeptide size (KDa)	Chromosome location	Approximate copy number	Function or homology	Related to	Sequenced
pTOM4 [82]	620 [82]			7 [52]			2A11[2]	yes[2]
pTOM5 [82]	1700 [82]	1.96 [62]	48 [82]	2/3 [52]		phytoene synthase [11]		yes [70]
pTOM6 [82]	1800 [82]	1.58 [62]	55 [82]	10 [52]	1 gene [10]	polygalacturonase [37]	F1, P1, C3, E41, pPG16	yes [37]
pTOM13 [82]	1400 [82]	1.4 [62]	35 [82]	7 [52]	3 genes [48]	ethylene-forming enzyme [41, 42]	E8, pTOM99	yes [47]
pTOM31 [82]	800 [82]			1 [52]				
pTOM36 [82]	1300 [82]	1.47 [62]	2 × 44 [82], 52 [36]	5/8/11 [52]				
pTOM75 [82]	950 [82]	1.2 [62]	28 [82]					yes[2]
pTOM96 [82]	700 [82]							
pTOM99 [82]	900 [82]	1.52 [62]	2 × 44 [82]	3/9 [52]			E8[3], pTOM13[3]	yes[3]
pTOM129 [82]	580 [82]			4 [52]				
pTOM137 [82]	1000 [82]	1.68 [62]	57 [82]					
pTOM25 [82]	900 [82]			2/6/6/9 [52]				
pTOM38 [82]	1100 [82]			1/3/4/5/6/6/9 [52]				
pTOM41 [82]	1550 [82]			1 [52]				
pTOM88 [82]	850 [82]			5/9/12 [52]				
pTOM92 [82]	1400 [82]			8 [52]				
pTOM94 [82]	600 [82]			8/10 [52]				
pTOM111 [82]	1000 [82]			3 [52]				
pTOM114 [82]	1400 [82]			1 [52]				
pTOM66 [68]	652 [28]	0.53 [19]				heat-shock protein [79]	Class 1 heat-shock protein [28]	yes [28]
pPE1 [71]	1665 [71]	1.6 [71]	42.5 [71]		≤3 copies [71]		pectin methylesterase [71]	yes [71]
22–2 [60]	1100 [60]	1.4 [60]			low copy number [60]			
24–3 [60]	700 [60]	0.7 + 0.75 [60]			low copy number [60]			
28–8 [60]	1400 [60]	1.6 [60]			low copy number [60]			
9–24 [60]	1800 [60]	1.8 [60]			multiple copies [60]			
2A11 [67]		0.7 [67]	10.6 [67]		single or low copy [67]		pTOM4[2]	yes [67]
J49 [55]		1.0 [55]	27 [55]		multiple copies [55] 5 genes [57]			
E4 [55]		0.88 [55]	25 [55]		single gene [57]			yes [16]
E8 [55]		1.6 [55]	42 [55]		3 genes [22]		pTOM13 [22] pTOM99[2]	yes [22]
E17 [55]		0.58 [55]			1 gene [22]	proteinase inhibitor [61]	pER1 [61]	yes [61]
E41 [55]		2.1 [55]	54 [55]			polygalacturonase [55]	pPG16, pTOM6. P1, C3, F1	yes [55]
D21 [24]								
C3 [80]	900 [80]	1.7 [80]					P1, F1	yes [80]

P1 [80]	1300 [80]					C3, F1	yes [80]
F1 [80]	1600 [80]		50.057 [80]		polygalacturonase [80]	pTOM6, P1, E41, C3, pPG16	yes [80]
pPG16 [23]	550 [6]	1.9 [23]	54 [23]	> 1 gene [23]	polygalacturonase [23]	pPG1.8, 1.9, F1, P1, C3, E41, pTOM6	yes [23]
pPG1.8 [6]	1800 [6]					pPG16, pTOM6, F1, P1, C3, E41	
pPG1.9 [6]	1900 [6]					pPG16, pTOM6, F1, P1, C3, E41	yes [25]
C14 [76]	1378 [76]	1.9 [76]		single copy [77]	thiol protease [76]	papain [77]	yes [76]
C17 [76]		1.2 [76]					yes [76]
C19 [76]		0.8 [76]					
pER1 [61]	524 [61]	0.6 [61]		multiple copies [61]		E17 [61]	yes [61]
pcVV4A [94]	1846 [94]	54.7 [94]			ACC synthase[4]	pcVV4B [94]	yes [94]
pcVV4B [94]	420 [94]					pcVV4A [94]	

[1] References in brackets.
[2] R. Fray, personal communication.
[3] J. Ray, personal communication.
[4] ACC = 1-amino cyclopropane-1-carboxylic acid.

cell wall degrading enzyme polygalacturonase (PG) [37]. Other cDNA clones have also been identified as encoding PG including E41 [55], pPG16, pPG1.8, pPG1.9 [23], C3, P1, F1 [80]. Pectinesterase, another cell wall degrading enzyme was shown to be encoded by pPE1, which is homologous to an mRNA which is present in green fruit and does not increase in amount during ripening [71]. pTOM 5 [70] has significant homology to a prokaryotic prephytoene pyrophosphate synthase gene suggesting it could be involved in carotenoid biosynthesis [2]. This has recently been confirmed by tomato antisense experiments (see section 6.3).

Other sequenced cDNAs include pTOM 66, which codes for a polypeptide related to a number of heat shock proteins [28]. E17 has homology to proteinase inhibitor I genes and may be involved in preventing insect attack [61]. 2A11, homologous to pTOM 4 (R. Fray, personal communication), encodes a peptide with slight homology to Bowman-Birk type proteinase inhibitors and storage proteins [67].

Transgenic experiments with antisense constructs (discussed in section 6.4) suggested that the pTOM 13 gene product is involved in ethylene synthesis [41] and it is now known to encode the ethylene-forming enzyme (EFE) [42, 87]. EFE has some amino acid homology with hydroxylase enzymes [41] and also has some homology to E8 [22], which is identical to pTOM 99 (J. Ray, personal communication). This suggests E8/pTOM 99 may also encode a hydroxylase, although its precise function is not known. Recently cDNA and genomic clones encoding tomato ACC synthase, another important enzyme in the ethylene biosynthesis pathway, have been identified and sequenced [73, 94].

4. Expression levels of ripening-related genes

4.1. Expression during ripening

The levels of the mRNAs homologous to ripening-related cDNAs have been studied during normal ripening and a number of other situ-

Table 2. mRNA levels in different organs and in response to environmental stimuli[1]

Clone	Wild-type green fruit	Wild-type ripening fruit	Other tissues	Mutant fruit			Temperature-treated fruit		Silver/NBD[2]-treated fruit	Leaf senescence	C_2H_4-related
				rin	*nor*	*nr*	35 °C	4 °C			
pTOM4	low [19]	increased [68]					reduced [68]	reduced [98]	silver-reduced [20]	no [19]	
pTOM5	very low [82] undetected [62]	increased [82, 62, 68]	ethylene-treated *rin* fruit [53]	reduced [53]		reduced [53]	reduced [68]	reduced [98]	silver-reduced [20]	no [19]	yes [62]
pTOM6	very low [82] undetected [62]	increased [82, 62, 68]		much detected [53]		reduced [53]	reduced [68]	reduced [98]	silver-reduced [20]	no [19]	yes [62]
pTOM13	very low [62]	increased [62, 68]	ethylene-treated *rin* fruit, [53] wounded leaves/ fruit [48]	reduced [53]		unaffected [53]	reduced [68]	increased [98]	silver-reduced [20, 19]	yes [19]	yes [62]
pTOM31	low [19]	increased [68]					reduced [68]	reduced [98]	silver-reduced [19, 20]	yes [19]	
pTOM36	low [82] undetected [62]	increased [82, 62, 68]	ethylene-treated *rin* fruit [53]	reduced [53]		unaffected [53]	reduced [68]	reduced [98]	silver-reduced [19, 20]	yes [19]	yes [62]
pTOM75	undetected [62]	low [62] increased [68]	roots/ leaves [62] leaves/ wounded leaves [48]	unaffected [53]		unaffected [53]	reduced [68]		silver-reduced [19, 20]	yes [19]	yes [62]
pTOM96		increased [68, 19]					little change [68]		silver-reduced [20]	no [19]	
pTOM99	undetected [62]	increased [62, 68, 19]	ethylene-treated *rin* fruit [53]	reduced [53]		unaffected [53]	reduced [68]		silver-reduced [20]	no [19]	yes [62]
pTOM129	low [19]	increased [68]					transient increase [68]	reduced [98]	silver-induced [19]	yes [19]	
pTOM137	present [62] low [19]	present [62] increased [68]	roots/ leaves [62]	increased [53]		unaffected [53]	reduced [68]		silver-reduced [19] increase [20]	yes [19]	yes [62]
pTOM66	low [19]	increased [68]					transient increase [68]	reduced [98]	silver-reduced [19]	yes [19]	
pPE1	high [71]	reduced [71]		similar [71]							
22–2	undetected [60]	increased [60]	leaf [60]	much reduced [60]	reduced [60]	unaffected [60]					
24–3	low [60]	increased [60]	leaf [60]	reduced [60]	reduced [60]	reduced [60]					
28–8	very low [60]	increased [60]		much reduced [60]	unaffected [60]	reduced [60]					
9–24	low [60]	increased [60]	leaf [60]	reduced [60]	reduced [60]	reduced [60]					
2A11	low [67]	increased [67]	ovary [67]								
J49	low [55]	increased [55]	leaves, ethylene-treated leaves, ethylene-treated fruit [61]	reduced [56]					NBD-reduced [55]		yes [55, 61, 57]

E4	low [55]	increased [55]	ethylene-treated *rin* fruit [56]	much reduced [56]					NBD-reduced [55]	yes [55, 57]
E8	low [55]	increased [55]	ethylene-treated *rin* fruit [56]	much reduced [56]					NBD-reduced [55]	yes [55, 57]
E17	low [55]	increased [55]	ethylene-treated *rin* fruit [56]	much reduced [56]					NBD-reduced [55]	yes [55, 57]
E41	undetected [55]	increased [55]								no [55, 57]
D21	present [57]	present [24, 57]		similar [24]	similar [24]	similar [24]				
C3	undetected [80]	increased [80]		undetected [80]						
pPG16	very low [23] undetected [6]	increased [23]								
pPG1.9	undetected [24]	increased [24]		much reduced [24]	almost undetected [24]	reduced [24]				
C14	undetected [76]	reduced [76]	chilled fruit [76]				increase [77]	increase [76]		
C17	undetected [76]	reduced [76]	chilled fruit [76]					increase [76]		
C19	detected [76]	increased [76]	chilled fruit [76]					increase [76]		
pER1	undetected [61]	increased [61]	ethylene-treated leaves [61], fruit/leaves [61]							yes [55, 57]
pcVV4a	undetected [94]	increased [94]	wounded fruit [94]							

[1] References in brackets.

[2] NBD = norbornadiene.

ations. This information is summarised in Table 2. Virtually all the ripening-related mRNAs have been shown to be absent or at a low level in immature fruit, increase dramatically during fruit ripening and decline as ripening progresses. Several mRNAs increase both during ripening and during senesence of leaves, which bears some physiological resemblences to ripening [19]. These mRNAs may be involved in events which are common to ripening and senescence such as ethylene synthesis and metabolism, or degradation of starch and chlorophyll. A number of the ripening-related mRNAs have also been detected in other organs such as leaves and roots but several, including the fruit PG mRNA, appear to be ripening-specific.

4.2. Ethylene and gene expression

Application of exogenous ethylene, which is known to induce ripening, causes increases in the level of a number of ripening-related mRNAs, but interestingly does not affect the level of PG mRNA [55]. Application of silver ions, which are thought to interfere with ethylene perception or action, reduces the level of a number of ripening-related mRNAs in tomato fruit, although not all mRNAs are affected. Paradoxically, silver reduces the level of PG mRNA [18] (see section 5.4).

Some of the mRNAs homologous to ripening-related cDNAs have been found to be present at a reduced level in the fruit of tomato ripening mutants such as *Nr* and *rin* [60, 24, 53]. In *rin* fruit, the amount of some but not all of these mRNAs can be restored by addition of ethylene, showing that *rin* fruit must have functional ethylene receptors [56, 53]. However ethylene treatment of *rin* fruit does not restore normal ripening characteristics of the fruit such as pigment accumulation, indicating either that a second ripening-specific receptor is missing or, alternatively, there is a lesion in the signal transduction chain that links ethylene perception to the ripening response.

4.3. Expression in response to environmental stimuli

mRNA homologous to pTOM 13 accumulates within 20 minutes after wounding of leaves [83]. This result first suggested to us that the pTOM 13 protein may play a role in ethylene synthesis as it is detected during fruit ripening and other ethylene regulated processes, such as wounding [47] and leaf senescence [19]. Other stimuli, such as salt stress and high and low temperatures affect accumulation of other fruit mRNAs. For example, many of the ripening-related mRNAs were found at a reduced level in fruit stored at 35 °C or 4 °C, when several aspects of ripening are also impaired [68, 98] and the reduced levels of at least one mRNA in the *nor* mutant fruit were increased in response to NaCl [21]. A chill-induced cDNA (C14) has been identified as encoding a thiol protease [76].

5. Control of gene expression in fruit

Several fruit-specific or ripening-related cDNAs have been used to identify corresponding genomic clones for PG [10], E4 and E8 [22, 16], EFE [47, 48, 54], 2A11 [67] and pTOM5 [72]. Regions of DNA flanking some of these ripening-related genes have been studied *in vitro* and *in vivo* in order to identify *cis*-acting regions of DNA with regulatory functions.

5.1. The E4 and E8 genes

The E8 (pTOM99) and E4 genes of tomato display a temporal and tissue-specific mode of expression; their mRNAs are abundant in ripening fruit, but are not detectable in leaf, root, stem or unripe fruit. Both E8 and E4 gene expression is activated by ethylene, and nuclear run-on transcription assays indicate that both genes are regulated at the transcriptional level. A genomic clone of the E8 gene was obtained on a 4.4 kb fragment, together with 2 kb of 5′-flanking sequence and 0.5 kb of 3′-flanking sequence [22]. To determine whether sequences sufficient for the normal expression of the E8 gene are contained on the 4.4 kb fragment, it was reinserted into the tomato genome by transformation. So that expression of thc reintroduced E8 gene could be distinguished from that of the endogenous E8 gene, a tag consisting of a λ DNA fragment was inserted into the cloned E8 gene, and RNA detected by hybridization with labelled λ DNA. RNA was barely detectable in the unripe fruit of transformed plants, whereas it was present at a level similar to that of the endogenous E8 gene in ripening fruit. Furthermore, expression of the tagged gene was inducible by exogenous ethylene [22], showing that sequences required for the ethylene-responsive and developmentally regulated expression of the E8 gene are contained on the 4.4 kb E8 genomic fragment.

Mobility shift assays performed with sequences from the 2 kb upstream region of the E8 gene indicate that there are multiple protein recognition sites in the 5′-flanking sequences [22] and activation of the E8 gene is correlated with the accumulation of proteins that are able to interact with these. A constitutive binding activity which interacts at other sites in the upstream sequence is also detectable in unripe and ripening fruit extracts. One proteinaceous binding factor, the activity of which is particularly intense and appears to increase in ripening fruit, has been studied in more detail [22]. Binding occurs 920 to 936 bp upstream of the E8 transcription start site and cannot be out-competed by other upstream E8 sequences that are also involved in complex formation. This suggests that the sequence to which the factor binds is unique, and that distinct binding factors participate in the formation of other complexes. The factor, however, also appears able to interact with a sequence located 18 to 34 bp upstream of the E4 transcription start site which overlaps the putative TATA box of the E4 gene. This suggests that the factor may be involved in regulation of both the E8 and E4 genes which are coordinately transcribed during ripening [22]. The variation in distance between the E8 or E4 transcription start sites and this recognition sequence indicates that it may be enhancer-like [16]. A comparison of the binding sites which interact with the same factor upstream of the E8 and E4 genes revealed the following 17 bp consensus sequence; ATT–C–A–––A–AGAAA [16]. Methylation interference analyses showed that a pair of guanine residues which are separated by 8 bp on opposite DNA strands are an essential component of both the E8 and E4 binding sites. The DNA sequences surrounding these guanine residues are identical at 11 of 17 positions, with more similar sequences flanking a non-conserved A/T-rich core. The additional conserved sequences are also important for binding site recognition at both the E8 and E4 sites [16].

Outside of their common binding sites, the 5′-flanking sequences of the E8 and E4 genes have little DNA sequence identity. This is presumably related to their different regulation patterns. Although the genes are coordinately transcribed during ripening, there are certain differences in their patterns of expression. For example, exposing whole plants to exogenous ethylene results in E4 gene expression in many plant organs, whereas E8 gene expression is induced only in fruit [57].

5.2. The 2A11 gene

The 2A11 cDNA [67] encodes a polypeptide with homology to pTOM 4. Expression of the 2A11 gene is only detectable in fruit, but not in leaf, stem or root tissue. 2A11 expression, however, is not ripening-specific; the mRNA first appears in ovaries at anthesis, and rises to a maximum level in mature ripe fruit [67]. Tomato plants have been transformed with the β-glucuronidase (GUS) reporter gene, flanked by 5′ and 3′ regions of the 2A11 gene, in an attempt to delimit the 2A11 DNA regulatory elements. Sequences spanning 4 kb and 1.3 kb upstream of the 2A11 gene were able to confer high level, fruit-specific GUS expression [95]. A 1.8 kb upstream sequence, though, was unable to direct GUS activity in either fruit or leaf tissue, indicating that a negative regulatory element is located in the region between approximately 1.3 and 1.8 kb upstream of the 2A11 transcription start site. A positive fruit-specific regulatory element is also present more than 1.8 kb upstream. A 3′ region, spanning 2.3 kb downstream from the stop codon of the 2A11 gene, appeared to have little, if any, role in fruit-specific expression [96].

Short sequences from the 1.8 kb upstream region of the 2A11 gene were also tested for their ability to enhance expression of the GUS gene from a truncated CaMV 35S promoter in transgenic tomato plants. This revealed the presence of several positive regulatory elements; at least four were fruit-specific, and three were general. In addition three negative regulatory elements appeared to be present. DNA-binding studies also indicated that a number of proteins interact with sequences in the same 1.8 kb region upstream of the 2A11 gene. A sequence comparison of the 5′ region of the 2A11 gene with the upstream regions of the PG, E8 and E4 genes did not reveal the presence of any conserved sequences [96].

5.3. The polygalacturonase gene

Only one gene for the endo-acting PG responsible for metabolising cell wall pectin during fruit ripening was isolated from the tomato genome [10]. It is probable that the PG isoenzymes PG1, PG2a and PG2b are all products of this same gene. This has been demonstrated, at least for PG2a and 2b, by showing both are produced in transgenic tobacco expressing this gene [66]. Different genes with homology to PG are expressed in pollen [13] and possibly abscission zones [89]. 1.4 kb of DNA upstream of the PG coding sequence was used to direct expression of a reporter gene chloramphenicol acetyl transferase (CAT), in transgenic tomato plants. CAT activity was only found in ripe fruit and not in unripe fruit or roots, stems and leaves [10]. *Cis*-acting DNA sequences which regulate gene expression during ripening must therefore be present within this 1.4 kb DNA upstream of the PG gene. The level of expression of the reporter gene was significantly lower than the level of PG gene expression during ripening, implying that other regions of DNA also affect PG gene expression *in vivo* (C. Watson, C. Smith, D. Grierson and W. Schuch, unpublished results).

The PG mRNA first becomes detectable within one to two days after the rise in ethylene production. Run-on transcription assays indicate that this increase in PG mRNA results from activation of transcription [26]. This reaches a peak in fully orange fruit, where it is at least 50-fold greater than at the mature green stage, and continues until very late in the ripening process [27]. Steady state PG mRNA levels rise over 2000-fold during ripening, and closely parallel the increase in PG transcription. PG mRNA seems to be particularly stable, however, which also accounts at least partly for the high level of PG mRNA accumulation in ripe fruit [26] (see section 5.4).

A computer search has not revealed the presence of the E8/E4 consensus protein recognition site in the 1.45 kb PG promoter upstream of the transcription start site and there were no other significant sequence homologies apart from variable stretches of A/T-rich sequences between the flanking sequences of the E8 or E4 fruit ripening genes and the PG promoter (J. Shabbeer, unpublished results). Additionally, it has been shown that 5′-flanking sequences of the PG gene cannot out-compete binding by proteins to the E8/E4 consensus site [16].

5.4. Comparison of E8 and PG gene expression

In fruit producing maximum levels of ethylene, the transcription rate of the E8 gene is over six-fold greater than the maximum PG transcription rate. Despite this, the PG mRNA rises to a level six-fold greater than that of a constitutively expressed control gene D21, and 3.5-fold higher than the level of E8 mRNA [26]. This suggests that post-transcriptional processes, such as mRNA stability or cytoplasmic entry rates, also contribute to the relatively high accumulation of PG mRNA during ripening.

The E8 and PG genes display differences in their response to ethylene, which may reflect some fundamental difference in the control mechanisms of the two genes. For example, prolonged exposure (10 to 24 hours) of mature green fruit to exogenous ethylene induces the expression of a number of ripening genes, including E8 and PG. If, however, fruit are exposed to exogenous ethylene for less than eight hours, E8 gene expression is activated, whereas PG expression is not [55]. This suggests that relatively few events that are stimulated by ethylene may be required for rapid E8 gene expression, whereas a greater number of events following ethylene exposure may be needed for PG expression. Differences in signal transduction pathways may also reflect changes in expression of the two genes in *rin* fruit. Fruit with the *rin* mutation do not ripen normally, and do not produce elevated levels of ethylene. The maximum E8 transcription rate in *rin* fruit, though, is approximately 60 % that of the corresponding rate in wild-type fruit. In contrast, the PG gene shows very low transcriptional activity in *rin* fruit [26]. Furthermore, exogenous ethylene, although incapable of restoring normal ripening in *rin* fruit, can increase the expression of the E8 (pTOM 99) gene to a level similar to that in ripening wild-type fruit. Exogenous ethylene, however, does not induce up-regulation of the PG gene at all in *rin* fruit [53].

6. Manipulating ripening with antisense genes

Recent experiments have shown that it is possible to 'turn off' the expression of existing genes in transgenic plants by introducing a gene constructed to generate antisense RNA [93, 84, 80]. This has allowed expression of specific ripening-related genes to be curtailed, permitting their identification and assessment of their function in ripening [85, 41, 11].

Despite the importance of the antisense gene approach for causing directed mutations, the precise mechanism of action of antisense RNA is not clear. The general observation is that antisense genes are specific for the target gene and cause a dramatic reduction in the accumulation of the normal sense mRNA. It is widely assumed that transcription of antisense RNA is required for them to be effective. In cells where the endogenous gene is normally expressed, there is also a reduction in antisense RNA accumulation. It is possible that antisense RNA interferes with transcription of the target gene, although *in vitro* transcription studies with PG antisense fruit indicate this does not occur [81]. Alternatively, antisense RNA might interfere with processing or translation of the mRNA. It seems most probable, however, that the stability of the message is interfered with by the antisense RNA transcript base pairing specifically with the target mRNA. The resulting double-stranded mRNA duplex is presumed to be rapidly degraded although it is not yet clear whether this occurs in the nucleus, the cytosol, or both.

In some cases expression of sense genes in transgenic plants has also caused down regulation of both the target gene and transgene gene [51, 86]. Although this may involve a new mechanism for gene inactivation [51] it has been suggested that it could be due to the unexpected production of antisense RNA by transcriptional read-through from nearby promoters [40].

6.1. Antisense inhibition of polygalacturonase gene expression

The first experiments which utilised antisense RNA techniques to study fruit ripening involved the manipulation of polygalacturonase gene expression. We used a DNA construct consisting of a 730 bp fragment of the PG cDNA, which was inserted in the antisense orientation between the CaMV 35S RNA promoter and the 3′-end of the nopaline synthase gene, and transformed into tomato [84]. Similar experiments have been carried out by Sheehy *et al.*, using a full-length cDNA [81]. Transgenic plants expressing the antisense gene had reduced levels of both PG mRNA and PG enzyme activity in the fruit at all stages of ripening. Other aspects of ripening such as ethylene production and lycopene accumulation were not affected in the mutant plants. The antisense genes were stably inherited, and the progeny resulting from self-fertilization contained 0, 1 or 2 copies of the antisense gene [86]. Plants that did not inherit an antisense gene had normal PG ac-

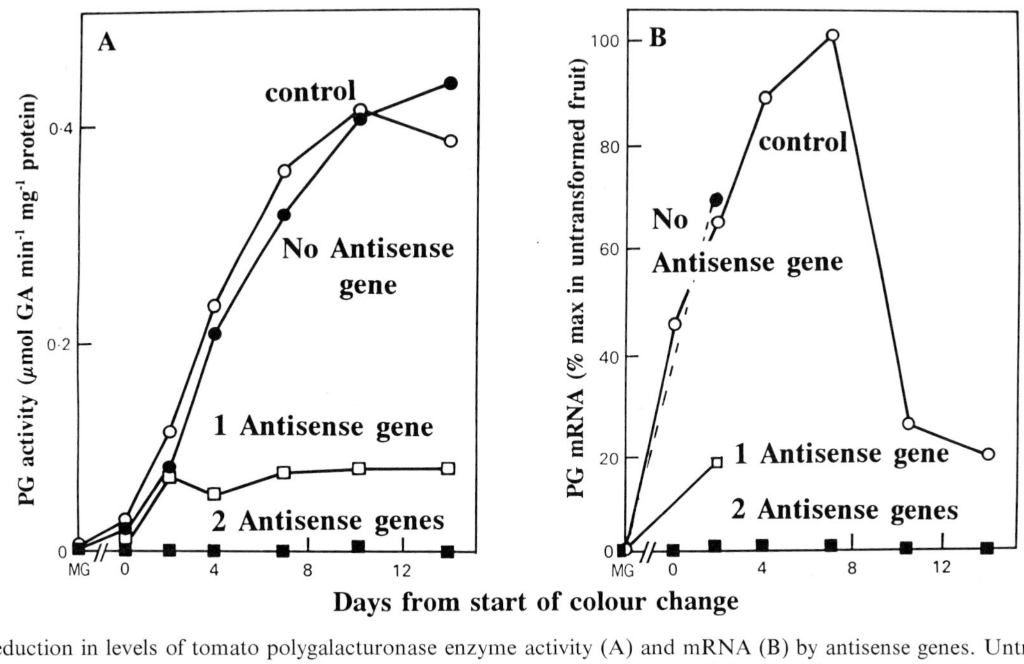

Fig. 1. Reduction in levels of tomato polygalacturonase enzyme activity (A) and mRNA (B) by antisense genes. Untransformed control tomatoes and fruit of transformed progeny containing 0, 1 or 2 antisense PG (pTOM 6) genes were picked and assayed at various stages during ripening. Redrawn from Smith *et al.* [86].

A B

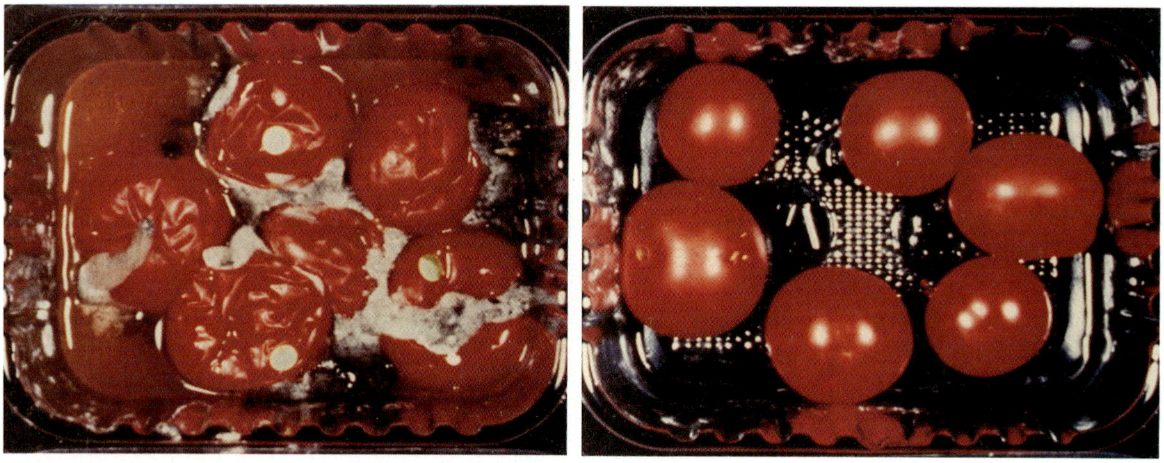

A B

A B C

tivity, showing that the antisense gene exerted no permanent effect on its target gene. Plants that inherited 2 copies of the antisense gene had PG activity reduced to ca. 1% of normal, illustrating that antisense inhibition can be an extremely powerful way of inhibiting expression of a specific gene (Fig. 1). All isoforms of fruit PG were greatly reduced by the antisense gene [86], supporting the conclusion that they are derived from a single gene. Despite this dramatic decrease in PG expression no significant change in the softening of the mutant fruit could be detected by compressibility measurements [84]. However, pectin degradation in the fruit cell walls was inhibited [86]. This result leads us to the conclusion that although PG is important for pectin degradation, it is not, as previously believed, the primary determinant of softening in tomato fruit. It is probable that other cell wall degrading enzymes make an important contribution to softening, as may other processes, such as redistribution of calcium ions. Another experiment supporting this conclusion involved transformation of an inducible PG gene into the *rin* tomato ripening mutant. This mutant normally expresses PG at a very low level [24, 53] and the fruit barely soften. When transformed with the PG gene driven by the ethylene-inducible E8 promoter the *rin* fruit still did not soften although increased pectin degradation was observed in ethylene-treated fruit [29].

It has been proposed that pectin fragments, released from cell walls by PG action, elicit the production of ethylene which in turn regulates expression of ripening genes [90, 3]. This does not fit well with measurements showing ethylene production precedes PG synthesis in ripening fruit [33]. Furthermore the antisense PG experiments do not support this view, since although pectin degradation is much reduced in these fruit, ethylene synthesis and ripening is apparently unaffected [85].

Although the antisense PG fruit appear to soften normally they do exhibit other characteristics which may give them commercial value. The PG antisense fruit have several advantages for processing and are more resistant to cracking or mechanical damage and secondary fungal infection (Fig. 2) [78]. These characteristics are probably related to the inhibition of pectin degradation. This may cause the cells in the pericarp to be bonded together more firmly and also alter the physical properties of tomato extracts.

6.2. Antisense inhibition of fruit pectinesterase

Pectinesterase (PE) removes methyl groups from the pectin component of cell walls. Enzyme activity is present in a range of organs and tissues, including ripening fruit, where it may play a role in cell wall metabolism. There are probably at least three PE genes [43]. A cDNA for a fruit pectinesterase has been identified [71] and used to construct an antisense gene controlled by a 35S promoter. Expression of this antisense gene in transgenic tomatoes greatly inhibited PE activity in fruit but had no effect on enzyme activity in leaves or roots (L. Hall, personal communication).

6.3. Antisense inhibition of a carotenoid biosynthesis gene

pTOM 5 has sequence homology with the *crtB* genes from *Erwinia herbicola* and *Rhodobacter capsulatus* [2]. These genes are believed to encode prephytoene pyrophosphate synthase, which ca-

Fig. 2. (opposite page, top). Ripe PG (pTOM 6) antisense tomato fruit have increased resistance to cracking and secondary fungal infections. Fruit of transformed progeny containing 0 (A), or 2 (B) antisense PG genes were picked 7 days after onset of colour change and stored at 12 °C for 21 days.

Fig. 3. (opposite page, middle). Phytoene synthase (pTOM 5) antisense tomato fruit ripen to a yellow colour. A, ripe untransformed fruit; B, ripe phytoene synthase antisense fruit.

Fig. 5. (opposite page, bottom). Ripening of EFE antisense (pTOM 13) tomato fruit. Untransformed control fruit (A) and homozygous pTOM 13 antisense fruit (B and C) were picked at mature green stage and incubated with (C) or without (A and B) the addition of 20μl/l ethylene. Fruit were photographed two weeks after the onset of colour change.

talyses the dimerisation of two molecules of geranylgeranyl pyrophosphate to form phytoene, the first C_{40} carotene in the carotenoid synthesis pathway. These observations suggested that the pTOM 5 gene product is involved in the production of carotenoid pigments during tomato fruit ripening. Expression of antisense pTOM 5 RNA in transgenic tomato plants confirmed this. The transformed plants had fruit which ripened to a yellow colour and also had pale yellow flowers (Fig. 3) [11]. Interestingly, this phenotype of ripe yellow fruit and pale yellow flowers is characteristic of the *yellow flesh* tomato mutant. Lycopene could not be detected in these fruit although other aspects of ripening such as PG mRNA levels remained unaffected. As prephytoene pyrophosphate, the product of the step catalysed by prephytoene pyrophosphate synthase, does not accumulate in these transgenic fruit it seems possible that the tomato enzyme catalyses two steps of the carotenoid biosynthesis pathway. Both *yellow flesh* and a pTOM 5 gene are located on chromosome 3 [52] and work is now in progress to determine whether the *yellow flesh* phenotype is a result of a mutation within the pTOM 5 gene. It is possible that if the pTOM 5 gene were to be overexpresssed in tomato plants the fruit may have enhanced colour, which may be valuable to tomato processors. Recently, a phytoene desaturase gene from soybean has been cloned by complementation of a mutant of *Rhodobacter capsulatus* [4]. A similar antisense or complementation strategy could now be used to identify and manipulate many other genes involved in pigment biosynthesis.

6.4. Identification of genes encoding the ethylene-forming enzyme

pTOM 13 mRNA levels increase both during ripening and rapidly after wounding of fruit and leaves, which suggested that the pTOM 13 gene product could be involved in ethylene synthesis [83]. To test this hypothesis antisense pTOM 13 mRNA was expressed in transgenic tomato plants, which consequently had reduced levels of the normal pTOM 13 mRNA both during fruit ripening and after wounding. In both these situations ethylene production was also reduced in the transgenic plants [41]. In common with the antisense PG experiments, the extent of antisense inhibition was dependent on gene dosage and progeny which did not inherit an antisense gene had a normal pattern of ethylene synthesis (Fig. 4). This suggested that pTOM 13 could encode one of the two enzymes involved in the ethylene biosynthesis pathway: ACC synthase, which catalyses the formation of ACC (1-aminocyclopropane-1-carboxylic acid) from S-adenosyl methionine, and the ethylene-forming enzyme (EFE or ACC oxidase), which catalyses the conversion of ACC to ethylene [100].

It was found that EFE activity was inhibited in pTOM 13 antisense transgenic plants in a gene dosage-dependent manner [41]. This suggested that pTOM 13 encodes at least part of the ethylene-forming enzyme. In further experiments pTOM 13 was expressed in *Saccharomyces cerevisiae*, after first correcting a number of cloning artefacts identified in the original cDNA sequence. Transformed yeast expressing the reconstructed pTOM 13 cDNA were able to convert ACC to ethylene in a stereospecific manner characteristic of the plant enzyme, whereas untrans-

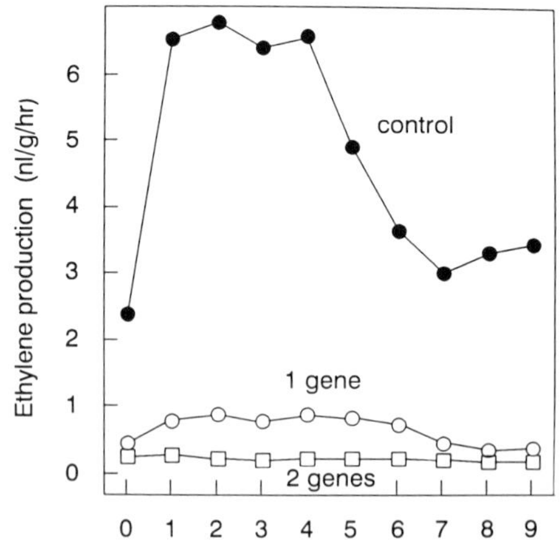

Fig. 4. Ethylene evolution of during ripening of detached tomato fruit. Daily measurements were taken from untransformed control fruit and fruit carrying 1 or 2 EFE (pTOM 13) antisense genes. Redrawn from Hamilton *et al.* [41].

formed yeast controls were unable to produce ethylene [42]. Recently, Spanu *et al.* [87] used pTOM13 to isolate an elicitor-induced EFE clone from tomato cell cultures and showed that RNA transcribed from this cDNA caused EFE synthesis when microinjected into *Xenopus* oocytes. Prior to these experiments EFE had not been well characterised and no molecular probes were available. The enzyme has been notoriously difficult to purify in an active form, and was previously believed to be membrane bound. The amino acid sequence of EFE, predicted from the pTOM 13 sequence [47] showed homology to the sequence of flavanone 3-hydroxylase [Prescott and Martin, cited in 41]. This was a significant observation as one of the possible mechanisms suggested for the conversion of ACC to ethylene requires a hydroxylation step [100]. EFE has now been extracted and solubilised in an active form using the conditions developed for flavanone 3-hydroxylase [97]. This development, and the ability to express the enzyme in yeast [42] should facilitate the characterisation of EFE. At present it seems unlikely to be located in a membrane and may be soluble.

6.5. Inhibition of ripening by reducing ethylene synthesis with antisense genes

Experiments to determine the phenotype of the pTOM 13 antisense plants showed that fruit left on the plant changed colour but over a period of weeks did not over-ripen or shrivel like the controls [41]. This would obviously be a desirable characteristic for prolonging the shelf-life of fresh fruit. When the fruit were picked at a mature green stage the reduced ethylene phenotype became more pronounced. The fruit changed to a yellow colour, turning orange after several weeks. When similar fruit were picked and ripened in the presence of ethylene these effects were overcome and the fruit ripened in a similar way to normal control fruit (Fig. 5). The inhibition of ethylene synthesis, with a consequent inhibition of ripening, has also been achieved using ACC synthase antisense genes in tomato [65]. Further experiments are underway to characterise more fully these pTOM 13 antisense plants. Since the effectiveness of an antisense gene varies in different plants, probably in relation to the site of insertion of the gene, it is likely that a range of transformants can be generated in which ethylene synthesis is inhibited to different extents. In addition, using organ-specific or developmentally regulated promoters to drive antisense gene expression, it should be possible to inhibit ethylene production in particular parts of a plant or at particular stages in the life cycle. Because ethylene is believed to play a role in regulating many developmental processes and responses to stress, such low ethylene plants should prove useful in studying a diverse range of ethylene mediated processes such as ripening, senescence, abscission, infection by pathogens, and responses to waterlogging.

6.6. Ethylene synthesis and action

cDNA clones for tomato and squash ACC synthases have recently been identified and sequenced [74, 94, 64, 75] and there are several tomato genes for this enzyme [73]. There are also three EFE genes homologous to pTOM 13 in the tomato genome, *eth1*, *eth2*, and *eth3* [47, 54]. It is possible that expression of these genes is regulated by different stimuli, and that different EFE and ACC synthase genes are responsible for ethylene production in, for example, the initiation of ripening, or during senescence or abscission, and in response to wounding. It is now important to study the expression of these different genes to help us to understand how ethylene synthesis is controlled during development and in response to stress. The next significant steps must be the identification of the ethylene receptor and the elucidation of the signal transduction chain that leads to changes in gene expression.

Acknowledgements

We thank Rupert Fray for his helpful suggestions during the preparation of this manuscript. Work in the authors' laboratory was funded by AFRC and SERC.

References

1. Abeles FB: Ethylene in Plant Biology. Academic Press, New York (1973).
2. Armstrong GA, Alberti M, Hearst JE: Conserved enzymes mediate the early reactions of carotenoid biosynthesis in nonphotosynthetic and photosynthetic prokaryotes. Proc Natl Acad Sci USA 87: 9975–9979 (1990).
3. Baldwin EA, Pressey R: Tomato polygalacturonase elicits ethylene production in tomato fruit. J Am Soc Hort Sci 113: 92–95 (1988).
4. Bartley GE, Viitanen PV, Pecker I, Chamouitz D, Hirschberg J, Scolnik PA: Molecular cloning and expression in photosynthetic bacteria of a soybean cDNA coding for phytoene desaturase, an enzyme of the carotenoid biosynthesis pathway. Proc Natl Acad Sci USA 88: 6532–6536 (1991).
5. Bathgate B, Purton ME, Grierson D, Goodenough PW: Plastid changes during the conversion of chloroplasts to chromoplasts in ripening tomatoes. Planta 165: 197–204 (1985).
6. Bennett AB, DellaPenna D: Polygalacturonase gene expression in ripening tomato fruit. In: Nevins DJ, Jones RA (eds) Tomato Biotechnology, pp. 299–308. Alan R Liss, New York (1987).
7. Bernatzky R, Tanksley SD: Towards a saturated linkage map in tomato based on isozymes and random cDNA sequences. Genetics 112: 887–898 (1986).
8. Biale JB, Young RE: Respiration and ripening in fruits – retrospect and prospect. In: Friend J, Rhodes MJC (eds) Recent Advances in the Biochemistry of Fruit and Vegetables, pp. 1–39. Academic Press, London (1981).
9. Biggs MS, Harriman RW, Handa AK: Changes in gene expression during tomato fruit ripening. Plant Physiol 81: 395–403 (1986).
10. Bird CR, Smith CJS, Ray JA, Moureau P, Bevan MW, Bird AS, Hughes S, Morris PC, Grierson D, Schuch W: The tomato polygalacturonase gene and ripening-specific expression in transgenic plants. Plant Mol Biol 11: 651–662 (1988).
11. Bird CR, Ray JA, Fletcher JD, Boniwell JM, Bird AS, Teulieres C, Blain I, Bramley PM, Schuch W: Using antisense RNA to study gene function: Inhibition of carotenoid biosynthesis in transgenic tomatoes. Bio/technology 9: 635–639 (1991).
12. Brady CF: Fruit ripening. Annu Rev Plant Physiol 38: 155–178 (1987).
13. Brown SM, Crouch ML: Characterization of a gene family abundantly expressed in *Oenothera organensis* pollen that shows sequence similarity to polygalacturonase. Plant Cell 2: 268–274 (1990).
14. Christoffersen RE, Warm E, Laties GG: Gene expression during fruit ripening in avocado. Planta 155: 52–57 (1982).
15. Christoffersen RE, Tucker ML, Laties GG: Cellulase gene expression in ripening avocado fruit: The accumulation of cellulase mRNA and protein as demonstrated by cDNA hybridization and immunodetection. Plant Mol Biol 3: 385–391 (1984).
16. Cordes S, Deikman J, Margossian LJ, Fischer RL: Interaction of a developmentally regulated DNA-binding factor with sites flanking two different fruit-ripening genes from tomato. Plant Cell 1: 1025–1034 (1989).
17. Darby LA, Ritchie DB, Taylor IB: Isogenic lines of the tomato 'Ailsa Craig.' Glasshouse Crops Res Annu Rep 1977: 168–184 (1978).
18. Davies KM, Hobson GE, Grierson D: Silver ions inhibit the ethylene-stimulated production of ripening-related mRNAs in tomato. Plant Cell Envir 11: 729–738 (1988).
19. Davies KM, Grierson D: Identification of cDNA clones for tomato (*Lycopersicon esculentum* Mill) mRNAs that accumulate during fruit ripening and leaf senescence in response to ethylene. Planta 179: 73–80 (1989).
20. Davies KM, Hobson GE, Grierson D: Differential effect of silver ions on the accumulation of ripening-related mRNAs in tomato fruit. J Plant Physiol 135: 708–713 (1990).
21. Davies K, Grierson D, Edwards R, Hobson G: Salt-stress induces partial ripening of the *nor* tomato mutant but expression of only some ripening-related genes. J Plant Physiol 96: 545–550 (1991).
22. Deikman J, Fischer RL: Interaction of a DNA binding factor with the 5′-flanking region of an ethylene-responsive fruit ripening gene from tomato. EMBO J 7: 3315–3320 (1988).
23. DellaPenna D, Alexander DC, Bennett AB: Molecular cloning of tomato fruit polygalacturonase: analysis of polygalacturonase mRNA levels during ripening. Proc Natl Acad Sci USA 83: 6420–6424 (1986).
24. DellaPenna D, Kates DS, Bennett AB: Polygalacturonase gene expression in Rutgers, *rin*, *nor* and *Nr* tomato fruits. Plant Physiol 85: 502–507 (1987).
25. DellaPenna D, Bennett AB: *In vitro* synthesis and processing of tomato fruit polygalacturonase. Plant Physiol 86: 1057–1063 (1988).
26. DellaPenna D, Lincoln JE, Fischer RL, Bennett AB: Transcriptional analysis of polygalacturonase and other ripening associated genes in Rutgers, *rin*, *nor* and *Nr* tomato fruit. Plant Physiol 90: 1372–1377 (1989).
27. DellaPenna D, Giovannoni JJ: Regulation of gene expression in ripening tomatoes. In: Grierson D (ed) Developmental Regulation of Plant Gene Expression, pp. 182–216 Blackie, Glasgow/London (1991).
28. Fray RG, Lycett GW, Grierson D: Nucleotide sequence of a heat-shock and ripening-related cDNA from tomato. Nucl Acids Res 18: 7148 (1990).
29. Giovannoni JJ, DellaPenna D, Bennett AB, Fischer RL: Expression of a chimeric polygalacturonase gene in transgenic *rin* (*ripening inhibitor*) tomato fruit results in polyuronide degradation but not fruit softening. Plant Cell 1: 53–63 (1989).

30. Given NK, Venis MA, Grierson D: Purification and properties of phenylalanine ammonia-lyase from strawberry fruit and its synthesis during ripening. J Plant Physiol 133: 31–37 (1988).
31. Goodenough PW, Thomas TH: Biochemical changes in tomatoes stored in modified gas atmospheres. I. Sugars and acids. Ann Appl Biol 98: 507–515 (1981).
32. Goodwin TW: The Biochemistry of the Carotenoids, vol I, 2nd ed. Chapman and Hall, London (1980).
33. Grierson D, Tucker GA: Timing of ethylene and polygalacturonase synthesis in relation to the control of tomato ripening. Planta 157: 174–179 (1983).
34. Grierson D, Slater A, Maunders M, Crookes P, Tucker GA, Schuch W, Edwards K: Regulation of the expression of tomato fruit ripening genes: The involvement of ethylene. In: Roberts JA, Tucker GA (eds) Ethylene and Plant Development, pp. 147–161. Butterworths, London (1984).
35. Grierson D, Slater A, Speirs J, Tucker GA: The appearance of polygalacturonase mRNA in tomatoes: one of a series of changes in gene expression during development and ripening. Planta 163: 263–271 (1985).
36. Grierson D, Maunders MJ, Slater A, Ray J, Bird CR, Schuch W, Holdsworth MJ, Tucker GA, Knapp JE: Gene expression during tomato ripening. Phil Trans R Soc Lond B 314: 399–410 (1986).
37. Grierson D, Tucker GA, Keen J, Ray J, Bird CR, Schuch W: Sequencing and identification of a cDNA clone for tomato polygalacturonase. Nucl Acids Res 14: 8595–8603 (1986).
38. Grierson D, Kader AA: Fruit ripening and quality. In: Atherton JG, Rudich J (eds) The Tomato Crop, pp. 241–280, Chapman and Hall, London (1986).
39. Grierson D, Purton ME, Knapp JE, Bathgate B: Tomato ripening mutants. In: Thomas H, Grierson D (eds) Developmental mutants in higher plants, pp. 73–94. Cambridge University Press, Cambridge (1987).
40. Grierson D, Fray RG, Hamilton AJ, Smith CJ, Watson CF: Does co-suppression of sense genes in transgenic plants involve antisense RNA? Trends Biotechnol 9: 122–123 (1991).
41. Hamilton AJ, Lycett GW, Grierson D: Antisense gene that inhibits synthesis of the hormone ethylene in transgenic plants. Nature 346: 284–287 (1990).
42. Hamilton AJ, Bouzayen M, Grierson D: Identification of a tomato gene for the ethylene forming enzyme by expression in yeast. Proc Natl Acad Sci USA, in press (1991).
43. Harriman RV, Handa AK: Identification and characterization of three pectinmethylesterase genes in tomato. Am Soc Plant Phys, book of abstracts (1990).
44. Harris WM, Spurr AR: Chromoplasts of tomato fruits. II. The red tomato. Am J Bot 56: 380–389 (1969).
45. Hobson GE: The firmness of tomato fruit in relation to polygalacturonase activity. J Hort Sci 40: 66–72 (1965).
46. Hobson GE: Electrophoretic investigation of enzymes from developing *Lycopersicon esculentum* fruit. Phytochemistry 13: 1383–1390 (1974).
47. Holdsworth MJ, Bird CR, Ray J, Schuch W, Grierson D: Structure and expression of an ethylene-related mRNA from tomato. Nucl Acids Res 15: 731–739 (1987).
48. Holdsworth MJ, Schuch W, Grierson D: Organisation and expression of a wound/ripening-related small multigene family from tomato. Plant Mol Biol 11: 81–88 (1988).
49. Horsch RB, Fry JE, Hoffman NL, Eiochholtz D, Rogers SG, Frayley RT: A simple and general method for transferring genes into plants. Science 227: 1229–1231 (1985).
50. Jeffery D, Smith C, Goodenough P, Prosser I, Grierson D: Ethylene-independent and ethylene-dependent biochemical changes in ripening tomatoes. Plant Physiol 74: 32–38 (1984).
51. Jorgensen R: Altered gene expression in plants due to *trans* interactions between homologous genes. Trends Biotechnol 8: 340–344 (1990).
52. Kinzer SM, Schwager SJ, Mutschler MA: Mapping of ripening-related or specific cDNA clones of tomato (*Lycopersicon esculentum*). Theor Appl Genet 79: 489–496 (1990).
53. Knapp J, Moureau P, Schuch W, Grierson D: Organization and expression of polygalacturonase and other ripening related genes in Ailsa Craig 'Neverripe' and 'Ripening inhibitor' tomato mutants. Plant Mol Biol 12: 105–116 (1989).
54. Köch M, Hamilton AJ, Grierson D: *eth* 1, a gene involved in ethylene synthesis in tomato. Plant Mol Biol 17: 141–142 (1991).
55. Lincoln JE, Cordes S, Read E, Fischer RL: Regulation of gene expression by ethylene during *Lycopersicon esculentum* (tomato) fruit development. Proc Natl Acad Sci USA 84: 2793–2797 (1987).
56. Lincoln JE, Fischer RL: Regulation of gene expression by ethylene in wild-type and *rin* tomato (*Lycopersicon esculentum*) fruit. Plant Physiol 88: 370–374 (1988).
57. Lincoln JE, Fischer RL: Diverse mechanisms for the regulation of ethylene-inducible gene expression. Mol Gen Genet 212: 71–75 (1988).
58. Lyons JM, Pratt HK: Effect of stage of maturity and ethylene treatment on respiration and ripening of tomato fruits. Proc Am Soc Hort Sci 84: 491–500 (1964).
59. McCormick S, Neidermeyer J, Fry J, Barnason A, Horsch A, Frayley R: Leaf disc transformation of cultivated tomato (*L. esculentum*) using *Agrobacterium tumefaciens*. Plant Cell Rep 5: 81 (1986).
60. Mansson PE, Hsu D, Stalker D: Characterisation of fruit specific cDNAs from tomato. Mol Gen Genet 200: 356–361 (1985).
61. Margossian LJ, Federman AD, Giovannoni JJ, Fischer RL: Ethylene-regulated expression of a tomato fruit rip-

ening gene encoding a proteinase inhibitor I with a glutamic residue at the reactive site. Proc Natl Acad Sci USA 85: 8012–8016 (1988).

62. Maunders MJ, Holdsworth MJ, Slater A, Knapp JE, Bird CR, Schuch W, Grierson D: Ethylene stimulates the accumulation of ripening-related mRNAs in tomatoes. Plant Cell Envir 10: 177–184 (1987).
63. Mutschler MA, Tanksley SD, Rick CM: Linkage maps of the tomato (*Lycopersicon esculentum*). Tomato Genet Coop Report No 37 (1987).
64. Nakijima N, Mori H, Yamazaki K, Imaseki H: Molecular cloning and sequence of a complementary DNA encoding 1-aminocyclopropane-1-carboxylic synthase induced by tissue wounding. Plant Cell Physiol 31: 1021–1029 (1990).
65. Oeller PW, Wong LM, Taylor LP, Pike DA, Theologis A: Reversible inhibition of tomato fruit senescence by antisense 1-aminocyclopropane-1-carboxylate synthase. Science, in press (1992).
66. Osteryoung KW, Toenjes K, Hall B, Winkler V, Bennett AB: Analysis of tomato polygalacturonase expression in transgenic tobacco. Plant Cell 2: 1239–1248 (1990).
67. Pear JR, Ridge N, Rasmussen R, Rose RE, Houck CM: Isolation and characterization of a fruit specific cDNA and the corresponding genomic clone from tomato. Plant Mol Biol 13: 639–651 (1989).
68. Picton S, Grierson D: Inhibition of expression of tomato-ripening genes at high temperature. Plant Cell Envir 11: 265–272 (1988).
69. Rattanapanone N, Spiers J, Grierson D: Evidence for changes in messenger RNA content related to tomato fruit ripening. Phytochemistry 17: 1485–1486 (1978).
70. Ray J, Bird C, Maunders M, Grierson D, Schuch W: Sequence of pTOM 5, a ripening related cDNA from tomato. Nucl Acids Res 15: 10587 (1987).
71. Ray J, Knapp J, Grierson D, Bird C, Schuch W: Identification and sequence determination of a cDNA clone for tomato pectinesterase. Eur J Biochem 174: 119–124 (1988).
72. Ray J, Moureau P, Bird C, Bird A, Grierson D, Maunders M, Fray R, Truesdale M, Bramley P, Schuch W: Cloning and characterisation of phytoene synthase gene from tomato. Plant Mol Biol, submitted.
73. Rottmann WH, Peter GF, Oeller PW, Keller JA, Shen NF, Nagy BP, Taylor LP, Campbell AD, Theologis A: 1-Aminocyclopropane-1-carboxylate synthase in tomato is encoded by a multigene family whose transcription is induced during fruit and floral senescence. J Mol Biol, in press (1991).
74. Sato T, Theologis A: Cloning the mRNA encoding 1-amino-cyclopropane-1-carboxylate synthase, the key enzyme for ethylene biosynthesis in plants. Proc Natl Acad Sci USA 86: 6621–6625 (1989).
75. Sato T, Oeller PW, Theologis A: The 1-aminocyclopropane-1-carboxylate synthase of *Cucurbita*. J Biol Chem 266: 3752–3759 (1991).
76. Schaffer MA, Fischer RL: Analysis of mRNAs that accumulate in response to low temperature identifies a thiol protease gene in tomato. Plant Physiol 87: 431–436 (1988).
77. Schaffer MA, Fischer RL: Transcriptional activation by heat and cold of a thiol protease gene in tomato. Plant Physiol 93: 1486–1491 (1990).
78. Schuch W, Hobson G, Kanczler J, Tucker G, Robertson D, Grierson D, Bright S, Bird C: Improvement of tomato fruit quality through genetic engineering. HortScience, in press (1991).
79. Schöffl F, Raschke E, Nagao RT: The DNA sequence analysis of soybean heat-shock genes and identification of possible regulatory promoter elements. EMBO J 3: 2491–2497 (1984).
80. Sheehy RE, Pearson J, Brady CJ, Hiatt WR: Molecular characterization of tomato fruit polygalacturonase. Mol Gen Genet 208: 30–36 (1987).
81. Sheehy RE, Kramer M, Hiatt WR: Reduction of polygalacturonase activity in tomato fruit by antisense RNA. Proc Natl Acad Sci USA 85: 8805–8809 (1988).
82. Slater A, Maunders MJ, Edwards K, Schuch W, Grierson D: Isolation and characterization of cDNA clones for tomato polygalacturonase and other ripening related proteins. Plant Mol Biol 5: 137–147 (1985).
83. Smith CJS, Slater A, Grierson D: Rapid appearance of a mRNA correlated with ethylene synthesis encoding a protein of MW 35,000. Planta 168: 94–100 (1986).
84. Smith CJS, Watson CF, Ray J, Bird CR, Morris PC, Schuch W, Grierson D: Antisense RNA inhibition of polygalacturonase gene expression in transgenic tomatoes. Nature 334: 724–726 (1988).
85. Smith CJS, Watson CF, Bird CR, Ray J, Schuch W, Grierson D: Expression of a truncated tomato polygalacturonase gene inhibits expression of the endogenous gene in transgenic plants. Mol Gen Genet 224: 477–481 (1990).
86. Smith CJS, Watson CF, Morris PC, Bird CR, Seymour GB, Gray JE, Arnold C, Tucker GA, Schuch W, Harding S, Grierson D: Inheritance and effect on ripening of antisense polygalacturonase genes in transgenic tomatoes. Plant Mol Biol 14: 369–379 (1990).
87. Spanu P, Reinhart D, Boller T: Analysis and cloning of the ethylene-forming enzyme from tomato by functional expression of its mRNA in *Xenopus laevis* oocytes. EMBO J 10: 2007–2013 (1991).
88. Speirs J, Brady CJ, Grierson D, Lee E: Changes in ribosome organisation and messenger RNA abundance in ripening tomato fruits. Aust Plant Physiol 11: 225–233 (1984).
89. Taylor JE, Tucker GA, Lasslett Y, Smith CJS, Arnold CM, Watson CF, Schuch W, Grierson D, Roberts JA: Polygalacturonase expression during leaf abscission of normal and transgenic tomato plants. Planta 183: 133–138 (1990).
90. Tigchelaar EC, McGlasson WB, Buescher RW: Genetic

regulation of tomato fruit ripening. J HortScience 17: 508–513 (1978).
91. Tucker GA, Grierson D: Synthesis of polygalacturonase during tomato fruit ripening. Planta 155: 64–67 (1982).
92. Tucker GA, Brady CJ: Silver ions interrupt tomato fruit ripening. J Plant Physiol 127: 165–169 (1987).
93. van der Krol AR, Lenting PE, Veenstra J, van der Meer IM, Koes RE, Gerats AGM, Mol JNM, Stuitje AR: An anti-sense chalcone synthase gene in transgenic plants inhibits flower pigmentation. Nature 333: 866–869 (1988).
94. van der Straeten D, van Wiemeersch L, Goodman HM, van Montagu M: Cloning and sequence of two different cDNAs encoding 1-aminocyclopropane-1-carboxylate synthase in tomato. Proc Natl Acad Sci USA 87: 4859–4863 (1990).
95. van Haaren MJJ, Houck CM: Strong negative and positive regulatory elements contribute to the high level fruit-specific expression of the tomato 2A11 gene. Plant Mol Biol, in press (1991).
96. van Haaren MJJ, Houck CM: A functional map, containing both enhancer and silencer elements, of the fruit-specific promoter of the tomato 2A11 gene. Plant Cell, submitted.
97. Ververidis P, John P: Complete recovery *in vitro* of ethylene forming enzyme activity. Phytochemistry 30: 725–727 (1991).
98. Watkins CB, Picton S, Grierson D: Stimulation and inhibition of expression of ripening-related mRNAs in tomatoes as influenced by chilling temperatures. J Plant Physiol 136: 318–323 (1990).
99. Yang SF, Hoffman NE: Ethylene biosynthesis and its regulation in higher plants. Annu Rev Plant Physiol 35: 155–189 (1984).
100. Yang SF: Biosynthesis and action of ethylene. HortScience 20: 41–45 (1985).

Plant Molecular Biology **19**: 89–107, 1992.
© 1992 *Kluwer Academic Publishers. Printed in Belgium.*

Developmental aspects of the Rhizobium-legume symbiosis

Henk J. Franssen, Irma Vijn, Wei Cai Yang and Ton Bisseling
Department of Molecular Biology, Agricultural University, Dreijenlaan 3, 6703 HA Wageningen, Netherlands

Key words: Rhibozobium-legume symbiosis, Nod factors, Nodulin genes, nodule differentiation, N_2-fixation

Introduction

In the plant kingdom a great variety of pathogenic, saprophytic and symbiotic interactions between plants and microorganisms occur, and several of these interactions have been the subject of intensive research. One of the best studied interactions is the symbiosis of *Rhizobium*, *Bradyrhizobium* or *Azorhizobium* bacteria with legume plants, which results in the formation of root nodules, in which bacteria are able to fix atmospheric nitrogen into ammonia. This process of symbiotic N_2 fixation is the major naturally occurring mechanism by which nitrogen is reduced and the ecological and agricultural importance of this process has been an important incentive to study this plant-microbe relation. The process of symbiotic N_2 fixation has been discussed in several recent reviews [33].

Among plant-microbe interactions the *Rhizobium*-legume symbiosis forms a unique system since this interaction results in the formation of a highly organized structure that can be considered a new plant organ, whereas most other plant-microbe interactions result in structures with a much lower level of organization.

The formation of a root nodule involves several developmental steps. In short, rhizobia attach to the root hairs of their host and cause deformation and curling of root hairs. Then the rhizobia invade the plant by newly formed tubular structures, the infection threads. These infection threads start in the curl of the root hair. Concomitantly with the infection process, cells in the root cortex start to divide and form the nodule primordium. Infection threads enter individual primordium cells and bacteria are released in the cytoplasm of the plant cells. Then the nodule primordium differentiates into a nodule [7, 63]. Like the formation of other plant organs root nodule formation also involves the expression of a set of organ specific genes, the so-called nodulin genes [92]. The nodulin genes that are markedly expressed before the onset of N_2 fixation are named early nodulin genes [29, 58]. These genes are supposedly involved in root hair curling, infection and the formation of the nodule structure. The nodulin genes expressed shortly before or concomitantly with the start of N_2 fixation are the late nodulin genes [29, 58].

Here we will discuss aspects of the *Rhizobium*-legume interaction related to the development of nodule formation with an emphasis on the bacterial signal molecules that induce plant developmental processes and the plant genes specifically involved in the different steps of nodule formation.

Bacterial genes and factors involved in inducing steps in nodule development

In several *Rhizobium* species genetic studies have revealed the presence of genes involved in establishing an effective symbiosis. In the fast-growing *Rhizobium* species these genes are located on a large plasmid (pSym), whereas in the slow-growing *Bradyrhizobium* species these genes are carried on the chromosome [54]. The rhizobial genes essential for infection and nodule formation include the nodulation (*nod*) genes, and several sets of genes involved in the structure of the bacterial surface such as genes concerned with the synthesis of exopolysaccharides (*exo* genes), lipopolysaccharides (*lps* genes) and β-glucans (*ndv* genes) [54].

Nod *Genes and Nod factors*

In free-living rhizobia the *nod* genes are not transcribed, with the exception of the *nodD* gene, which is constitutively expressed. The *nod* genes are induced by host plant secreted flavonoids [54, 66, 67]. In addition to these flavonoids, *nod* gene expression also requires the presence of the NodD protein, which probably acts as a transcriptional activator (Fig. 1) [54].

The *nod* genes are classified according to their ability to complement mutations of similar genes in other *Rhizobium* species. One set of *nod* genes, the common *nod* genes (*nodABCIJ*), are functionally interchangeable in all *Rhizobium* species, and share a high degree of sequence homology. Mutations in *nodABC* completely abolish nodulation; including root hair curling, cortical cell division and infection thread formation [54]. In contrast, mutations in *nodI* and *nodJ* cause only a slight delay of nodulation [44]. Members of the other set of *nod* genes, the host-specific nodulation genes are not functionally conserved since alleles from different *Rhizobium* species cannot substitute for one another's function on their respective host plants. Mutations of these genes generally cause delayed or less effective nodulation, moreover, they can alter the host range of the mutated rhizobia. For example, *R. meliloti* which has a mutation in the *nodFE* or *nodQ* genes is able to nodulate the original host with reduced efficiency, but also has an extended host range since it is able to deform root hairs on white clover and to induce infection thread formation in common vetch [20]. *NodH* mutants have the same phenotype as the *nodFE* and *nodQ* mutants on clover and common vetch, but they have lost their capacity to nodulate the natural host plant [20, 36].

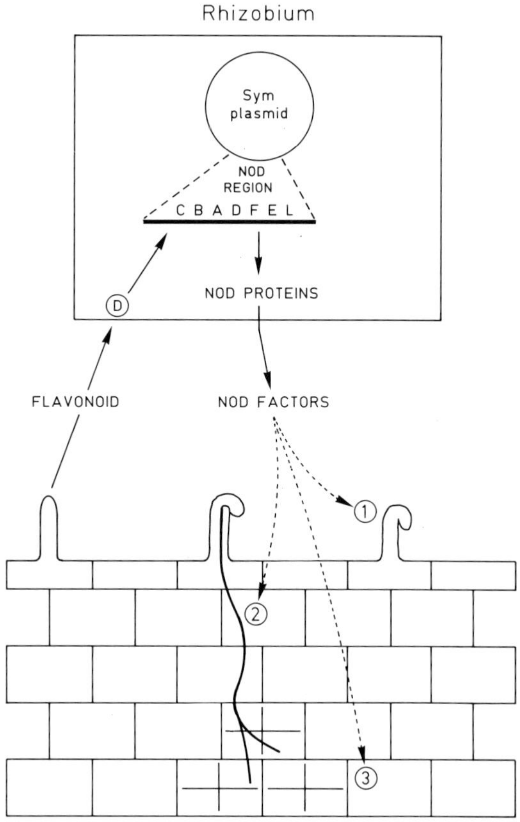

Fig. 1. Schematic representation of the interaction between *Rhizobium* and legume roots. Plant secreted flavonoids, in contact with the bacterial NodD protein regulate the transcription of bacterial *nod* genes located mostly in the sym plasmid. The bacterial *nod* gene products are involved in the synthesis of Nod factors. These Nod factors are able to induce the critical steps leading to nodule formation: 1) root hair deformation; 2) at least some steps of the infection process; 3) cortical cell division.

Upon induction of the *nod* genes, *Rhizobium* secretes a so-called Nod factor (Fig. 1). NodRm-1, the factor produced by *R. meliloti*, was the first Nod factor for which the structure was elucidated. NodRm-1 is a sulphated β, 1–4-linked tetrasaccharide of D-glucosamine in which three amino groups are acetylated. At the non-reducing end the sugar unit carries a C16 unsaturated fatty acid group, while the last unit of the reducing end contains a sulphate group (Fig. 2) [49]. Recently, the structures of several Nod factors of *R. leguminosarum* bv. *viciae* have been determined. *R. leguminosarum* bv. *viciae* secretes at least 5 different Nod

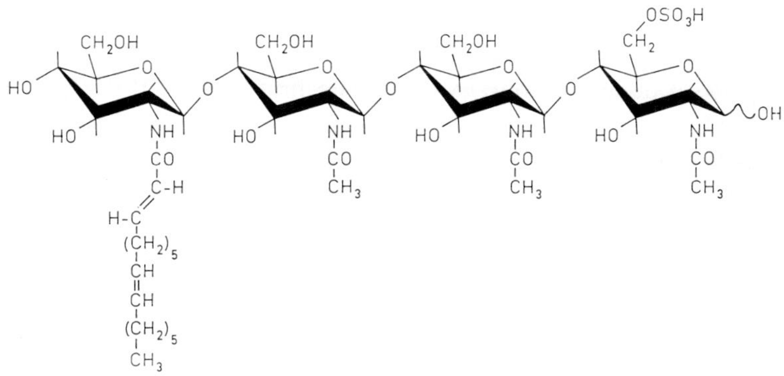

Fig. 2. Structure of NodRm-1, the Nod factor produced by *R. meliloti* as described by Lerouge *et al.* [49].

factors. Spaink *et al.* [80] showed that the production of these factors requires the presence of *nodABCEF* and *nodL* genes. Thus the other *nod* genes appeared not to be essential for the production of these factors. These NodRlv factors share overall structural similarities with NodRm-1, but there are also significant differences. Like NodRm-1 all NodRlv factors are oligomers of N-acetyl D-glucosamine but the chain length is 4 or 5 glucosamine residues. All *R. leguminosarum* bv. *viciae* Nod factors have an O-acetyl group at the non-reducing terminal sugar residue. This O-acetyl group is not present, however, in the absence of the *nodL* gene. NodRm-1 is produced by a *R. meliloti* strain lacking *nodL*, however, when a *nodL* gene is present in *R. meliloti* only a Nod factor containing an O-acetyl group, Ac-NodRm-1, is formed also [69; J. Dénarié, personal communication]. *R. leguminosarum* bv. *viciae* has the ability to form 2 different fatty acid chains, lipid 1 and lipid 2. Both lipids are different from the acyl group of NodRm-1 (H. Spaink, personal communication). The Nod factors containing lipid 1 are only formed when *Rhizobium* contains a functional *nodE* gene, while lipid 2 containing compounds are still formed by a *nodE* mutant. This observation clearly shows the involvement of *nodE* in the synthesis of the fatty acid chain. Recently it was shown that *R. meliloti* also produces Nod factors with a chain length of 5 glucosamines and furthermore Nod factors with an acyl group different from the one present in NodRm-1 have been found (J. Dénarié, personal communication). A major difference between the NodRlv factors and NodRm-1 is the absence of a sulphate group [80]. In *R. meliloti nodH*, *nodP* and *nodQ* are involved in the sulphurylation of NodRm-1 [70, 78] and these genes do not occur in *R. leguminosarum* bv. *viciae*.

Since the common *nod* genes are highly conserved in different *Rhizobium* species it is very likely that all *Rhizobium* species produce Nod factors with a similar structure as the NodRm and NodRlv factors. This conclusion is supported by preliminary studies on Nod factors secreted by *Rhizobium*. sp. NGR234 [6].

Rhizobium *genes encoding surface components*

Rhizobium extracellular polysaccharides include charged exopolysaccharides (EPS), lipopolysaccharides (LPS) and neutral β-glucans [10]. Rhizobia with mutations in genes involved in the synthesis of these surface compounds frequently form empty nodules in which abortive or no infection threads are visible [54].

The genes involved in EPS synthesis (*exo* genes) of *R. meliloti* have been studied in most detail. EPS are composed of repeated units, which consist of a sugar backbone and a side-chain of sugar residues with substitutions. The repeated unit of EPS I (succinoglucan) in wild-type *R. meliloti* is an octasaccharide, which contains glucose and galactose in its backbone, while its side-chain harbours glucoses with pyruvate, acetyl and succinyl modifications [3]. EPS I occurs in a high and low molecular weight form. In addition, *R. meliloti* has the capacity to produce a second exopolysaccharide named EPS II, which varies in its structure and chemical content from EPS I [26].

The majority of the *R. meliloti exo* genes are located on another megaplasmid than pSym. Mutations in the *exoA*, *B*, *F*, *L*, *M*, *P*, *Q* or *T* genes completely abolish the production of EPS I [48, 51]. These mutants form empty nodules on alfalfa, which lack infection threads as well as intracellular bacteria [21, 34, 90]. Mutations in the *exoH* gene affect the succinylation of EPSI and *exoH* mutants form aberrant nodules on alfalfa roots [47]. Another class of mutants such as *exoG* and *J* are characterized by a diminished production of the high-molecular-weight EPS I, but members of this class are still able to form N_2-fixing nodules, albeit with decreased efficiency [51]. In *exoG* mutants the production of a low-molecular-weight exopolysaccharide of yet unidentified structure has been detected [51].

The LPS consist of three components: lipid A, which is situated in the bilayer of the outer membrane; the core, whose basic structure is similar in several rhizobia; and the O-antigen, which is variable between different species [9]. Genetic studies have shown that mutants defective in the production of LPS O-antigenic saccharides or with diminished production of LPS, are not able to induce infection thread formation or the formed threads are aborted [8, 14].

The third surface compound essential for normal nodule formation, cyclic β-1,2 glucan, consists of 16–25 glucose units. *Rhizobium* strains with mutations involved in the production or secretion of cyclic β-1,2-glucan retain the ability to cause nodule formation; but, as with other *exo* and *lps* mutants, the nodules are developmentally defective and do not fix N_2. The genes *ndvA* and *B* which are responsible for cyclic β-1,2-glucan production are located on the chromosome and have high homology to the *chv* (chromosomal virulence) loci of *Agrobacterium* [19].

Root hair deformation

The first step in the interaction between *Rhizobium* and the plant is the attachment of bacteria to root hairs. *Rhizobium* interacts with almost all emerging root hairs, but only about 25 % of these young root hairs actually form a curl, which resembles a shepherd's crook and which is thought to be essential for successful interaction [42, 96, 97]. Computer simulation studies suggest that this curling is the result of a local stimulation of growth of the root hair tip, rather than an inhibition of growth [42].

Recently, changes in root hair gene expression after inoculation with *Rhizobium* have been studied. In pea root hairs one mRNA, RH-42, has been identified whose synthesis is induced by the interaction of the root hairs with *R.leguminosarum* bv. *viciae*. The RH-42 gene is not expressed in root hairs of uninoculated plants or in developing nodules. Therefore, it is likely that RH-42 is involved in curling or deformation of root hairs [27]. The expression of another pea gene, RH-44, is markedly stimulated in pea root hairs after inoculation with *R. leguminosarum* bv. *viciae* but this gene is already expressed at a low level in root hairs of uninoculated plants [27]. In *Vigna unguiculata* five root hair proteins were identified which appear within 24 h after inoculation of plants with *Rhizobium* sp. NGR234. Two of these proteins appear to be confined to root hairs, while the other three have been identified in nodules as well [6].

The purified Nod factors of *R. meliloti* and *R. leguminosarum* bv. *viciae* are able to induce root hair deformation in a host specific manner [49, 80]. Addition of purified NodRm-1 in a concentration range of $10^{-8}-10^{-11}$M to alfalfa roots causes root hair deformation on these plants but vetch root hairs are not susceptible to this factor. In contrast, the NodRlv factors deform vetch root hairs but not alfalfa root hairs.The Nod factor produced by a *R. meliloti nodH* mutant, NodRm-2, is apart from the absence of the sulphate group, identical to NodRm-1. The lack of the sulphate group in NodRm-2 is responsible for the incapability of this factor to deform alfalfa root hairs, but instead the NodRm-2 is able to deform vetch root hairs [70]. This demonstrates that at the level of root hair deformation the host specificity of the Nod factors is determined by the presence or absence of the sulphate group.

NodRlv factors, containing lipid 1 or lipid 2, are equally efficient in inducing root hair deformation on vetch [80] and NodRm-2 (having an acyl group which is different from lipid 1 or lipid 2 of the NodRlv factors; H. Spaink, personal communication) is also able to deform root hairs on vetch [71]. Therefore the structure of the lipid moiety appears to be of less importance than the presence of a sulphate group in determining the root hair deforming activity of a Nod factor on a particular host.

Since Nod factors are sufficient to induce root hair deformation it is likely that these compounds also induce (some of) the changes in plant gene expression occurring in root hairs after inoculation with *Rhizobium*. In pea root hairs gene expression was studied after application of exudates of *R. leguminosarum* bv. *viciae* containing Nod factors [27]. It was shown that the expression of the RH-44 gene is stimulated by these exudates, however, the transcription of the RH-42 gene was not detected [27]. This indicates that Nod factors cause changes in gene expression in root hairs but most likely other *Rhizobium* components will also be involved in the induction of expression of particular plant genes in root hairs. The involvement of additional *Rhizobium* components in the interaction with root hairs is already shown by preliminary studies of Battisti *et al.* [5], which showed that the low-molecular-weight form of *R. meliloti* EPS I is able to induce root hair curling on alfalfa in the absence of bacteria. Hence these data indicate that during the first contact between plant and bacterium already several signal molecules are involved.

The infection process

The cytological aspects of the infection process have been studied extensively among others by Newcomb [61, 62, 63], Bakhuizen [4] and Turgeon [88]. They have shown that after entrapment of rhizobia in a curled root hair a local lesion of the root hair cell wall is formed by hydrolysis of the cell wall. The mechanism of hydrolysis of the cell wall is not known. Either the bacteria may induce hydrolytic enzymes that are responsible for cell wall dissolution or the bacteria may exploit endogenous plant mechanisms such as those used when epidermal cells form root hairs [42]. Rhizobia enter the roots at the sites where root hair cell walls are hydrolysed. The penetration occurs by invagination of the plasma membrane. Around the invaginated membrane the plant forms an infection thread by depositing cell wall-like material [42, 54, 96]. Concomitantly with infection thread formation, particular cortical cells divide to form a nodule primordium, and the infection threads grow towards these primordia. The root cortical cells through which infection threads will pass on their way to the nodule primordia change markedly before they will be penetrated by an infection thread. Detailed cytological analyses have shown that microtubules rearrange, the nucleus migrates to the cell centre and an additional cell wall is formed. On the basis of these cytological changes it has been suggested that the cortical cells become prepared for infection thread penetration. These morphological changes resemble changes observed in the G2 phase of the cell cycle, indicating that these cortical cells start a cell division but become arrested in the G2 phase [4, 43].

Two early nodulin genes have been identified that are involved in the infection process in pea. The early nodulin cDNA clones pPsENOD5 and pPsENOD12 were isolated from a pea nodule cDNA library by differential screening [75, 76]. *In situ* hybridization studies revealed that the PsENOD12 gene is induced in root hairs, root cortical cells and nodule cells containing growing infection threads. Interestingly, this early nodulin gene is expressed not only in cells containing growing infection threads, but also within several cell layers in front of the infection thread tip, where cells are preparing for infection thread penetration [75].

In alfalfa nodules several nodulin genes have

been identified encoding proteins in which the first 24 amino acids at the N-terminus are identical to the N-terminus of the pea encoded PsENOD12 protein [35; O.G. Barker, personal communication].

The spatial distribution of the PsENOD5 mRNA in infected pea roots is strikingly different from that of the PsENOD12 transcript. While the PsENOD12 mRNA is present within several root cortical cell layers in front of the infection thread tip, the PsENOD5 gene is expressed only in cells containing the growing infection thread [76].

The PsENOD5 early nodulin also is a proline-rich protein, but it does not show the repetitive structure of PsENOD12. PsENOD5 is rich in Pro, Ala, Gly and Ser residues, and therefore this protein may be related to arabinogalactan proteins some of which are plasma membrane components [43]. Thus PsENOD5 may be a component of the infection thread plasma membrane.

Besides nodulins, other plant proteins play an important role in the infection process. By immunocytology it was shown that some pea glycoproteins are located in the infection thread [89]. Furthermore, it was demonstrated that lectin is involved in the infection process [16]. *R. leguminosarum* bv. *viciae* normally infects and nodulates pea and vetch, but not white clover. However, *R. leguminosarum* bv. *viciae* is able to induce root hair deformation on white clover roots, although neither infection threads are formed nor cortical cell divisions are induced [16]. Introduction of a pea lectin gene in white clover is sufficient to allow infection and nodule formation by *R. leguminosarum* bv. *viciae* to occur [16]. Apparently, the presence of pea lectin in white clover roots is sufficient to make clover susceptible to *R. leguminosarum* bv. *viciae* signal molecules eliciting infection and cortical cell division.

Genetic analysis of *Rhizobium* has led to the identification of several genes that are required for the infection process. In all *Rhizobium* species studied it was shown that the *nod* genes are essential. In addition to the *nod* genes, several genes involved in the production of bacterial surface components are involved in the infection process. Examples are the *exo*, *lps* and *ndv* (involved in β-1,2-glucan production) genes (see above: '*Rhizobium* genes encoding surface components'). These bacterial genes might all be involved in the production of signal molecules inducing the infection process [18, 68]. However, at present it has only been demonstrated that the *nod* genes are involved in the production of signal molecules eliciting plant gene expression involved in infection, since purified Nod factors of *R. leguminosarum* bv. *viciae* are able to induce the PsENOD12 gene in pea root hairs [59; B. Horváth, personal communication]. The induction of this infection-related early nodulin gene by the Nod factors shows that these Nod factors trigger at least certain steps of the infection process despite the fact that these factors cannot elicit the formation of infection threads.

Genetic studies on *Rhizobium* indicate that especially the lipid moiety of the Nod factor plays a critical role in the induction of the infection process. It has been demonstrated that in *R. leguminosarum* bv. *viciae* the *nodE* gene is the main determinant of host specificity. *R. leguminosarum* bv. *trifolii* nodulates clover but not vetch. By exchanging in this species the *nodE* gene for a chimaeric *nodE* gene, existing of parts of *R. leguminosarum* bv. *viciae nodE* and *R. leguminosarum* bv. *trifolii nodE* genes, *R. leguminosarum* bv. *trifolii* is now able to nodulate vetch [81]. Since *nodE* in *R. leguminosarum* bv. *viciae* is involved in the production of lipid 1 of the Nod factors it is very likely that the lipid moiety of the Nod factor determines whether the infection process can be induced in a particular host plant.

The mechanism by which the Nod factors induce steps in the *Rhizobium*-plant interaction is unknown. However, most likely it starts with perception of the Nod factor by a receptor located in the plasma membrane of the root hair. At present no such receptor has been identified but for several reasons it has been postulated [49, 52] that lectins are receptors or are part of receptor complexes: as described above pea lectin is the only molecule required to make clover susceptible to signals of *R. leguminosarum* bv. *viciae* eliciting infection [16]; lectins are sugar binding proteins;

and lectins are specifically located in the region of the root that is susceptible to infection by *Rhizobium* [43].

Lectins of legumes lack a transmembrane domain that occurs in all surface receptors of animal systems [1]. Therefore we think that it is unlikely that lectin by itself is a receptor of the Nod factor, but it is still possible that lectins are part of a complex that forms the Nod factor receptor.

Nodule primordium and meristem

Nodule organogenesis starts with *Rhizobium*-induced cell divisions in the fully differentiated root cortex, which result in the formation of a nodule primordium. Which root cortex cells divide depends upon the type of nodule a particular host plant forms. In general, in temperate legumes, such as pea, vetch and alfalfa the primordium is formed from cells of the root inner cortex. These legumes form indeterminate nodules and have a persistent apical meristem [62]. On the other hand, in most tropical legumes, such as soybean and French bean, a primordium is formed in the root outer cortex and the formed nodules have a determinate growth pattern. In these determinate nodules mitotic activity ceases during development and cell expansion, rather than cell division, is responsible for the increase in nodule size [61].

In the development of both types of nodules, infection threads grow towards the nodule primordia and bacteria are released into the plant cells by an endocytotic process by which bacteria become enclosed by a plant membrane. This membrane is named the peribacteroid membrane or the symbiosome membrane [71]. At this stage of development of the indeterminate nodule type a meristem is formed at the distal site of the primordium. The meristematic cells can be distinguished at a morphological and molecular level from the primordium cells. These meristematic cells are smaller in size and contain smaller vacuoles. Moreover, *in situ* hybridization studies have shown that the infection-related PsENOD12 gene is expressed in all cells of the nodule primordium, whereas this gene is not expressed in the nodule meristem. [75].

The Nod factors of *Rhizobium* play a pivotal role in the induction of root cortical cell division since both NodRm-1 and NodRlv factors can induce cortical cell division. Application of NodRm-1 to alfalfa seedlings is sufficient to induce the formation of genuine nodules [87]. The *R. leguminosarum* bv. *viciae* factors elicit primordium formation in vetch roots, but these primordia do not develop into nodules [80]. In both systems the structure of the fatty acid chain appears to be of crucial importance. When unsaturated bounds of the acyl group of NodRm-1 are hydrogenated, it loses the ability to induce cortical cell division [87] and only *R. leguminosarum* bv. *viciae* Nod factors containing lipid 1 can induce cell division [80]. So in addition to the role of the lipid moiety in inducing the infection process, it also has a critical role in induction of cortical cell division.

Biological activity of Nod factors

Application of Nod factors to legume roots is sufficient to induce root hair deformation, cortical cell division and at least some steps of the infection process (see previous three sections). Thus a single molecule or a few very closely related oligosaccharides of *Rhizobium* are able to induce a wide range of responses in legume roots. At present it is unknown how Nod factors induce three apparently different processes. However, it becomes clear that the induction of distinct steps in nodule development demands different structural requirements of the Nod factors; e.g. whereas only molecules with a specific fatty acid chain can induce root cortical cell division, there are no strict requirements to the acyl group with respect to induction of root hair deformation. Therefore we think that the Nod factors induce root hair deformation, cortical cell division and infection by different mechanisms.

An often applied strategy to obtain clues to the mechanism by which a particular signal molecule induces a process, is to search for other compounds that mimick the effect of such a signal molecule. It is thought that such compounds trigger the same signal transduction pathway as the signal molecule. Two types of experiments along that line indicate that the Nod factors induce cell division by changing the phytohormone balance in the root. First, the zeatin gene, involved in cytokinin synthesis, can partly complement Nod mutants of *R. meliloti* [53]. Second, it was shown that compounds that are supposed to block the polar transport of auxin, namely naphtylphthalamic acid (NPA) and triiodobenzoic acid (TIBA) induce the formation of nodule-like structures on the roots of leguminous plants. In the early 1950s, Allan *et al.* [2] showed that especially legumes are susceptible to these auxin transport inhibitors (ATIs). More recently it was shown that these ATI-induced nodule-like structures resemble *Rhizobium*-induced nodules since early nodulin genes are expressed in these structures at positions similar to those in normal root nodules. Thus the Nod factor may accomplish changes in the balance of phytohormones in the root [34, 90], which secondarily result in nodule formation. One way in which the Nod factor might change the hormone balance is through the synthesis of endogenous ATIs. Flavonoids are most likely endogenous ATIs [37] and hence it is interesting that the expression of the chalcone synthase (CHS) gene is stimulated in nodule primordia [100]. Since CHS is a key enzyme in the synthesis of flavonoids [32], it is possible that the increased expression level of CHS genes leads to a local accumulation of flavonoids. Based on this observation, we hypothesize that Nod factors mediate cortical cell division by stimulating the local production of flavonoids which causes a change in the phytohormone balance subsequently leading to cortical cell division.

Root nodule differentiation

During normal nodule development the nodule meristem first differentiates into a root nodule after the release of bacteria from the infection thread. The determinate and indeterminate nodule types have a similar organization of their tissues, a central tissue and several peripheral tissues (Fig. 3). The peripheral tissues include the nodule cortex, the nodule endodermis and the nodule parenchyma [91] ('inner cortex'), which contains the vascular bundles connecting the nodule with the root stele.

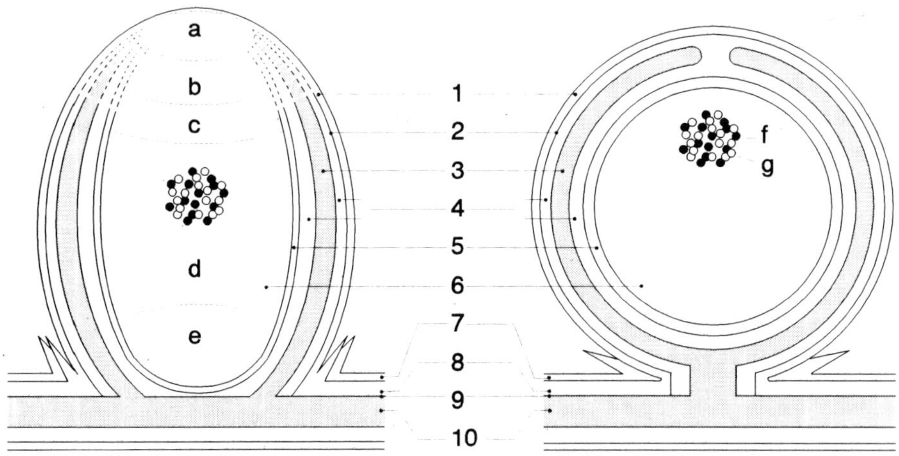

Fig. 3. Schematic representation of the determinate and indeterminate nodule with all tissues. The zonation indicated in the indeterminate nodule is as suggested by Vasse *et al.* [93]. a, meristem (zone I); b, infection zone II; c, interzone II–III; d, nitrogen-fixing zone III; e, senescent zone IV; f, infected cells; g, uninfected cells. Tissues are indicated by numbers: 1, nodule cortex; 2, nodule endodermis; 3, nodule vascular bundle; 4, nodule parenchyma; 5, boundary layer; 6, central tissue; 7, root epidermis; 8, root cortex; 9, root endodermis; 10, root central cylinder.

The central tissue

The nodule tissue in which the bacteria are hosted is named the central tissue. This tissue consists of two cell types, infected and uninfected cells. The infected cells are fully packed with bacteria, whereas the uninfected cells contain no rhizobia. The uninfected cells of determinate nodules are specialized for the assimilation of fixed N_2 to ureides [60, 64], whereas it is unknown whether the uninfected cell type of the indeterminate nodules has a specific function in ammonium assimilation.

In indeterminate nodules the cells of the central tissue are of graded age, the youngest cells near the meristem and the oldest cells in the proximal part of the nodule. Thus a zonation of consecutive stages of development is established in the central tissue [62]. Going in a proximal direction from the apically located persistent meristem, Newcomb [62] described an infection zone in which infection threads penetrate plant cells and release rhizobia into the cells. Then an early symbiotic zone can be discerned in which the rhizobia proliferate until the infected cells are almost completely filled with bacteroids (the symbiotic form of the rhizobia), while the infected and uninfected cells themselves enlarge considerably. In the late symbiotic zone the bacteroids are able to fix N_2. In older nodules the late symbiotic zone is followed by a senescent zone at the base of the nodule, in which the plant cells, together with the bacteroids, are degraded.

This indeterminate growth pattern has allowed a detailed study of morphological and molecular changes of both symbionts during the different steps in the development of indeterminate nodules. Pea and alfalfa nodules have been most closely studied in this respect. Recently, Vasse *et al.* [93] studied bacterial differentiation of *R. meliloti* in alfalfa nodules. Their ultrastructural studies revealed that 5 different bacteroid types can be discerned in successive zones of the nodule. On the basis of these ultrastructural characteristics Vasse *et al.* have proposed a new nomenclature to distinguish different zones of the indeterminate nodule (see Fig. 5). The bacterium-free meristematic region at the apex is *zone I*. In the *infection zone II* infection occurs and cell differentiation begins. In the *nitrogen-fixing zone III* the plant cells have reached their maximum size and bacteroids start to fix N_2. In between zone II and III the so-called *interzone II–III* is located. This zone is characterized by bacteroids with a specific morphology and the start of amyloplast deposition. In older nodules a *senescent zone IV* is present.

In some cases the transition of one bacteroid type into another occurs in adjacent cell layers. An example is given in Fig. 4 where the morphological changes of bacteroids at the transition of infection zone II into interzone II–III is shown. This figure clearly demonstrates the cell-to-cell changes of bacteroid morphology which occur during nodule development. Ultrastructural characteristics of bacteroids and amyloplast formation offer better tools to distinguish developmental zones of an indeterminate nodule than do the cytological criteria used by Newcomb [62]. Moreover, some molecular changes coincide with the transition of zones as proposed by Vasse *et al.* [93] (see below).

Here we will summarize the molecular changes observed in the central tissue of the pea nodule during development and will use the nomencla-

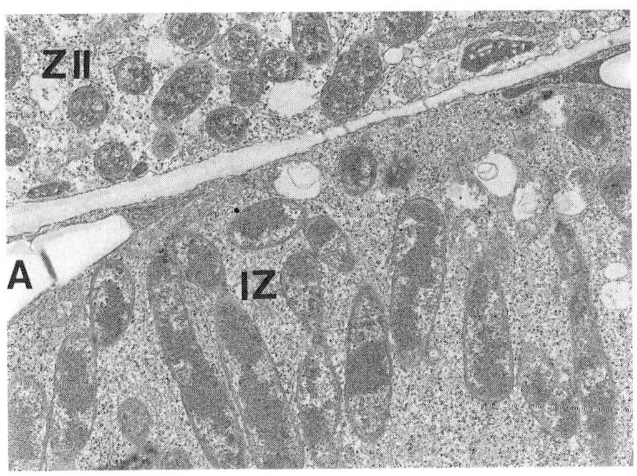

Fig. 4. Ultrastructural differentiation of bacteroids in 4-week-old alfalfa nodules showing the cell-to-cell ultrastructural changes that occur at the transition from infection zone II (ZII) to interzone II–III (IZ). A, amyloplast (according to Vasse *et al.* [93]).

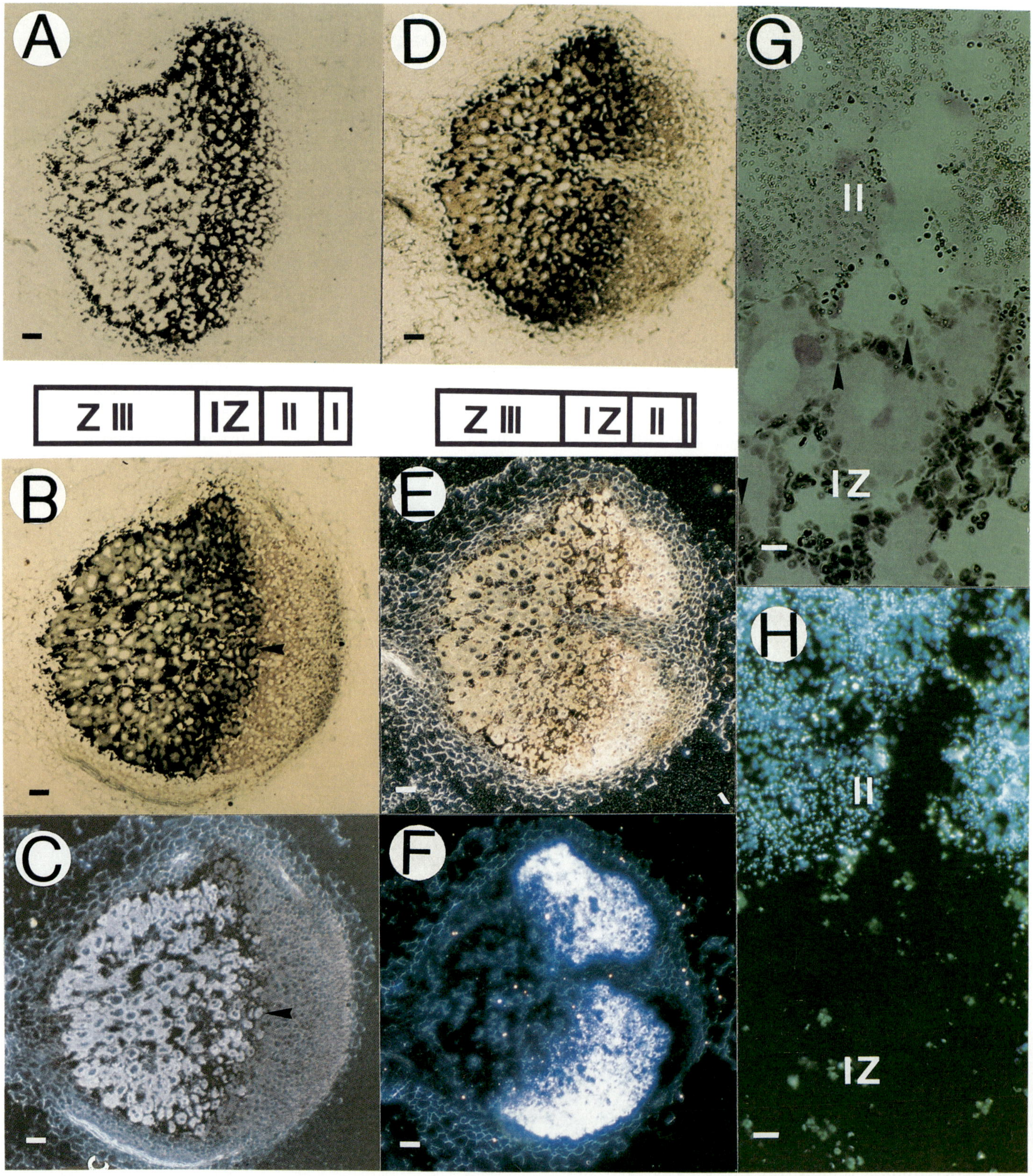
A
D
G
Z III
IZ
II
I
Z III
IZ
II
B
E
II
IZ
C
F
H
II
IZ

ture of Vasse *et al.* [93] to describe the different zones of this tissue.

No ultrastructural studies on bacteroid development in pea nodules are available, hence amyloplast formation is used to identify the different developmental zones. Amyloplasts first accumulate in nodules of 13-day-old pea plants and at day 16 the cells of the proximal half of the central tissue contain prominent amyloplasts (Fig. 5). At this stage of pea nodule development the zone containing amyloplasts consists of about 15 cell layers. At day 18 the zone containing amyloplasts starts to decrease in size, since amyloplasts disappear in the proximal region of the nodule. At day 20 only 6–8 cell layers contain prominent amyloplasts (Fig. 5) and at day 28 this zone is even reduced to two cell layers. We will name the zone of the central tissue of the pea nodule containing amyloplasts the interzone II–III. The zone between the apical meristem (zone I) and the cell layer where amyloplast accumulation starts is the infection zone II. The nitrogen-fixing zone III starts where the number of amyloplasts decreases.

Using *in situ* hybridization, the expression of early and late nodulin genes was studied in adjacent longitudinal sections of pea nodules and it was shown that these genes are expressed in subsequent zones of the central tissue. The early nodulin gene PsENOD12 is only expressed in the distal part of the infection zone II, which is consistent with a function of PsENOD12 in the infection process [75]. The PsENOD5 gene is expressed in the whole infection zone II of the central tissue and PsENOD5 gene expression reaches its highest level in the proximal part of the infection zone II (Fig. 5) [76]. In this region, infection threads do not grow, but bacteria proliferate, and these bacteria become surrounded by a plant-derived peribacteroid membrane. The relatively high expression level of the PsENOD5 gene at this stage of development suggests that the arabinogalactan-like protein ENOD5 may not be only a part of the plasma membrane of the infection thread, but also part of the peribacteroid membrane. At the transition of infection zone II into interzone II–III there is a sudden drop in the level of PsENOD5 mRNA (Fig. 5).

In the proximal part of infection zone II and in the interzone II–III, the expression of two other early nodulin genes, PsENOD3 and PsENOD14, is found in the infected cells (Fig. 5) [76]. Like the PsENOD12 and PsENOD5 genes the PsENOD3 and PsENOD14 genes are transiently expressed during pea nodule development. In nodules older than 20 days the decrease of the level of PsENOD3 and PsENOD14 mRNA matches with the disappearance of amyloplasts (Fig. 5). In younger nodules the level of these mRNAs drops markedly before the amyloplasts disappear. Sequence analyses of the PsENOD3 and PsENOD14 cDNA clones show that the mRNAs code for 6 kDa polypeptides that are 55 % homologous. They have a putative signal peptide at the amino terminal end and contain 4 cysteine

Fig. 5. The correlation between early nodulin/*nif* gene expression and amyloplast accumulation in the central tissue of pea nodules. Longitudinal sections of 20-day-old pea nodules are hybridized with antisense RNA of pea early nodulin mRNAs or *nifH* mRNA. To visualize the amyloplasts sections have been stained with 0.1 M aqueous KI-I_2. The zonation of the nodules shown in A to C and D to F is indicated below A and D, respectively. I, meristem; II, infection zone; IZ, interzone; ZIII, nitrogen-fixing zone. A to C. Localization of *nifH* mRNA. A and B are bright field micrographs of adjacent sections. Section A is stained with KI-I_2 for a shorter period than section B. Amyloplasts are the dark stained material. In section A the interzone (IZ) is more clearly visible than in B. C is the corresponding dark-field micrograph in B, to visualize *nifH* mRNA. The *nifH* mRNA is present in the infected cells of IZ and ZIII. The arrowhead indicates the same cell in B and C. Bar = 100 μm. D to F. Localization of PsENOD5 (D and E) and PsENOD3 mRNA (F) in adjacent sections. D is the bright-field micrograph corresponding to the dark-field micrograph shown in E. PsENOD5 mRNA is present in the interzone II–III (E). F is a bright-field micrograph showing the localization of PsENOD3 mRNA. PsENOD3 mRNA starts to accumulate in the infection zone (IZ) and its concentration markedly drops at the transition of 12 into ZIII. Bar = 100 μm. G and H. Magnification of the central tissue of a PsENOD5 hybridized section at the transition of infection zone into interzone. G is the bright-field micrograph corresponding to the epipolarization micrograph shown in H. Note when amyloplasts accumulate (arrowheads). PsENOD5 mRNA concentration decreases. Bar = 10 μm.

residues with a spatial distribution resembling that of metal-binding proteins [76].

Another molecular change that coincides with the transition of infection zone II into the interzone II–III is the induction of the *nif* genes of *Rhizobium* [99]. Using *in situ* hybridization it was shown that both *nifA* and *nifH* genes are induced in the first cell layer of the interzone (Fig. 5).

So at the transition of infection zone II into interzone II–III several prominent developmental changes occur. In *Rhizobium* the *nif* genes are first expressed and bacteroid morphology changes [93]. In the plant, amyloplast accumulation begins and the concentration of the early nodulin transcript PsENOD5 drops markedly. Furthermore, the decrease of the level of PsENOD3 mRNA matches with the end of the interzone. Therefore, as in alfalfa [93], amyloplast accumulation is a very useful criterion to define developmental zones of the pea nodule.

The *in situ* location of only one late nodulin mRNA leghaemoglobin (Lb) has been studied to date. Lb mRNA is first detectable in the proximal part of the infection zone II of the pea nodule reaching a maximum concentration in the nitrogen-fixing zone III [76]. In alfalfa nodules the accumulation pattern of the Lb transcript is different from that observed in pea since alfalfa Lb mRNA first appears in the interzone II–III [11].

Several late nodulins appear to be involved in O (e.g. Lb), N (e.g. uricase) or C metabolism (e.g. sucrose synthetase) of the nodule [58]. Therefore we expect that most late nodulin genes are expressed in the central tissue of the nodule. Since the expression of late nodulin genes occurs concomitantly with or shortly after the induction of Lb genes [28, 55, 95], their expression patterns will most likely coincide with that of the Lb genes.

Late nodulin gene expression has been demonstrated in a large variety of legume species. These genes and their products have been extensively reviewed [28, 58], and therefore the molecular characteristics of this group of nodulins will not be discussed in detail in this review. A remarkable group of late nodulins are late nodulins associated with the peribacteroid membrane [23, 24, 58], the membrane interface between bacterium and plant. The precise biochemical function of these peribacteroid membrane nodulins is as yet unknown. The best studied is the soybean nodulin Ngm-26, which probably is a transmembrane protein spanning the full peribacteroid membrane [23]. Its amino acid sequence is homologous to the *Escherichia coli* glycerol facilitator protein, a pore-type protein in the *E. coli* cytoplasmic membrane, which functions in transport of small molecules [73]. This suggests that Ngm-26 might be an ion channel involved in translocation of small compounds across the peribacteroid membrane [94].

Peripheral tissues

The majority of the nodulin genes studied sofar, are expressed in the central tissue. However, some early nodulin genes have been characterized, that are transcribed in the peripheral tissues. The peripheral tissue consists of nodule cortex and nodule parenchyma separated by the nodule endodermis (Fig. 3). Within the nodule parenchyma vascular bundles are located.

The early nodulin gene ENOD2 is transcribed in nodule parenchyma of both indeterminate and determinate nodules [90], and in this tissue also the GmENOD13 gene is expressed (H.J. Franssen, unpublished results). In the pericycle of the vascular tissue of soybean nodules the soybean GmENOD40 gene is expressed (H.J. Franssen, unpublished results). The specific expression of early nodulin genes in the peripheral tissues strongly suggests that these tissues have an important role in this symbiosis.

A special function for the nodule parenchyma has also been indicated by physiological studies. The groups of Witty [98] and Tjepkema [86] have shown that the nodule parenchyma regulates free O_2 concentration in the nodule. They showed that the low O_2 concentration in the nodule is achieved by the high O_2 consumption rate of the rhizobia in concert with an O_2 diffusion barrier in the nodule parenchyma [98]. The few and small intracellular spaces in the nodule parenchyma contribute to the formation of the O_2 diffusion barrier [86, 98]. The early nodulin ENOD2 is composed of two repeating pentapeptides containing two proline residues each, suggesting that also this early nodulin is a cell wall component. The soybean early nodulin protein GmENOD13 is 50 % homologous with GmENOD2 and has a similar repetitive nature [25]. Since the cell wall is a major factor in determining cell morphology, it is likely that early nodulins like ENOD2 and GmENOD13 contribute to the special morphology of the nodule parenchyma and consequently to the formation of the O_2 diffusion barrier.

The molecular and physiological studies described in this section strongly indicate that, in addition to the central tissue the peripheral tissues will have a specific function in this symbiosis. Some of the characteristics of the peripheral tissues may even be major determinants of the efficiency of the N_2 fixation process.

Bacterial factors involved in nodule differentiation

As described above, it has become clear that Nod factors play a pivotal role in the induction of root hair deformation, infection and cortical cell division. Furthermore, NodRm-1 induces the formation of empty nodules on alfalfa roots, showing that at least in some plant species Nod factors are sufficient to induce the first steps of nodule differentiation.

Nodulin gene expression has not been studied in the Nod factor-induced alfalfa nodules [87]. However, as to cell structure these empty nodules are very similar to those formed by, for example, *R. meliloti exo* mutants [21]. In the latter empty nodules some early nodulin genes are expressed but transcription of late nodulin genes has never been detected [17, 65]. Hence, we expect that also in Nod factor-induced nodules no late nodulin genes are expressed, which implies that Nod factors are most likely not involved in the induction of late nodulin gene expression. *In situ* localization of *nod* gene activity *in planta* also indicates that Nod factors are not involved in induction of late nodulin gene expression, since *nod* genes are only transcribed in the distal part of the infection zone of the nodule [77, 79]. It is far more likely that other bacterial signal molecules are required for the induction of expression of late nodulin genes. Alfalfa nodules formed by *exo* or *ndv* mutants of *R. meliloti* do not contain late nodulin mRNAs [e.g. 17, 65]. However, the fact that these mutants do not infect the roots (see above) means that development is already blocked at a stage preceding the induction of late nodulin gene expression. Therefore it cannot be concluded that these *exo* and *ndv* mutants fail to produce a signal, triggering the expression of late nodulin genes.

Mutants of *R. leguminosarum* bv. *viciae* that do not produce LPS or β-1,2-glucans form nodules in which only a small number of cells are infected [14]. In these infected cells the expression of late nodulin genes is induced [100]. Therefore neither LPS nor β-1,2-glucan are absolutely required for the induction of pea late nodulin genes. Therefore, at present it is unclear which signal molecules are actually involved in the induction of expression of late nodulin genes.

Regulation of nodulin gene expression

In both plant and animal systems studies of the molecular basis for regulation of gene expression have shown the complexity of this process [30, 50]. Several specific DNA-binding sequences (*cis*-acting elements) are present in the promoter region of these genes, and gene expression is then controlled by the binding of specific proteins, so-called *trans*-acting factors, to these specific sequences.

Nodulin genes are expressed in a time- and tissue-specific manner, thus *cis*- and *trans*-acting elements will also be part of the ultimate control mechanism of nodulin gene expression. Here we shall give a brief outline of the available data on the *cis*- and *trans*-acting elements involved in nodulin gene expression.

Cis-*regulatory elements*

Of the nodulin genes, the late nodulin genes are the best studied with respect to the *cis*-regulatory elements involved in regulation of gene expression. The promoter regions of the Lb genes of soybean (Lbc3) [83, 84] and *Sesbania rostrata* (Srglb) [12, 56, 85], the soybean late nodulin N23 gene [41, 72] and the nodule enhanced expression of a *Phaseolus* glutamine synthetase (*gln*) gene [22] have been studied in detail.

Chimaeric genes, consisting of the promoter of a nodulin gene and a reporter gene, were studied in transgenic leguminous plants such as *Lotus corniculatus* and alfalfa. The *lbc3* promoter has at least four different regions which are required for efficient nodule specific expression; a strong positive element (SPE), an organ-specific element (OSE), a negative element (NE) and a 'minimal promoter' containing the CAT and TATA boxes [82, 84]. Moreover, an interdependence of at least two of these elements (SPE and OSE) or their *trans*-acting factors is involved in nodule specific expression [82]. Comparison of the promoter regions of two Lb genes of the tropical legume *S. rostrata* with the soybean *lbc3* promoter revealed analogous *cis*-regulatory elements [85] and showed that an interdependent interaction of the CAT and TATA box region with another *cis*-regulatory element located further upstream is required for nodule specific expression [12]. In both the *lbc3* gene and the Srglb genes two highly conserved DNA sequences, AAAGAT and CTCTT, are found in the OSE element [56, 84].

Another late nodulin gene of which the promoter has been studied is the N23 gene of soybean [41], whose gene product is associated with the peribacteroid membrane [72]. Also in the promoter region of this gene an SPE, a positive element (PE) and an OSE, containing the highly conserved CTCTT and AAAGAT sequences, have been found. In contrast to the Srglb and *lbc3* promoter regions, the *cis*-regulatory elements in the promoter region of the N23 gene are located relatively close to the start of transcription (within 500 bp). In the *lbc3* and Srglb promoter regions the *cis*-elements are located as far as 1.5 kb from the start of transcription.

Trans-*regulatory elements*

In mammalian systems many DNA binding proteins have been identified and the genes for several of these proteins have been characterized [57]. There is not that much known for plants, but here also the specific interaction between DNA-binding proteins and their corresponding *cis*-regulatory elements have been described [12, 45, 74] and the first genes encoding *trans*-acting factors have been cloned [e.g. 13, 31].

Jensen *et al.* [39, 40] have shown that a nodule-specific *trans*-acting factor binds with two distinct A/T-rich DNA elements in the 5′ upstream region of the soybean *lbc3* promoter. This factor also binds to analogous sites, which are repeated several times in the *S. rostrata* Srglb promoter regions. With respect to the N23 promoter region, two distinct nodule factors (NAT1 and NAT2) and one leaf factor (LAT1) have been found to bind with A/T-rich sites in the 5′ upstream region of this promoter [38]. NAT1 and LAT1 are related to the high mobility group I (HMGI) protein of eukaryotic chromatin.

Our knowledge about the interaction of *cis*- and *trans*-regulatory elements of nodulin genes is still far from complete. This is very well illustrated by the fact that no *trans*-acting factors have been identified which interact with *cis*-regulatory elements controlling nodule specific expression.

Concluding remarks

During the past decade our insight into the role of the host plant in root nodule formation has markedly increased especially by studies on nodulin genes. Research on nodulins started with the discovery of the soybean nodulin N-35 [46] in 1979 and at present many nodulins of several legumes have been cloned [15]. Sequence analysis of the available nodulin clones has revealed interesting characteristics of these proteins. For example, most early nodulins are proline-rich proteins and show a high degree of homology to certain cell wall proteins. Biochemical studies on these early nodulins can add to our understanding why plants need that many different cell wall proteins. Other nodulins like N-26 [72, 73, 94] are membrane components and it has been shown that they are located in the peribacteroid membrane. Since this membrane is the interface between bacterium and plant it is probable that some of these nodulins have interesting functions related to exchange of metabolites and signal molecules between the two partners. However, the exact function of most nodulins remains at present unknown. Elucidation of these functions, which is a major challenge for the coming years, will undoubtedly increase our perception of biochemical and physiological processes controlling nodule formation and functioning.

Despite the fact that root nodule formation is the result of a mutual interaction of both partners, the research programs on plant and bacterial symbiotic genes have developed almost independently. The recent discovery that Nod factors trigger the expression of certain early nodulin genes provides a basis to integrate the achievements of both bacterial genetics and molecular studies on host plant genes.

Nod factors play a pivotal role in the induction of all three developmental processes – namely root hair deformation, infection and nodule formation – induced by *Rhizobium* in legume roots. Therefore, it is of major importance to elucidate the mechanisms by which the Nod factors induce these processes. The availability of purified Nod factors as well as cloned plant genes involved in the different developmental processes, that can be induced by these factors, will enormously facilitate studies on the molecular mechanisms underlying root nodule formation. The major question concerning these mechanisms that have to be studied is the way the Nod factors are perceived and transduced by the plant and how a single Nod factor, or a few very closely related factors, can induce three apparently different developmental processes in the plant.

Acknowledgements

We thank G. Truchet, H. Spaink and J. Brady for critical reading of the manuscript and stimulating discussions, G. Truchet for making Fig. 4 available, G. Heitkönig and M.J. van Iersel for secretarial assistance, and J. Spee and P. Madern for drawings.

I.V. and H.J.F. are supported by a grant from the Netherlands Organization for the Advancement of Scientific Research (NWO).

References

1. Alberts B, Bray D, Lewis J, Raff M, Roberts K, Watson JD: Molecular Biology of the Cell, 2nd ed. Garland, New York/London (1989).
2. Allen EK, Allen ON, Newman AS: Pseudonodulation of leguminous plants induced by 2-bromo-3,5-dichlorobenzoic acid. Am J Bot 40: 429–435 (1953).
3. Aman P, McNeil M, Franzen L-E, Darvill AG, Albersheim P: Structural elucidation, using HPLC-MS and GLC-MS, of the acidic polysaccharide secreted by *Rhizobium meliloti* strain 1021. Carbohydr Res 95: 263–282 (1981).
4. Bakhuizen R: The plant cytoskeleton in the *Rhizobium*-legume symbiosis. Ph.D. Thesis, Leiden University, Netherlands (1988).
5. Battisti L, Lee CC, Leigh JA: Specific forms of exopolysaccharides from *Rhizobium meliloti* restore symbiotic effectiveness to *R. meliloti exo* mutants. In: Abstracts 5th International Symposium on Molecular Genetics of Plant-Microbe Interactions, Interlaken, Switzerland p. 164, (1990).
6. Broughton WJ, Krause A, Lewin A, Perret X, Price NPJ, Relic B, Rochepeau P, Wong C-H, Pueppke SG, Brenner S: Signal exchange mediates host-specific nodulation of tropical legumes by the broad host-range *Rhizobium* species NGR234. In: Hennecke H, Verma DPS (eds), Advances in Molecular Genetics of Plant-Microbe Interactions, vol 1, pp. 162–167. Kluwer Academic Publishers, Dordrecht/Boston/London (1991).
7. Calvert HE, Pence MK, Pierce M, Malik NSA, Bauer WD: Anatomical analysis of the development and distribution of *Rhizobium* infections in soyabean roots. Can J Bot 62: 2375–2384 (1984).
8. Carlson RW, Kalembasa S, Tunoroski D, Packori P, Noel KD: Characterization of the lipopolysaccharide from a *Rhizobium phaseoli* mutant that is defective in infection thread development. J Bact 169: 4923–4928 (1987).
9. Carlson RW: Heterogenicity of *Rhizobium* lipopolysaccharides. J Bact 158: 1012–1017 (1984).
10. Carlson RW: Surface chemistry. In: Broughton WJ (ed), Nitrogen Fixation, vol 2, *Rhizobium*, pp. 199–234. Clarendon Press, Oxford (1982).
11. De Billy F, Barker OG, Gallusci P, Truchet G: Leghaemoglobin gene transcription is triggered in a single cell layer in the indeterminate nitrogen fixing root nodule of alfalfa. Plant J 1: 27–37 (1991).

12. De Bruijn FJ, Felix G, Grunenberg B, Hoffman H-J, Metz B, Ratet P, Simons-Schreier A, Szabados L, Welters P, Schell J: Regulation of plant genes specifically induced in nitrogen fixing nodules: Role of *cis*-acting elements and *trans*-acting factors in leghemoglobin gene expression. Plant Mol Biol 13: 319–325 (1989).
13. Dehesh K, Bruce WB, Quail PH: A trans-acting factor that binds to a GT-motif in a phytochrome gene promotor. Science 250: 1397–1399 (1990).
14. de Maagd RA, van Rossum C, Lugtenberg BJJ: Recognition of individual strains of fast-growing rhizobia by using profiles of membrane proteins and lipopolysaccharides. J Bact 170: 3782–3785 (1988).
15. Delauney A, Verma DPS: Cloned nodulin genes for symbiotic nitrogen fixation. Plant Mol Biol Rep 6: 279–285 (1988).
16. Diaz CL, Melchers LS, Hooykaas PJJ, Lugtenberg BJJ, Kijne JW: Root lectins as a determinant of host-plant specificity in the *Rhizobium*-legume symbiosis. Nature 338: 579–581 (1989).
17. Dickstein R, Bisseling T, Reinhold VN, Ausubel FM: Expression of nodule-specific genes in alfalfa root nodules blocked at an early stage of development: Genes Devel 2: 677–687 (1988).
18. Djordjevic S, Chen H, Batley M, Redmond JW, Rolfe B: Nitrogen fixation ability of exopolysaccharide synthesis mutants of *Rhizobium* strain NGR234 and *Rhizobium trifolii* is restored by the addition of homologous exopolysaccharides. J Bact 169: 53–60 (1987).
19. Dylan T: *Rhizobium meliloti* genes required for nodule development are related to chromosomal virulence genes in *Agrobacterium tumefaciens*. Proc Natl Acad Sci USA 83: 4403–4407 (1986).
20. Faucher C, Camut S, Dénarié J, Truchet G: The *nodH* and *nodQ* host range genes of *Rhizobium meliloti* behave as avirulence genes in *R. leguminosarum* bv. *viciae* and determine changes in the production of plant-specific extracellular signals. Mol Plant-Microbe Inter 2: 291–300 (1989).
21. Finan TM, Hirsch AM, Leigh JA, Johansen E, Kuldau GA, Deegan S, Walker GC, Signer ER: Symbiotic mutants of *Rhizobium meliloti* that uncouple plant from bacterial differentiation. Cell 40: 869–877 (1985).
22. Forde BG, Day HM, Turton JF, Wen-Jun S, Cullimore JV, Oliver JE: Two glutamine synthase genes from *Phaseolus vulgaris* L. display contrasting developmental and spatial patterns of expression in transgenic *Lotus corniculatus* plants. Plant Cell 1: 391–401 (1989).
23. Fortin MG, Morrison NA, Verma DPS: Nodulin 26 a peribacteroid membrane nodulin is expressed independently of the development of the peribacteroid compartment. Nucl Acids Res 15: 813–824 (1987).
24. Fortin MG, Zelechowska M, Verma DPS: Specific targetting of membrane nodulins to the bacteroid-enclosing compartment in soybean nodules. EMBO J 4: 3041–3046 (1985).
25. Franssen HJ, Scheres B, Van de Wiel C, Bisseling T: Characterization of soybean (hydroxy)proline-rich proteins. In: Palacios R, Verma DPS (eds) Molecular Genetics of Plant-Microbe Interaction, pp. 321–326. APS Press, St. Paul, MN (1988).
26. Glazebrook J, Walker GC: A novel exopolysaccharide can function in place of the Calcofluor-binding exopolysaccharide in nodulation of alfalfa by *Rhizobium meliloti*. Cell 56: 661–672 (1989).
27. Gloudemans T, Bhuvaneswari TV, Moerman M, van Brussel AAN, van Kammen A, Bisseling T: Involvement of *Rhizobium leguminosarum* nodulation genes on gene expression in pea root hairs. Plant Mol Biol 12: 157–167 (1989).
28. Govers F, Nap JP, Moerman M, Franssen HJ, van Kammen A, Bisseling T: cDNA cloning and developmental expression of pea nodulin genes. Plant Mol Biol 8: 425–435 (1987).
29. Govers F, Nap JP, van Kammen A, Bisseling T: Nodulins in the developing root nodule. *Plant Physiol Biochem* 25: 309–322 (1987).
30. Grierson D, Covey SN: Plant Molecular Biology, 2nd ed. Blackie, London (1988).
31. Guiltinan MJ, Marcotte WR, Quatrano RS: A plant leucine zipper protein that recognizes an abscisic acid response element. Science 250: 267–271 (1990).
32. Hahlbrock K, Scheel D: Physiology and molecular biology of phenylpropenoid metabolism. Annu Rev Plant Physiol Plant Mol Biol 40: 347–369 (1989).
33. Hennecke H: Nitrogen fixation genes involved in the *Bradyrhizobium japonicum*-soybean symbiosis. FEBS Lett 268: 422–426 (1990).
34. Hirsch AM, Bhuvaneswari TV, Torrey J-G, Bisseling T: Early nodulin genes are induced in alfalfa root outgrowths elicited by auxin transport inhibitors. Proc Natl Acad Sci USA 86: 1244–1248 (1989).
35. Hirsch AM, Bochenek B, Löbler M, McKhann HI, Reddy A, Li H-H, Ong M, Wong J: Patterns of nodule development and nodulin gene expression in alfalfa and afghanistan pea. In: Hennecke H, Verma DPS (eds) Advances in Molecular Genetics of Plant-Microbe Interactions, vol 1, pp. 317–324. Kluwer Academic Publishers, Dordrecht/Boston/London (1991).
36. Horváth B, Kondorosi E, John M, Schmidt J, Torok I, Gyorgypal Z, Barabas I, Wieneke U, Schell J, Kondorosi A: Organization, structure and symbiotic function of *Rhizobium meliloti* nodulation genes determining host specificity for alfalfa. Cell 46: 335–343 (1986).
37. Jacobs M, Rubery HP: Naturally occurring auxin transport vegetators. Science 241: 346–349 (1988).
38. Jacobsen K, Laursen NB, Jensen EØ, Marcker A, Poulsen C, Marcker KA: HMG1-like proteins from leaf and nodule nuclei interact with different AT motifs in soybean nodulin promoters. Plant Cell 2: 85–94 (1990).
39. Jensen EØ, Stougaard J, Jørgensen J-E, Sandal NN, de Bruijn FJ, Schell J, Marcker KA: Regulation of nodule-

specific plant genes. In: Bothe H, de Bruijn FJ, Newton WE (eds) Nitrogen Fixation: Hundred Years After, pp. 351–356. Gustav Fischer, New York (1988).

40. Jensen EØ, Marcker KA, Schell J, De Bruijn FJ: Interaction of a nodule specific, trans-acting factor with distinct DNA elements in the soybean leghaemoglobin Lbc3 5′ upstream region. EMBO J 7: 1265–1271 (1988).
41. Jørgensen J-E, Stougaard J, Marcker A, Marcker KA: Root nodule specific gene regulation: Analysis of the soybean nodulin N23 gene promoter in heterologous symbiotic systems. Nucl Acids Res 16: 39–50 (1988).
42. Kijne J: In: Stacey G, Burris R, Evans H (eds) Biological Nitrogen Fixation. Chapman and Hall, New York (in press).
43. Knox JP, Day S, Roberts KA: A set of surface glycoproteins forms an early marker of cell position, but not cell type, in the root apical meristem of *Daucus carota* L. Development 106: 47–56 (1989).
44. Kondorosi A, Horváth B, Göttfert M, Putnoky P, Rostas K, Györgypal Z, Kondorosi E, Török I, Bachem C, John M, Schmidt J, Schell J: Identification and organization of *Rhizobium meliloti* genes relevant to the initiation and development of nodules. In: Evans HJ, Bottomley PJ, Newton WE (eds) Nitrogen Fixation Research Progress, pp. 73–79. Martinus Nijhoff, Dordrecht/Boston/Lancaster (1985).
45. Kuhlemeyer C, Green PJ, Chua N-H: Regulation of gene expression in higher plants. Annu Rev Plant Physiol 38: 221–257 (1987).
46. Legocki RP, Verma DPS: A nodule specific plant protein (Nodulin-35) from soybean. Science 205: 190–193 (1979).
47. Leigh JA, Reed JW, Hanks JF, Hirsch AM, Walker GC: *Rhizobium meliloti* mutants that fail to succinylate their Calcofluor-binding exopolysaccharide are defective in nodule infection. Cell 51: 579–587 (1987).
48. Leigh JA, Signer ER, Walker GC: Exopolysaccharide-deficient mutants of *Rhizobium meliloti* that form ineffective nodules. Proc Natl Acad Sci USA 82: 6231–6235 (1985).
49. Lerouge P, Roche P, Faucher C, Maillet F, Truchet G, Promé JC, Dénarié J: Symbiotic host-specificity of *Rhizobium meliloti* is determined by a sulphated and acylated glucosamine oligosaccharide signal. Nature 344: 781–784 (1990).
50. Lewin B: Genes IV. Wiley, New York (1989).
51. Long S, Reed JW, Himawan JW, Walker GC: Genetic analysis of a cluster of genes required for the synthesis of the Calcofluor-binding exopolysaccharide of *Rhizobium meliloti*. J Bact 170: 4239–4248 (1988).
52. Long SR, Atkinson EM: *Rhizobium* sweet-talking. Nature 344: 712–713 (1990).
53. Long SR, Cooper J: Overview of symbiosis. In: Palacios R, Verma DPS (eds) Molecular Genetics of Plant-Microbe Interactions, pp. 163–178. APS Press, St. Paul, MN (1988).
54. Long, SR: *Rhizobium*-legume nodulation: life together in the underground. Cell 56: 203–214 (1989).
55. Marcker A, Lund M, Jensen E, Marcker KA: Transcription of the soybean leghemoglobin genes during nodule development. EMBO J 3: 1691–1695 (1984).
56. Metz BA, Welters P, Hoffman HJ, Jensen EØ, Schell J, De Bruijn F: Primary structure and promoter analysis of leghemoglobin genes of the stem-nodulated tropical legume *Sesbania rostrata*: conserved coding sequences *cis*-elements and *trans*-acting factors. Mol Gen Genet 214: 181–191 (1988).
57. Mitchell PJ, Tjian R: Transcriptional regulation in mammalian cells by sequence-specific DNA-binding proteins. Science 245: 371–378 (1989).
58. Nap JP, Bisseling T: Nodulin function and nodulin gene regulation in root nodule development. In: Gresshoff PM (ed) The Molecular Biology of Symbiotic Nitrogen Fixation, pp. 181–229. CRC Press, Boca Raton, FL (1989).
59. Nap JP, Bisseling T: Developmental biology of a plant-prokaryote symbiosis: The legume root nodule. Science 250: 948–954 (1990).
60. Newcomb EH, Tandon SR: Uninfected cells of soybean root nodules: ultrastructure suggests key role in ureide production. Science 212: 1394–1396 (1981).
61. Newcomb W, Sippel D, Peterson RL: The early morphogenesis of *Glycine max* and *Pisum sativum* root nodules. Can J Bot 57: 2603–2616 (1979).
62. Newcomb W: A correlated light electron microscopic study of symbiotic growth and differentiation in *Pisum sativum* root nodules. Can J Bot 54: 2163–2186 (1976).
63. Newcomb W: Nodule morphogenesis and differentiation. Int Rev Cytol (Suppl 13): 247–297 (1981).
64. Nguyen T, Zelechowska M, Foster V, Bergmann H, Verma DPS: Primary structure of the soybean nodulin-35 gene encoding uricase II localized in the peroxisomes of uninfected cells of nodules. Proc Natl Acad Sci USA 82: 5040–5044 (1985).
65. Norris JH, Marcel LA, Hirsch AM: Nodulin gene expression in effective alfalfa nodules and in nodules arrested at three different stages of development. Plant Physiol 88: 321–328 (1988)
66. Peters NK, Frost JW, Long SR: A plant flavone, luteolin, induces expression of *Rhizobium meliloti* nodulation genes. Science 233: 977 (1986).
67. Redmond JW, Batley M, Djordjevic MA, Innes RW, Kuempel PL, Rolfe BG: Flavones induce expression of nodulation genes in *Rhizobium*. Nature 323: 632–633 (1986).
68. Reuber TL, Reed JW, Glazebrook J, Urzainqui A, Walker GC: Analyses of the roles of *R. meliloti* exopolysaccharides in nodulation. In: Hennecke H, Verma DPS (eds) Advances in Molecular Genetics of Plant-Microbe Interactions, vol 1, pp. 181–188. Kluwer Academic Publishers, Dordrecht / Boston / London (1991).

69. Roche P, Lerouge P, Ponthus C, Promé JC: Structural determination of bacterial nodulation factors involved in the *Rhizobium meliloti*-alfalfa symbiosis. J Biol Chem (in press).
70. Roche P, Lerouge P, Promé JC, Faucher C, Vasse J, Maillet F, Camut S, De Billy F, Dénarié J, Truchet G: NodRm-1, a sulphated lipo-oligosaccharide signal of *Rhizobium meliloti* elicits hair deformation, cortical cell division and nodule organogenesis on alfalfa roots. In: Hennecke H, Verma DPS (eds) Advances in Molecular Genetics of Plant-Microbe Interactions, vol 1, pp. 119–126. Kluwer Academic Publishers, Dordrecht/Boston/London (1991).
71. Roth E, Jean K, Stacey G: Homology in endosymbiotic systems: the term 'symbiosome'. In: Palacios R, Verma DPS (eds) Molecular Genetics of Plant-Microbe Interactions, pp. 220–225. APS Press, St. Paul, MN (1988).
72. Sandal NN, Bojsen K, Marcker KA: A small family of nodule specific genes from soybean. Nucl Acids Res 15: 1507–1519 (1987).
73. Sandal NN, Marcker KA: Soybean nodulin-26 is homologous to the major intrinsic protein of the bovine lens fiber membrane. Nucl Acids Res 19: 9347 (1988).
74. Schell JS: Transgenic plants as tools to study the molecular organisation of plant genes. Science 237: 1176–1183 (1987).
75. Scheres B, van de Wiel C, Zalensky A, Horváth B, Spaink H, van Eck H, Zwartkruis F, Wolters AM, Gloudemans T, van Kammen A, Bisseling T: The ENOD12 gene product is involved in the infection process during pea-*Rhizobium* interaction. Cell 60: 281–294 (1990).
76. Scheres B, van Engelen F, van der Knaap E, van de Wiel C, van Kammen A, Bisseling T: Sequential induction of nodulin gene expression in the developing pea nodule. Plant Cell 8: 687–700 (1990).
77. Schlaman HRM, Horváth B, Vijgenboom E, Okker R, Lugtenberg BJJ: Evidence for a new negative regulation mechanism involved in the suppression of nodulation gene expression of *Rhizobium leguminosarum* bv. *viciae*. J Bact (in press).
78. Schwedock J, Long SR: ATP sulphurylase activity of the *NodP* and *NodQ* gene products of *Rhizobium meliloti*. Nature 348: 644–646 (1990).
79. Sharma BS, Signer ER: Temporal and spatial regulation of the symbiotic genes of *Rhizobium meliloti* in plants revealed by transposon Tn*5-gusA*. Genes Devel 4: 344–356 (1990).
80. Spaink HP, Geiger O, Sheeley DM, van Brussel AAN, York WS, Reinhold VN, Lugtenberg BJJ, Kennedy EP: The biochemical function of the *Rhizobium leguminosarum* proteins involved in the production of host specific signal molecules. In: Hennecke H, Verma DPS (eds) Advances in Molecular Genetics of Plant-Microbe Interactions, vol 1, pp. 142–149. Kluwer Academic Publishers, Dordrecht/Boston/London (1991).
81. Spaink HP, Weinman J, Djordjevic MA, Wijffelman CA, Okker RJH, Lugtenberg BJJ: Genetic analysis and cellular localization of the *Rhizobium* host specificity-determining NodE protein. EMBO 8: 2811–2818 (1989).
82. Stougaard J, Joergensen JE, Christensen T, Kühle A, Marcker KA: Interdependence and nodule-specificity of cis-acting regulatory elements in soybean leghemoglobin lbc3 and N23 gene promoters. Mol Gen Genet 220: 353–360 (1990).
83. Stougaard J, Petersen TE, Marcker KA: Expression of a complete soybean leghemoglobin gene in root nodules of transgenic *Lotus corniculatus*. Proc Natl Acad Sci USA 84: 5754–5757 (1987).
84. Stougaard J, Sandal NN, Grøn A, Köhle A, Marcker KA: 5′ analysis of the soybean leghemoglobin lbc3 gene: regulatory elements required for promoter activity and organ specificity. EMBO J 6: 3565–3569 (1987).
85. Szabados L, Ratet P, Grunenberg B, de Bruijn FJ: Functional analysis of the *Sesbania rostrata* leghemoglobin glb gene 5′ upstream region in transgenic *Lotus corniculatus* and *Nicotiana tabacum* plants. Plant Cell 2: 973–986 (1990).
86. Tjepkema JD, Yocum CS: Measurement of oxygen partial pressure within soybean nodules by oxygen microelectrodes. Planta 119: 351–360 (1974).
87. Truchet G, Roche P, Lerouge P, Vasse J, Camut S, De Billy F, Promé J-C, Dénarié J: Sulphated lipo-oligosaccharide signals of *R. meliloti* elicit root nodule organogenesis and alfalfa. Nature 351: 670–673 (1991).
88. Turgeon BG, Bauer WD: Early events in the infection of soybean by *Rhizobium japonicum*. Time course and cytology of the initial infection process. Can J Bot 60: 152–161 (1982).
89. Van den Bosch KA, Bradley DJ, Knox JP, Perotto S, Butcher GW, Brewin NJ: Common components of the infection thread matrix and the intercellular space identified by immunochemical analysis of pea nodules and uninfected roots. EMBO J 8: 335–342 (1989).
90. van de Wiel C, Norris J, Bochenek B, Bisseling T, Hirsch AM: Nodulin gene expression and ENOD2 localization in effective, nitrogen-fixing and ineffective, bacteria-free nodules of alfalfa. Plant Cell 2: 1009–1017 (1990).
91. van de Wiel C, Scheres B, Franssen H, van Lierop MJ, van Lammeren A, van Kammen A, Bisseling T: The early nodulin transcript ENOD2 is located in the nodule parenchyma (inner cortex) of pea and soybean root nodules. EMBO J 9: 1–7 (1990).
92. van Kammen A: Suggested nomenclature for plant genes involved in nodulation and symbiosis. Plant Mol Biol Rep 2: 43–45 (1984).
93. Vasse J, De Billy F, Camut S, Truchet G: Correlation between ultrastructural differentiation of bacteroids and nitrogen fixation in alfalfa nodules. J Bact 172: 4295–4306 (1990).
94. Verma DPS: Endosymbiosis of *Rhizobium*: Internaliza-

tion of the 'extracellular compartment' and metabolites exchange. In: Gresshoff P, Roth E, Stacey G, Newton WB (eds) Nitrogen Fixation: Achievements and Objectives, pp. 235–237. Chapman Hall, New York/London (1990).

95. Verma DPS, Legocki RP, Auger S: Expression of nodule-specific host genes in soybean. In: Gibson AH, Newton WE (eds) Current Perspectives in Nitrogen Fixation, pp. 205–209. Australian Academy of Science, Canberra (1981).
96. Verma DPS, Long SR: The molecular biology of *Rhizobium*-legume symbiosis. Int Rev Cytol (Suppl 14): 211–245 (1983).
97. Vincent JM: Factors controlling the legume-*Rhizobium* symbiosis. In: Newton WE, Orme-Johnson WH (eds) Nitrogen Fixation II, pp. 103–129. University Park Press, Baltimore (1980).
98. Witty JF, Minchin FR, Shøt L, Sheehy JE: Nitrogen fixation and oxygen in legume root nodules. Oxford Surv Plant Cel Biol 3: 275–315 (1986).
99. Yang WC, Horváth B, Hontelez J, van Kammen A, Bisseling T: *In situ* hybridization of *Rhizobium* mRNAs in pea root nodules; *nifA* and *nifH* mRNA localization. Mol Plant-Micr Inter (in press) (1991).
100. Yang WC, Canter Cremers HCJ, Hogendijk P, Batley M, Wijffelman CA, van Kammen A, Bisseling T: The role of chalcone synthase in pea nodules formed by wild type and β(1–2) glucan mutants of *Rhizobium leguminosarum* bv. *viciae*. Submitted.

Plant Molecular Biology **19**: 109–122, 1992.
© 1992 Kluwer Academic Publishers. Printed in Belgium.

The molecular biology of disease resistance

N.T. Keen
Department of Plant Pathology, University of California, Riverside, CA 92521, USA

Key words: Pathogenicity and virulence mechanisms, hypersensitive reaction (HR), elicitors, disease resistance genes, gene-for-gene complementarity

Introduction

Plants have evolved with pathogens and insect pests for millions of years. It is therefore not surprising that a particular plant is resistant to most of them. When certain environmental conditions exist, however, virulent pathogens can cause damage to plant tissues both in the field and in storage. Indeed, modern intensive agriculture has often exacerbated the occurrence of disease outbreaks by presenting the pest with an intense monoculture of genetically identical plants grown in close proximity under high water and fertilizer regimes. Plant pathologists and entomologists are faced with the job of preventing economic losses in these contrived situations. Historically, the most desirable and effective strategy has been the incorporation of disease resistance genes into commercially acceptable cultivars. In the past ten years, the tools of molecular biology have been applied to the study of the mechanisms accounting for disease resistance. This paper summarizes progress and notes areas in which greater strides toward understanding and improving disease resistance can be anticipated. Because of its broad nature, this paper is not comprehensive but focuses mainly on selected recent work and heavily cites review papers.

An overview of disease resistance in plants

Disease and pest resistance in plants is multifaceted, involving structural and chemical components, some of which are preformed while others are inducibly produced in response to pathogen infection. If the environment is conducive and a virulent pathogen is present, defense systems are necessary or disease results. Even with such a system in place, the pathogen may produce virulence factors that destroy or compromise components of plant defense. Similarly, if the pathogen overcomes static defenses and does not display factors that elicit active defense systems, disease may occur. On the other hand, the pathogen that fails to evade one or more components of the plant defense gauntlet will be unsuccessful.

Mechanisms of pathogen virulence and pathogenicity and how they relate to disease resistance

Parasites must possess the machinery required for growth in the environment of their hosts and must also be equipped to obtain nutrients and overcome host defenses. Not surprisingly, parasites have devised many approaches to solving these problems. Pathogens frequently possess pathogenicity and virulence genes that are tightly regulated and generally expressed only when the parasite is growing in the host [84, 86, 117]. Certain bacterial plant pathogens contain large clusters of plant-regulated genes, called *hrp* genes, which are required for growth and pathogenicity in plants [5, 75, 83, 115]. The *hrp* genes are also required to elicit hypersensitive defense reactions on incompatible host plants, but it is not yet clear whether this results from failure of *hrp* mutants to grow well in the plant or for more complex reasons. Huynh *et al.* [54], for example, have shown that *hrp* gene regulation is required for the expression of avirulence gene *B* in *Pseudomonas syringae* pv. *glycinea*. Also of considerable interest, a cloned *hrp* cluster from *Pseudomonas syringae* pv. *syringae* caused the hypersensitive reaction when introduced into other pathogenic or saprophytic bacteria, including *Escherichia coli* [51]. The biochemical functions of *hrp* genes are poorly understood, but some evidence suggests they may be involved in the uptake and/or secretion of nutrients and other factors when bacteria enter host tissues [52, 83].

General virulence mechanisms

Plant pathogens usually express several virulence mechanisms that increase their ability to colonize and damage host plant tissues. Thus, virulence in the pathogen and resistance in a plant host are reciprocal concepts and a discussion of plant resistance necessarily entails a consideration of virulence mechanisms in pathogens. Some of these are general mechanisms such as the production of enzymes, plant growth regulators or toxins. These agents damage plant cells and cause the leakage of nutrients as well as provide an optimal environment for the pathogen and reduce the capacity of plant cells for defense. For example, pathogen-produced pectic enzymes degrade the pectic fraction of the plant cell wall, resulting in tissue disintegration and nutrient release that facilitate growth of the pathogen.

The production of pectic enzymes is an important component of virulence in soft-rotting *Erwinia* sp. [69]. The most compelling evidence is from genetic experiments in which mutation of particular *pel* genes resulted in a marked decrease in virulence on a plant host [8, 91]. These experiments were only possible following the cloning of several structural genes (see [69]). Unlike the usual case in procaryotes, *Erwinia* sp. contain multi-gene families of tandem, highly homologous, *pel* genes encoding pectate lyase isoenzymes. Furthermore, these genes are independently regulated, each possessing its own promoter as well as a transcriptional terminator following the coding region [89, 106]. An important clue rationalizing the production of multiple pectate lyase isozymes resulted from work by Preston *et al.* [88]. Of the four isozymes produced by an *E. chrysanthemi* strain, two produced similar patterns of product release from the substrate, polygalacturonic acid. However, the other two enzymes exhibited patterns of product release that were different from each other and distinct from the former two enzymes. Thus, the pectate lyase isozymes exhibit considerably different catalytic properties despite their substantial amino acid homology. The importance of this complexity is thought to lie in the range of plant species attacked by strains of *Erwinia*. Thus, certain members of the *pel* multi-gene family may facilitate virulence on a particular plant while other of the isozymes may be more important in virulence on a different plant species.

Toxins (e.g. [7, 84]) and extracellular polysaccharides (e.g. [26]) have also been implicated in the virulence of bacterial pathogens by genetic experiments. For example, Kinscherf *et al.* [63] recently generated mutants of *Pseudomonas syringae* BR2 deficient in production of the toxin, tabtoxin, as well as the ability to produce disease symptoms on bean plants. The mutant strains multiplied as well in bean tissue as the wild-type bacterium but did not produce disease symptoms. This is an important observation since it implies that host plants would be resistant to disease if they could be made insensitive to the bacterial toxin. The prediction has been tested and Anzai *et al.* [2] reported that transgenic tobacco plants expressing a gene conferring resistance to tabtoxin were in fact also resistant to the disease caused by a toxigenic bacterial pathogen. This finding therefore has important implications for disease resistance in cases where the pathogen produces a toxin that is required for high virulence.

Several plant pathogens incite imbalances in the levels of plant growth regulators. Some of the classic work in this field was done by the group of Kosuge with *Pseudomonas savastanoi*, the causal agent of a gall disease on olive and oleander plants. These bacteria multiply to large numbers in the favorable environment of the gall tissues, which are incited due to production of IAA by the pathogen (e.g. [18]). The IAA biosynthetic genes can therefore be considered as important bacterial virulence factors. The work with *P. savastanoi* IAA biosynthetic genes is also of historical importance because it facilitated subsequent understanding of the role of *Agrobacterium tumefaciens onc* genes in causation of crown gall tumors (e.g. [1]).

Specific virulence mechanisms

Many pathogens produce virulence factors that allow pathogenesis only on certain cultivars or species of plant. These factors accordingly have importance in determining plant disease resistance and pathogen host range. For example, certain fungal pathogens produce low-molecular-weight toxins that damage only specific cultivars of a host plant species [19, 97]. The pathogen only causes disease on these precise cultivars. Such host-selective toxins therefore constitute virulence mechanisms which exhibit a high degree of specificity since susceptible plant cultivars frequently harbor a single gene accounting for their toxin sensitivity. Mechanisms accounting for the host-selective toxicity have been investigated, the best understood case being T toxin from *Helminthosporium maydis* race T, described in this issue by Levings [73].

Rhizobium sp. [78] and bacterial pathogens have been shown by mutation experiments to possess positive-acting genetic determinants required for nodulation of particular species and/or cultivars of plants. Such determinants may also function to define the range of host species attacked by a pathogen. For instance, Mellano and Cooksey [81] obtained mutants of *Xanthomonas campestris* pv. *translucens* that were reduced in the number of grass host species they could attack. The bacterial strain employed normally attacked four different grass species, but the several mutants obtained had lost virulence on one or more of the plant species while remaining fully virulent on the others. Waney *et al.* [112] confirmed these findings and were able to complement some of the mutations with cloned DNA from the wild-type strain. This should permit deduction of the various gene functions that compromise the resistance of particular plant species.

Swarup *et al.* [104] showed that a cloned DNA fragment from *Xanthomonas citri* permitted another citrus pathogen, *X. campestris* pv. *citrumelo*, to form cankerlike lesions on citrus leaves which are normally unique to the disease caused by *X. citri*. Thus, the cloned DNA appeared to confer a specific virulence factor associated with *X. citri*. More recently, the same authors [105] showed that this gene, called *pthA*, has high homology with a bacterial avirulence gene called *avrBs3*, to be discussed later. Of considerable interest, introduction of *pthA* into other bacterial pathogens, such as *X. phaseoli*, and *X. citri* pv. *malvacearum*, caused them to induce defensive reactions on their normal host plants. These results show that *pthA* not only confers a specific virulence function in *X. citri* but also determines host recognition in other bacterial pathogens leading to plant defense reactions. Another possible example of this evolutionarily appealing plant strategy for defensive pathogen surveillance involves *avrBs2*, an avirulence gene cloned from *X. campestris* pv. *vesicatoria* [55]. When *avrBs2* was mutated in the bacterium, the resulting strain was reduced in virulence on pepper plants, regardless of whether they harbored the cognate resistance gene, *Bs2*. Thus, *Bs2* may target a bacterial metabolite specified by *avrBs2* that is important for bacterial virulence.

Mechanisms of disease resistance

Plants are equipped with an array of mechanisms for resistance to pathogens. Some of these are preformed barriers such as cell walls, cuticle and the root pericycle. An impressive proof of the role of the cuticle in preventing ingress of certain pathogens is supplied by the work of Dickman *et al.* [30]. These workers transformed a cutinase gene cloned from *Fusarium* (which normally directly penetrates the cuticle and cell wall of its host plant, pea) into *Mycosphaerella*, which normally only infects its host, papaya, through wounds in the cuticle/epidermis. However, *Mycosphaerella* transformants expressing the cutinase gene were able to directly penetrate papaya fruit through intact cuticle. This experiment therefore elegantly demonstrates that the cuticle is normally an effective resistance barrier to *Mycosphaerella* and that cutinase is an important virulence factor which breaches the cutin barrier. In addition to such structural features, preformed chemical barriers may also be important in disease resistance. For example, a genetic experiment was utilized to test the role of a preformed steroidal glycoalkaloid, tomatine, in the resistance of green tomato fruits to infection by the fungus *Fusarium* [24]. Mutant fungal strains were selected that grew in culture on high levels of tomatine and their virulence was also greater than the wild-type, tomatine sensitive pathogen. This indicated that tomatine is an important resistance factor against *Fusarium* infection. Inducible chemical and structural resistance mechanisms also occur in plants. For example, many induced structural wound-healing responses such as callose deposition, periderm formation, wound gums, lignin and suberization may confer resistance against certain pathogens [11].

Another important active defense system in plants is called the hypersensitive response (HR). The HR is characterized by the death of plant cells in the vicinity of an invading pathogen shortly after infection occurs. While plant cell necrosis *per se* may directly contribute to resistance, the actual curtailment of pathogen development and spread often occurs later. This resistance frequently correlates closely with the formation of antibiotic chemicals, called phytoalexins, or the erection of induced structural barriers such as lignin in the infection site. The expression of many plant genes is activated during the HR, including those encoding enzymes of the phenolic pathway, peroxidases, glucanases and chitinases, as well as hydroxyproline-rich glycoproteins (see [34, 46]). It should be emphasized that resistance modulated by HR is a cellular and tissue response, and accordingly the temporal and spatial expression patterns of defense response genes may be strategically important (see [45, 98]). Combinations of defense responses may also act synergistically to contain an invading pathogen. Curiously, the physiological role of most of the putative defense responses in inhibiting pathogen development has not been critically tested. The only exception thus far is phytoalexins.

The occurrence of phytoalexins in plant defense was inferred by the classic work of Müller 50 years ago (for review see [50]). The role of phytoalexins in disease defense has been studied extensively for several years. Although much of the available evidence is correlative, it cumulatively makes a strong case for the role of phytoalexins in the resistance of many plants to an array of pathogens [37, 56]. Some of the major experimental evidence emanates from the use of specific inhibitors on biosynthetic enzymes (e.g. [111]) and critical studies on the localization, quantitation and timing of phytoalexin production by plant tissues in relation to pathogen development [47, 53, 87, 100, 121]. Direct proof for the role of phytoalexins in disease resistance has been supplied by genetic experiments utilizing pathogen strains that vary in their virulence and ability to degrade a particular plant phytoalexin [27, 108]. Schäfer *et al.* [95] also employed a phytoalexin detoxifying gene cloned from the fungal pathogen *Fusarium solani* f. sp. *pisi* which detoxified the pea phytoalexin, pisatin. This gene was transformed into *Cochliobolus heterostrophus*, a pathogen of maize but not pea plants. Surprisingly, significant growth of the transformed fungus occurred in pea plants. Thus, the ability to defeat a single component of pea defense allowed *C. heterostrophus* to become a *quasi* pathogen. In all of the genetic experiments above, the ability to degrade a host phytoalexin was required for high virulence. The corollary conclusion is that unless a particular pathogen strain does not elicit or, alternatively, degrades or otherwise tolerates a host phytoalexin, it constitutes an effective defense barrier.

Recently, considerable progress has occurred

in isolating enzymes involved in phytoalexin biosynthetic pathways. For example, the laboratory of Prof. Hans Grisebach has made monumental strides in understanding the isoflavonoid pathway in soybean which accounts for production of pterocarpanoid phytoalexins called glyceollins (see [39] for a review). This and other groups have also succeeded in cloning genes encoding some of the key terminal enzymes in phytoalexin biosynthetic pathways from legumes and other plants (e.g. W. Barz and R. Dixon, personal communication; J. Chappell, personal communication; [48, 113]). These are important experiments because the biosynthetic genes can be transformed into foreign plants, permitting them to make new phytoalexins [48]. As will be discussed later, there is good reason to think that pathogens of such transgenic plants might have difficulty coping with the presence of phytoalexins not previously encountered.

Rapid reactions to elicitors and activation of defense response genes

Plant tissues inoculated with incompatible pathogens or treated with elicitors of defense responses undergo the rapid generation of activated oxygen species (e.g. [3]) and alterations in K^+ efflux (e.g. [62]). Associated with or following closely after these changes are alterations in cellular phosphorylation patterns (e.g. [31]) and, in at least some cases, in calcium flux [70, 101]. While all these changes may be involved in the signal transduction patterns by which elicitors are perceived by certain plant cells, it is not yet known how the observed derepression of defense response genes occurs. New developments in the sensitive monitoring of intracellular levels of signal molecules (e.g. [64]) should aid progress in this area.

Defense response genes encode enzymes involved in the biosynthesis of phytoalexins, others such as glucanases and chitinases that are antagonistic to pathogens and enzymes leading to the deposition of structural polymers such as lignin and hydroxyproline-rich glycoproteins. Since much of the work in this field has been recently reviewed [22, 33, 34, 96], coverage will be minimal here. The elegant use of *in situ* mRNA probing has revealed precise temporal and spatial expression patterns for defense response genes during a resistance reaction (e.g. [98]). A major interest in this field has been the definition of conserved promoter and exonic elements in several defense response genes that are associated with their specific induction during HR [22, 35, 36, 76, 80]. By identifying the cognate *trans*-acting plant proteins, it might be possible to elucidate the signalling mechanisms by which pathogen infection results in defense response gene activation.

Plant recognitional mechanisms leading to active resistance

Like the immune systems of vertebrates, plants perceive the presence of potential pathogens and invoke active defense systems such as the HR. Plants respond to a varied array of pathogen metabolites, called elicitors. In some cases, elaboration of an elicitor is directly determined by a single pathogen allele, called an avirulence gene. Unfortunately, the putative plant receptors that perceive these elicitors have not yet been characterized, nor have the genes encoding them been cloned. However, plants have been shown genetically to possess substantial batteries of disease resistance genes that confer resistance to some but not all biotypes of a particular pathogen species. These genes are proposed to encode receptors for pathogen elicitors that are specified by avirulence genes [28, 43, 60].

Pathogen-produced elicitors of plant defense

Cruickshank and Perrin [20] first demonstrated that a fungal plant pathogen elaborated a soluble factor which led to the induction of phytoalexin synthesis in pea plants. This elicitor was identified as a ca. 8 kDa peptide. Several elicitors of pathogen origin have subsequently been reported. Recent reviews (e.g. [33, 34, 60]) provide greater detail than is possible in this paper. As described in the next section, three different pathogen elicitors have been characterized that are associated with avirulence gene activity. It has been proposed that these elicitors are perceived by specific receptors in plants carrying the complementary disease resistance genes [28]. Unfortunately, there have not yet been critical tests for occurrence of these plant receptors. However, evidence obtained with other elicitors points to the occurrence of cognate receptors in plants.

Sharp *et al.* [99] elucidated the structure of a B-linked heptaglucan elicitor isolated from cell wall hydrolyzates of the fungus, *Phytophthora megasperma* f.sp. *glycinea*. This elicitor is highly active in soybean and a few additional plants but relatively inactive in others. Several groups have demonstrated the occurrence in soybean plasma membranes of specific binding sites for *Phytophthora* glucan elicitors [16, 38, 120]. Isolation of the putative receptor protein(s) should permit cloning of the encoding gene(s).

Victorin is a cyclic peptide of novel structure [118] that was initially regarded as a host-selective toxin in oat plants containing the disease resistance gene, *Pc-2*. Mayama *et al.* [79], however, demonstrated that victorin is in fact an elicitor of defense responses modulated by the *Pc-2* resistance gene. Wolpert and Macko [119] presented evidence for a specific 100 kDa victorin receptor in *Pc-2* oat cells by applying labeled victorin to leaf tissues and electrophoresing leaf proteins. However, incubation of victorin with soluble leaf extracts resulted in labelling of a 100 kDa membrane protein from both the *Pc-2* and *pc-2* oat genotypes. This discrepancy is not yet understood.

Pathogen avirulence genes

It has been known for 50 years [41] that a single plant disease resistance gene has a matching or complementary gene in the pathogen, called an avirulence gene. This has led to the concept of gene-for-gene complementarity [42, 43, 57]. As noted earlier, a functional avirulence allele must occur in a pathogen before the cognate disease resistance gene in a plant perceives the pathogen and invokes active defense reactions. Although many avirulence genes have been identified by classical genetic studies, only recently have they been cloned and characterized. The first avirulence gene cloned was *avrA* from race 6 of *Pseudomonas syringae* pv. *glycinea* [102]. The first cloned fungal avirulence gene was *avr9* from *Cladosporium fulvum* [28, 109] and the first proven viral avirulence gene was the coat protein gene of tobacco mosaic virus (TMV) [21]. The study of avirulence genes has greatly increased our understanding of how gene-for-gene systems function biochemically. They have also provided insight for the improved deployment of plant disease resistance genes to more effectively control pathogens and other pests.

In addition to their presence in pathogens, avirulence genes also appear to occur in *Rhizobium* sp. (for review, see [43]). For example, Lewis-Henderson and Djordjevic [74] recently reported that nodulation of the clover cultivar Woogenellup by *R. leguminosarum* bv. *trifolii* is blocked by the presence of two bacterial genes, *nodM*, and *csn-1*. Adding to the complexity, the function of these genes was suppressed by the presence of a functional *nodT* gene. Since both *nodM* and *csn-1* were required to give the avirulence phenotype, it was proposed that they contributed to a single biochemical pathway, resulting in an elicitor which in turn interacted with a resistance gene product in cv. Woogenellup. Future work with these genes should be of great interest since *nodM* is believed to encode D-glucosamine synthetase [4], a precursor of the signal molecules which stimulate mitosis in plant cells [72, 93]. The work with *nodM* and *csn-1* is of importance because it is the first case in which two different avirulence genes complement one plant disease resistance gene and the system should permit identification of the putative elicitor molecule. It will be interesting to see if other gene-for-gene interactions are discovered with this degree of complexity.

All of the currently known avirulence genes encode single protein products of various sizes. With the exception of *nodM* and the coat protein gene of TMV, discussed above, sequencing of avirulence genes has not yet resulted in determination of their functions in the pathogen. However, avirulence genes with distinct specificities may exhibit homology to each other (e.g. [107]). The most striking example of homologous avirulence genes is provided by the *avrBs3* family in pathovars of *Xanthomonas campestris*. Bonas *et al.* [10] cloned *avrBs3* from *X. campestris* pv. *vesicatoria* and showed that it encoded a large protein product which was unusual in containing several repeated amino acid sequences. Recently, other avirulence genes with unique specificities but highly homologous to *avrBs3* have been discovered in several *X. campestris* pathovars (D. Gabriel, personal communication; J. Leach, personal communication; U. Bonas and R. Stall, personal communications; [25, 105]).

The cloning of pathogen avirulence genes has provided definitive genetic proof of their role in recognition of pathogens by plants carrying the complementary resistance genes [43, 57]. Recent results have indicated that avirulence genes may also be associated with higher level determination of microbial host ranges, such as at the plant species level [14, 67, 114]. These studies have further raised the possibility that resistance genes from one plant may be expected to function if transformed into other, distantly related plant species.

Elicitors specified by avirulence genes

Recently, several avirulence genes have been associated with the production of discrete elicitors by the pathogens which harbor them. For instance, Knogge *et al.* [65] observed that an isolate of *Rynchosporium secalis* produced a linear peptide elicitor, called NIP1, which elicited active defense reactions in a barley cultivar carrying the *Rrs1* resistance gene. It will accordingly be of interest to see if the NIP1 peptide is the product of an avirulence gene in *R. secalis*. Ricci *et al.* [90] also showed that two non-pathogenic *Phytophthora* sp. produced peptides which elicited defense reactions in tobacco plants. However, the tobacco pathogen, *P. parasitica*, did not produce this peptide. While not yet proven, the genes encoding the different *Phytophthora* peptides could be regarded as avirulence gene alleles.

Pierre De Wit and colleagues have done elegant work on the characterization of *avr9* in *Cladosporium fulvum* and the mechanism by which this avirulence gene interacts with tomato plants expressing resistance gene *Cf9* (see [109]). The key initial experiment was the isolation of a small peptide produced in infected tomato leaves only by fungal isolates carrying *avr9* [29]. This peptide functioned as a specific elicitor of the hypersensitive response only in tomato plants carrying *Cf9*. Following sequencing of the peptide, an oligonucleotide probe was constructed and used to select cDNA and genomic clones from libraries of an *avr9 C. fulvum* biotype. It has also recently been shown that transformation of the cloned *avr9* gene into a fungal isolate lacking the gene converted it to the *avr9* phenotype (P. De Wit, personal communication). This result therefore proves the identity of *avr9* as an avirulence gene determining gene-for-gene specificity.

Dawson and colleagues showed that the coat protein gene of certain isolates of TMV behaves as an avirulence gene in tobacco plants carrying the N′ resistance gene. Comparison of the sequences of coat protein genes from avirulent and virulent viral strains of N′ plants led to the discovery that single amino acid changes in the coat protein modulated the avirulence phenotype [66]. More recently, transgenic tobacco plants have been constructed carrying the coat protein genes from virulent or avirulent TMV isolates, but no other viral sequences [21]. Plants transformed with the coat protein genes from virulent viral strains did not exhibit detectable symptoms, but transformants with the coat genes from avirulent isolates showed pronounced necrotic leaf symptoms which enlarged and eventually led to the death of entire leaves. Furthermore, the necrotic leaves accumulated defense response proteins symptomatic of the HR. These results with TMV are important because they constitute the only case to date in which an avirulence gene has been identified that has a known function in the pathogen. The coat protein therefore functions *per se* as an elicitor of the HR in N′ tobacco plants and recent results indicate that the structural organization of protein subunits cues plant recognition

(J. Culver and W. Dawson, personal communication).

Avirulence gene D from *Pseudomonas syringae* pv. *tomato* [68] is the only bacterial gene for which information is available on how avirulence genes interact with resistant plants. The cloned *avrD* gene functioned in *P. syringae* pv. *glycinea* cells to elicit the HR in only those cultivars of soybean carrying the disease resistance gene, *Rpg4* [59]. The *avrD* gene encodes a 34 kDa protein of unknown function in the bacteria. However, its expression causes several Gram negative bacteria, including *E. coli*, to extracellularly secrete a low molecular weight elicitor molecule that initiates the HR only in *Rpg4* soybean plants [61]. It is therefore suspected that *avrD* encodes an enzymatically active protein which converts a normal bacterial metabolite to the *avrD* elicitor or a precursor of the elicitor. The elicitor activity has been associated with two closely related fatty acid derivatives with several oxygen atoms at one end of the molecules (J. Sims, M. Stayton, S. Midland, M. Smith, D. Kobayashi, S. Kawasaki and N. Keen, unpublished observations). Thus, the elicitor dictated by *avrD* is quite different from the peptide gene products of *avr9* and the coat protein gene of TMV. It is probable, however, that elicitors specified by avirulence genes will be diverse in structure, just as is the case with antigens in vertebrates.

Plant disease resistance genes

Despite the accumulation of substantial classical genetic data on the behavior of plant disease resistance genes, none have yet been cloned and characterized. Several strategies are currently being pursued to attempt the cloning of these genes [40, 58]. The major difficulty is that disease resistance genes do not have identified protein products and can therefore only be detected by occurrence of the HR following inoculation of a plant with a pathogen strain carrying the complementary avirulence gene. Transposon tagging and chromosome walking from closely linked marker genes have thus far been the preferred approaches, but neither has yet produced a cloned resistance gene. The recent fine mapping of a disease resistance gene in *Arabidopsis thaliana* [23] and the inherent advantages of this plant for molecular genetic experiments portend the imminent cloning of this resistance gene by chromosome walking.

Technical advances have also permitted the application of two alternative strategies to the cloning of disease resistance genes. The first is subtractive cloning [103], in which a resistance gene that is absent in a deletion cultivar can be enriched from a cultivar possessing the gene by hybridization and PCR methods. The second new approach is a functional cloning strategy in which pools of DNA clones from a resistant plant are transformed with a particle acceleration device into suitable tissue of susceptible plants. The transformed areas are then screened by applying one of the specific elicitors discussed above or by inoculating with an avirulent pathogen strain in order to eventually select a single clone giving the hypersensitive resistant phenotype [58]. All of these approaches should facilitate the cloning of several disease resistance genes in the next few years.

Novel means of engineering disease resistance

There have been experimental successes with the transformation of various foreign genes into plants to effect disease resistance [116]. With virus diseases, for instance, considerable success has resulted from the transformation of viral coat protein or replicase genes into plant hosts [6, 44]. Similarly, introduction of a microbial chitinase gene into tobacco plants [13, 92] has shown promise for the control of certain fungal pathogens. If several naturally occurring plant disease resistance genes are eventually cloned, their transformation into various plant species should also have considerable impact on disease control [58].

Hain *et al.* [48] demonstrated that introduction of the key biosynthetic gene for a peanut stilbene phytoalexin into tobacco plants enabled them to produce the peanut phytoalexin, resveratrol. As noted earlier, it is predicted that plants able to make large amounts of foreign phytoalexins would be resistant to pathogens that could not detoxify the new chemical structures. Similarly, transgenic plants expressing protease inhibitor [94] or Bt toxin genes [85] offer promise for improved resistance to insects.

There are additional novel avenues that can be explored in order to develop resistant plants. One approach is to introduce genes that detoxify pathogen toxins [2] or inhibit essential pathogen enzymes [15]. Particularly appealing is the prospect that genes encoding antimicrobial peptides may be introduced into plants. Several such genes have been described from plants themselves [12, 17, 110] and animals [9, 32, 71, 122]. Salicylic acid has recently been identified as a natural signal that triggers resistance responses in certain plants [77, 82]. It might accordingly be possible to manipulate the production of salicylic acid in plants to improve resistance. Finally, antibody genes encoding heavy and light immunoglobulin chains have recently been expressed in plants [49]. This important development opens the possibility of transforming plants with genes encoding antibodies directed to important pathogen functions such as viral replicases or microbial toxins and cell wall degrading enzymes.

Summary and future directions

The application of molecular genetic approaches has revolutionized the study of plant-pathogen interactions by providing decisive tests of hypotheses and allowing the elucidation of novel biochemical mechanisms which underly these systems. This has resulted in the relatively rapid accumulation of information concerning mechanisms underlying virulence in pathogens and resistance in plants. Because of the large number of plants and pathogens involved and the highly precise specificities which determine pathogen virulence and plant resistance, the study of plant-pathogen interactions has also contributed greatly to our knowledge of basic plant biology. The study of enzymes, hormones and toxins specified by pathogens has particularly aided plant science, the most famous single case being *Agrobacterium tumefaciens*. With the recent acceleration of activity in study of plant-pathogen systems, we can expect continued contributions to plant biology from this field, particularly as concerns secondary plant metabolism, cell surface receptors, signal transduction mechanisms, gene regulation and cell differentiation.

While considerable information is known about the mechanisms determining plant disease resistance, greater understanding will occur when disease resistance genes are finally cloned and characterized from several higher plants. This imminent development will permit additional dissection of the molecular events involved in the hypersensitive reaction. The availability of cloned disease resistance genes will also set the stage for dramatically improved disease control in the field through their systematic transformation into various crop plants.

References

1. Akiyoshi DE, Morris RO, Hinz R, Mischke BS, Kosuge T, Garfinkel DJ, Gordon MP, Nester EW: Cytokinin-auxin balance in crown gall tumors is regulated by specific loci in the T-DNA. Proc Natl Acad Sci USA 80: 407–411 (1983).
2. Anzai H, Yoneyama K, Yamaguchi I: Transgenic tobacco resistant to a bacterial disease by the detoxification of a pathogenic toxin. Mol Gen Genet 219: 492–494 (1989).
3. Apostol I, Heinstein PF, Low PS: Rapid stimulation of an oxidative burst during elicitation of cultured plant cells. Role in defense and signal transduction. Plant Physiol 90: 109–116 (1989).
4. Baev N, Endre G, Petrovics G, Banfalvi Z, Kondorosi A: Six nodulation genes of *nod* box locus 4 in *Rhizobium meliloti* are involved in nodulation signal production: *nodM* codes for D-glucosamine synthetase. Mol Gen Genet 228: 113–124 (1991).
5. Bauer DW, Beer SV: Further characterization of an *hrp* gene cluster of *Erwinia amylovora*. Mol Plant-Microbe Inter 4: 493–499 (1991).
6. Beachy RN, Loesch-Fries S, Tumer NE: Coat protein-mediated resistance against virus infection. Annu Rev Phytopath 28: 451–474 (1990).

7. Bender CL, Stone HE, Sims JJ, Cooksey DA: Reduced pathogen fitness of *Pseudomonas syringae* pv. *tomato* Tn*5* mutants defective in coronatine production. Physiol Mol Plant Path 30: 273–283 (1987).
8. Boccara M, Diolez A, Rouve M, Kotoujansky A: The role of individual pectate lyases of *Erwinia chrysanthemi* strain 3937 in pathogenicity on saintpaulia plants. Physiol Mol Plant Path 33: 95–104 (1988).
9. Boman HG, Hultmark D: Cell-free immunity in insects. Annu Rev Microbiol 41: 103–126 (1987).
10. Bonas U, Stall RE, Staskawicz B: Genetic and structural characterization of the avirulence gene *avrBs3* from *Xanthomonas campestris* pv. *vesicatoria*. Mol Gen Genet 218: 127–136 (1989).
11. Bostock RM, Stermer BA: Perspectives on wound healing in resistance to pathogens. Annu Rev Phytopath 27: 343–371 (1989).
12. Broekert W, Lee H-I, Kush A, Chua N-H, Raikhel N: Wound-induced accumulation of mRNA containing a hevein sequence in laticifers of rubber tree (*Hevea brasiliensis*) Proc Natl Acad Sci USA 87: 7633–7637 (1990).
13. Broglie K, Chet I, Holliday M, Cressman R, Biddle P, Knowlton S, Mauvais CJ, Broglie R: Transgenic plants with enhanced resistance to the fungal pathogen *Rhizoctonia solani*. Science 254: 1194–1197 (1991).
14. Carney BF, Denny TP: A cloned avirulence gene from *Pseudomonas solanacearum* determines incompatibility on *Nicotiana tabacum* at the host species level. J Bact 172: 4836–4843 (1990).
15. Cervone F, Delorenzo G, Degra L, Salvi G, Bergami M: Purification and characterization of a polygalacturonase-inhibiting protein from *Phaseolus vulgaris* L. Plant Physiol 85: 631–637 (1987).
16. Cheong JJ, Hahn MG: A specific, high-affinity binding site for the hepta-B-glucoside elicitor exists in soybean membranes. Plant Cell 3: 137–147 (1991).
17. Chrispeels MJ, Raikhel NV: Lectins, lectin genes, and their role in plant defense. Plant Cell 3: 1–9 (1991).
18. Comai L, Kosuge T: Cloning and characterization of *iaaM*, a virulence determinant of *Pseudomonas syringae* pv. *savastanoi*. J. Bact 149: 40–46 (1982).
19. Crombie L: Natural resistance and susceptibility of plant hosts to fungal attack: the chemical base. In: Casida J (ed) Pesticides and Alternatives: Innovative Chemical and Biological Approaches to Pest Control, pp. 497–512. Elsevier, Amsterdam (1990).
20. Cruickshank IAM, Perrin DR: The isolation and partial characterization of monilicolin A, a polypeptide with phaseolin-inducing activity from *Monilinia fructicola*. Life Sci 7: 449–458 (1968).
21. Culver JN, Dawson WO: Tobacco mosaic virus elicitor coat protein genes produce a hypersensitive phenotype in transgenic *Nicotiana sylvestris* plants. Mol Plant-Microbe Inter 4: 458–463 (1991).
22. Dangl JL: Regulatory elements controlling developmental and stress-induced expression of phenylpropanoid genes. In: Boller T, Meins F (eds) Plant Gene Research, vol. 8. Genes Involved in Plant Defense. Springer-Verlag, New York, in press (1992).
23. Debener T, Lehnackers H, Arnold M, Dangl JL: Identification and molecular mapping of a single *Arabidopsis thaliana* locus determining resistance to a phytopathogenic *Pseudomonas syringae* isolate. Plant J 1: 289–302 (1991).
24. DeFago G, Kern H, Sedlar L: Genetic analysis of tomatine insensitivity, sterol content and pathogenicity for green tomato fruits in mutants of *Fusarium solani*. Physiol Plant Path 22: 39–43 (1983).
25. De Feyter R, Gabriel DW: At least six avirulence genes are clustered on a 90-kilobase plasmid in *Xanthomonas campestris* pv. *malvacearum*. Mol Plant-Microbe Inter 4: 423–432 (1991).
26. Denny TP, Baek SR: Genetic evidence that extracellular polysaccharide is a virulence factor of *Pseudomonas solanacearum*. Mol Plant-Microbe Inter 4: 198–206 (1991).
27. Desjardins AE, Gardner HW: Genetic analysis in *Gibberella pulicaris*: rishitin tolerance, rishitin metabolism, and virulence on potato tubers. Mol Plant-Microbe Inter 2: 26–34 (1989).
28. De Wit PJGM: Functional models to explain gene-for-gene relationships in plant pathogen interactions. In: Boller T, Meins F (eds.) Plant Gene Research, vol. 8. Genes Involved in Plant Defense. Springer-Verlag, New York, in press (1991).
29. De Wit PJGM, Hofman AE, Velthuis GCM, Kuc J: Isolation and characterization of an elicitor of necrosis isolated from intercellular fluids of compatible interactions of *Cladosporium fulvum* (syn. *Fulvia fulva*) and tomato. Plant Physiol 717: 642–647 (1985).
30. Dickman MB, Podila GK, Kolattukudy PE: Insertion of cutinase gene into a wound pathogen enables it to infect intact host. Nature 342: 446–448 (1989).
31. Dietrich A, Mayer JE, Hahlbrock K: Fungal elicitor triggers rapid, transient, and specific protein phosphorylation in parsley cell suspension cultures. J Biol Chem 265: 6360–6368 (1990).
32. Dimarcq J-L, Zachary D, Hoffmann JA, Hoffmann D, Reichart J-M: Insect immunity: expression of the two major inducible antibacterial peptides, defensin and diptericin, in *Phormia terranovae*. EMBO J 9: 2507–2515 (1990).
33. Dixon RA, Harrison MJ: Activation, structure, and organization of genes involved in microbial defense in plants. Adv. Genet 28: 165–234 (1990).
34. Dixon RA, Lamb CJ: Molecular communication in interactions between plants and microbial pathogens. Annu Rev Plant Physiol Plant Mol Biol 41: 339–367 (1990).
35. Douglas CJ, Hauffe KD, Itesmorales ED, Ellard M, Paszkowki U, Hahlbrock K, Dangl J: Exonic sequences

are required for elicitor and light activation of a plant defense gene, but promoter sequences are sufficient for tissue specific expression. EMBO J 10: 1767–1776 (1991).
36. Dron M, Clouse SD, Dixon RA, Lawton MA, Lamb CJ: Glutathione and fungal elicitor regulation of a plant defense gene promoter in electroporated protoplasts. Proc Natl Acad Sci USA 85: 6738–6742 (1988).
37. Ebel J: Phytoalexin synthesis: the biochemical analysis of the induction process. Annu Rev Phytopath 24: 235–264 (1986).
38. Ebel J, Cosio EG, Frey T: Perception of pathogen-derived elicitor and signal transduction in host defenses. In: Hennecke H, Verma DPS (eds) Advances in Molecular Genetics of Plant-Microbe Interactions, vol 1, pp. 421–427. Kluwer Academic Publishers, Dordrecht, Netherlands (1991).
39. Ebel J, Grisebach H: Defense strategies of soybean against the fungus *Phytophthora megasperma* f.sp. *glycinea*: a molecular analysis. Trends Biochem Sci 13: 23–27 (1988).
40. Ellis JG, Lawrence GJ, Peacock WJ, Pryor AJ: Approaches to cloning plant genes conferring resistance to fungal pathogens. Annu Rev Phytopath 26: 245–263 (1988).
41. Flor HH: Inheritance of pathogenicity in *Melampsora lini*. Phytopathology 32: 653–669 (1942).
42. Flor HH: The complementary genic systems in flax and flax rust. Adv Genet 8: 29–54 (1956).
43. Gabriel DW, Rolfe BG: Working models of specific recognition in plant-microbe interactions. Annu Rev Phytopath 28: 365–391 (1990).
44. Golemboski DB, Lomonossoff GP, Zaitlin M: Plants transformed with a tobacco mosaic virus nonstructural gene sequence are resistant to the virus. Proc Natl Acad Sci USA 87: 6311–6315 (1990).
45. Graham TL, Graham MY: Cellular coordination of molecular responses in plant defense. Mol Plant-Microbe Inter 4: 415–422 (1991).
46. Hahlbrock K, Scheel D: Physiology and molecular biology of phenylpropanoid metabolism. Annu Rev Plant Physiol Plant Mol Biol 40: 347–369 (1989).
47. Hahn MG, Bonhoff A, Grisebach H: Quantitative localization of the phytoalexin glyceollin I in relation to fungal hyphae in soybean roots infected with *Phytophthora megasperma* f.sp. *glycinea*. Plant Physiol 77: 591–601 (1985).
48. Hain R, Bieseler B, Kindl H, Schroder G, Stocker R: Expression of a stilbene synthase gene in *Nicotiana tabacum* results in synthesis of the phytoalexin resveratrol. Plant Mol Biol 15: 325–335 (1990).
49. Hiatt A, Cafferkey R, Bowdish K: Production of antibodies in transgenic plants. Nature 342: 76–78 (1989).
50. Hoxtermann E: Karl Otto Muller (1897-1978) und die Entdeckungsgeschichte der Phytoalexine (Anlasslich der Entdeckung eines neuen Abswehrprinzips der Pflanzen vor 50 Jahren). J Phytopath 132: 161–167 (1991).
51. Huang HC, Schuurink R, Denny TP, Atkinson MM, Baker CJ, Yucel I, Hutcheson SW, Collmer A: Molecular cloning of a *Pseudomonas syringae* pv. *syringae* gene cluster that enables *Pseudomonas fluorescens* to elicit the hypersensitive response in tobacco plants. J Bact 170: 4748–4756 (1988).
52. Huang HC, Hutcheson SW, Collmer A: Characterization of the *hrp* cluster from *Pseudomonas syringae* pv. *syringae* 61 and Tn*phoA* tagging of genes encoding exported or membrane-spanning Hrp proteins. Mol Plant-Microbe Inter 4: 469–476 (1991).
53. Huang JS, Barker KR: Glyceollin I in soybean-cyst nematode interactions–spatial and temporal distribution in roots of resistant and susceptible soybeans. Plant Physiol 96: 1302–1307 (1991).
54. Huynh TV, Dahlbeck D, Staskawicz BJ: Bacterial blight of soybean: Regulation of a pathogen gene determining host cultivar specificity. Science 245: 1374–1377 (1989).
55. Kearney B, Staskawicz BJ: Widespread distribution and fitness contribution of *Xanthomonas campestris* avirulence gene *avrBs2*. Nature 346: 385–386 (1990).
56. Keen NT: Phytoalexins and their elicitors. In: Microbes and Microbial Products as Herbicides, p. 114–129. ACS Symposium Series 439. American Chemical Society, Washington, DC (1990).
57. Keen NT: Gene-for-gene complementarity in plant-pathogen interactions. Annu Rev Genet 24: 447–463 (1990).
58. Keen NT, Bent A, Staskawicz B: Plant disease resistance genes: interactions with pathogens and their improved utilization to control plant diseases. In: Chet I (ed) Biotechnological Prospects for Plant Pathogen Control. Ecological and Applied Microbiology Series. J. Wiley New York, in press (1992).
59. Keen NT, Buzzell RI: New disease resistance genes in soybean against *Pseudomonas syringae* pv. *glycinea*: evidence that one of them interacts with a bacterial elicitor. Theor Appl Genet 81: 133–138 (1991).
60. Keen NT, Dawson WO: Pathogen avirulence genes and elicitors of plant defense. In: Boller T, Meins F (eds) Genes Involved in Plant Defense, vol. 8. Plant Gene Research, Springer-Verlag, New York, in press (1992).
61. Keen NT, Tamaki S, Kobayashi D, Gerhold D, Stayton M, Shen H, Gold S, Lorang J, Thordal-Christensen H, Dahlbeck D, Staskawicz B: Bacteria expressing avirulence gene D produce a specific elicitor of the soybean hypersensitive reaction. Mol Plant-Microbe Inter 3: 112–121 (1990).
62. Keppler LD, Baker CJ, Atkinson MM: Active oxygen production during a bacteria-induced hypersensitive reaction in tobacco suspension cells. Phytopathology 79: 974–978 (1989).
63. Kinscherf TG, Coleman RH, Barta TM, Willis DK: Cloning and expression of the tabtoxin biosynthetic re-

gion from *Pseudomonas syringae*. J Bact 173: 4124–4132 (1991).

64. Knight MR, Campbell AK, Smith SM, Trewavas AJ: Transgenic plant aequorin reports the effects of touch and cold-shock and elicitors on cytoplasmic calcium. Nature 352: 524–526 (1991).
65. Knogge W, Hahn M, Lehnackers H, Rupping E, Wevelsiep L: Fungal signals involved in the specificity of the interaction between barley and *Rhynchosporium secalis*. In: Hennecke H, Verma DPS (eds) Advances in Molecular Genetics of Plant-Microbe Interactions, vol. 1, pp. 250–253. Kluwer Academic Publishers Dordrecht, Netherlands (1991).
66. Knorr DA, Dawson WO: A point mutation in the tobacco mosaic virus capsid protein gene induces hypersensitivity in *Nicotiana sylvestris*. Proc Natl Acad Sci USA 85: 170–174 (1988).
67. Kobayashi DY, Tamaki SJ, Keen NT: Cloned avirulence genes from the tomato pathogen *Pseudomonas syringae* pv. *tomato* confer cultivar specificity on soybean. Proc Natl Acad Sci USA 86: 157–161 (1989).
68. Kobayashi DY, Tamaki SJ, Keen NT: Molecular characterization of avirulence gene D from *Pseudomonas syringae* pv. *tomato*. Mol Plant-Microbe Inter 3: 94–102 (1990).
69. Kotoujansky A: Molecular genetics of pathogenesis by soft-rot Erwinias. Annu Rev Phytopath 25: 405–430 (1987).
70. Kurosaki F, Tsurusawa Y, Nishi A: The elicitation of phytoalexins by Ca^{2+} and cyclic AMP in carrot cells. Phytochemistry 26: 1919–1923 (1987).
71. Lehrer RI, Ganz T, Selsted ME: Defensins: endogenous antibiotic peptides of animal cells. Cell 64: 229–230 (1991).
72. Lerouge P, Roche P, Faucher C, Maillet F, Truchet G, Prome JC, Denarie J: Symbiotic host-specificity of *Rhizobium meliloti* is determined by a sulphated and acylated glucosamine oligosaccharide signal. Nature 344: 781–784 (1990).
73. Levings CS: CMS, the Texas cytoplasm and URF-13. Plant Mol Biol, in press (1992).
74. Lewis-Henderson W, Djordjevic MA: A cultivar-specific interaction between *Rhizobium leguminosarum* bv. *trifolii* and subterranean clover is controlled by *nodM*, other bacterial cultivar specificity genes, and a single recessive host gene. J Bact 173: 2791–2799 (1991).
75. Lindgren PB, Peet RC, Panopoulos NJ: Gene cluster of *Pseudomonas syringae* pv. '*phaseolicola*' controls pathogenicity on bean plants and hypersensitivity on non-host plants. J Bact 168: 512–522 (1986).
76. Lois R, Dietrich A, Hahlbrock K: A phenylalanine ammonialyase gene from parsley: structure, regulation and identification of elicitor and light responsive *cis*-acting elements. EMBO J 8: 1641–1648 (1989).
77. Malamy J, Carr JP, Klessig DF, Raskin I: Salicylic acid: a likely endogenous signal in the resistance response of tobacco to viral infection. Science 250: 1002–1004 (1990).
78. Martinez E, Romero D, Palacios R: The *Rhizobium* genome. Crit Rev Plant Sci 9: 59–93 (1990).
79. Mayama S, Tani T, Ueno T, Midland SL, Sims JJ, Keen NT: The purification of victorin and its phytoalexin elicitor activity in oat leaves. Physiol Mol Plant Path 29: 1–18 (1986).
80. Meier I, Hahlbrock K, Somssich IE: Elicitor-inducible and constitutive *in vivo* DNA footprints indicate novel *cis*-acting elements in the promoter of a parsley gene encoding pathogenesis-related protein 1. Plant Cell 3: 313–315 (1991).
81. Mellano VJ, Cooksey DA: Development of host range mutants of *Xanthomonas campestris* pv. *translucens*. Appl Environ Microbiol 54: 884–889 (1988).
82. Metreaux P, Signer H, Ryals J, Ward E, Wyss-Benz M, Gaudin J, Raschdorf K, Schmid E, Blum W, Invernardi B: Increase in salicylic acid at the onset of systemic acquired resistance in cucumber. Science 250: 1004–1006 (1990).
83. Mills D, Mukhopadhyay P, Zhao Y, Romantschuk M: Organization and function of pathogenicity genes of *Pseudomonas syringae* pathovars *phaseolicola* and *syringae*. In: Patil S *et al.* (eds) Molecular Strategies of Pathogens and Host Plants, pp. 69–81. Springer-Verlag, Berlin (1991).
84. Mo YY, Gross DC: Expression *in vitro* and during plant pathogenesis of the *syrB* gene required for syringomycin production by *pseudomonas syringae* pv. *syringae*. Mol Plant-Microbe Inter 4: 28–36 (1991).
85. Perlak FJ, Fuchs RL, Dean DA, McPherson SL, Fischoff DA: Modification of the coding sequence enhances plant expression of insect control protein genes. Proc Natl Acad Sci USA 88: 3324–3328 (1991).
86. Peters NK, Verma DPS: Phenolic compounds as regulators of gene expression in plant-microbe interactions. Mol Plant-Microbe Inter 3: 4–8 (1990).
87. Pierce M, Essenberg M: Localization of phytoalexins in fluorescent mesophyll cells isolated from bacterial blight-infected cotton cotyledons and separated from other cells by fluorescence-activated cell sorting. Physiol Mol Plant Path 31: 273–290 (1987).
88. Preston JF, Rice JD, Chow MC, Brown BJ: Microbial strategies for the depolymerization of plant and algal polyuronates. Leatham GF, Himmel ME (eds) Enzymes in Biomass Conversion, pp. 450–466. ACS Symposium Series 460, American Chemical Society, Washington, DC (1991).
89. Reverchon S, Van Gijsegem F, Rouve M, Kotoujansky A, Robert-Baudouy J: Organization of a pectate lyase gene family in *Erwinia chrysanthemi*. Gene 49: 215–224 (1986).
90. Ricci P, Bonnet P, Huet JC, Sallantin M, Beauvais-Cante F, Bruneteau M, Billard V, Michel G, Pernollet JC: Structure and activity of proteins from pathogenic

fungi *Phytophthora* eliciting necrosis and acquired resistance in tobacco. Eur J Biochem 183: 555–563 (1989).

91. Ried JL, Collmer A: Construction and characterization of an *Erwinia chrysanthemi* mutant with directed deletions in all of the pectate lyase structural genes. Mol Plant-Microbe Inter 1: 32–38 (1988).
92. Roby D, Broglie K, Cressman R, Biddle P, Chet I, Broglie R: Activation of a bean chitinase promoter in transgenic tobacco plants by phytopathogenic fungi. Plant Cell 2: 999–1007 (1990).
93. Roche P, Lerouge P, Prome JC, Faucher C, Vasse J, Maillet F, Camut S, De Billy F, Denarie J, Truchet G: NodRM-1, a sulphated lipo-oligosaccharide signal of *Rhizobium meliloti* elicits hair deformation, cortical cell division and nodule organogenesis on alfalfa roots. In: Hennecke H, Verma DPS (eds) Advances in Molecular Genetics of Plant-Microbe Interactions, pp. 119–126. Kluwer Academic Publishers, Dordrecht, Netherlands (1991).
94. Ryan CA: Protease inhibitors in plants: genes for improving defenses against insects and pathogens. Annu Rev Phytopath 28: 425–449 (1990).
95. Schafer W, Straney D, Ciuffetti L, VanEtten HD, Yoder OC: One enzyme makes a fungal pathogen, but not a saprophyte, virulent on a new host plant. Science 246: 247–249 (1989).
96. Scheel D, Parker JE: Elicitor recognition and signal transduction in plant defense gene activation. Z Naturforsch 45c: 569–575 (1990).
97. Scheffer RP: Role of toxins in evolution and ecology of plant pathogenic fungi. Experientia 47: 804–810 (1991).
98. Schmelzer E, Kruger-Lebus S, Hahlbrock K: Temporal and spatial patterns of gene expression around sites of attempted fungal infection in parsley leaves. Plant Cell 1: 993–1001 (1989).
99. Sharp JK, McNeil M, Albersheim P: The primary structures of one elicitor-active and seven elicitor-inactive hexa (B-D-glucopyranosyl)-D-glucitols isolated from the mycelial walls of *Phytophthora megasperma* f.sp. glycinea. J. Biol Chem 259: 11321–11336 (1984).
100. Snyder BA, Nicholson RL: Synthesis of phytoalexins in sorghum as a site-specific response to fungal ingress. Science 248: 1637–1639 (1990).
101. Stab MR, Ebel J: Effects of Ca^{2+} on phytoalexin induction by fungal elicitor in soybean cells. Arch Biochem Biophys 257: 416–423 (1987).
102. Staskawicz BJ, Dahlbeck D, Keen NT: Cloned avirulence gene of *Pseudomonas syringae* pv. *glycinea* determines race-specific incompatibility on *Glycine max* (L.) Merr. Proc Natl Acad Sci USA 81: 6024–6028 (1984).
103. Straus D, Ausubel F: Genomic subtraction for cloning DNA corresponding to deletion mutations. Proc Natl Acad Sci USA 87: 1889–1893 (1990).
104. Swarup S, DeFeyter R, Brlansky RH, Gabriel DW: A pathogenicity locus from *Xanthomonas citri* enables strains from several pathovars of *X. campestris* to elicit cankerlike lesions on citrus. Phytopathology 81: 802–809 (1991).
105. Swarup S, Yang Y, Gabriel DW: A *Xanthomonas citri* pathogenicity gene, *pthA*, pleiotropically encodes gratuitous avirulence on non-hosts. Mol Plant-Microbe Inter, Okasset, still in press (1991).
106. Tamaki SJ, Gold S, Robeson M, Manulis S, Keen NT: Structure and organization of the *pel* genes from *Erwinia chrysanthemi* EC16. J Bact 170: 3468–3478 (1988).
107. Tamaki SJ, Kobayashi DY, Keen NT: Sequence domains required for the activity of avirulence genes *avrB* and *avrC* from *Pseudomonas syringae* pv. *glycinea*. J Bact 173: 301–307 (1991).
108. VanEtten HD, Matthews DE, Matthews PS: Phytoalexin detoxification: importance for pathogenicity and practical implications. Annu Rev Phytopath 27: 143–164 (1989).
109. van Kan JAL, van den Ackerveken GFJM, de Wit PJGM: Cloning and characterization of cDNA of avirulence gene *avr9* of the fungal pathogen *Cladosporium fulvum*, causal agent of tomato leaf mold. Mol Plant-Microbe Inter 4: 52–59 (1991).
110. Vigers AJ, Roberts WK, Selitrennikoff CP: A new family of plant antifungal proteins. Mol Plant-Microbe Inter 4: 315–323 (1991).
111. Waldmuller T, Grisebach H: Effects of R-(1-amino-2-phenylethyl)phosphonic acid on glyceollin accumulation and expression of resistance to *Phytophthora megasperma* f.sp. *glycinea* in soybean. Planta 172: 424–430 (1987).
112. Waney VR, Kingsley MT, Gabriel DW: *Xanthomonas campestris* pv. *translucens* genes determining host specific virulence and general virulence on cereals identified by Tn*5*-*gus*A insertion mutagenesis. Mol Plant-Microbe Inter 4: 623–627 (1991).
113. Welle R, Schroder G, Schiltz E, Grisebach H, Schroder J: Induced plant responses to pathogen attack. Analysis and heterologous expression of the key enzyme in the biosynthesis of phytoalexins in soybean (*Glycine max* L. Merr. cv. Harosoy 63). Eur J Biochem 196: 423–430 (1991).
114. Whalen MC, Stall RE, Staskawicz BJ: Characterization of a gene from a tomato pathogen determining hypersensitive resistance in non-host species and genetic analysis of this resistance in bean. Proc Natl Acad Sci USA 85: 6743–6747 (1988).
115. Willis DK, Rich JJ, Hrabak EM: *hrp* genes of phytopathogenic bacteria. Mol Plant-Microbe Inter 4: 132–138 (1991).
116. Willmitzer L: Genetic engineering for plant disease resistance. In: Chet I (ed) Innovative Approaches to Plant Disease Control, pp. 353–364, John Wiley, New York (1987).
117. Woloshuk CP, Kolattukudy PE: Mechanism by which contact with plant cuticle triggers cutinase gene expression in the spores of *Fusarium solani* f.sp. *pisi*. Proc Natl Acad Sci USA 83: 1704–1708 (1986).

118. Wolpert TJ, Macko V, Acklin W, Juan B, Seibl J, Meili J, Arigoni D: Structure of victorin C, the major host-selective toxin from *Cochliobolus victoriae*. Experientia 41: 1524–1529 (1985).
119. Wolpert TJ, Macko V: Specific binding of victorin to a 100-kDa protein from oats. Proc Natl Acad Sci USA 86: 4092–4096 (1989).
120. Yoshikawa M, Keen NT, Wang MC: A receptor on soybean membranes for a fungal elicitor of phytoalexin accumulation. Plant Physiol 73: 497–506 (1983).
121. Yoshikawa M, Yamauchi K, Masago H: Glyceollin: its role in restricting fungal growth in resistant soybean hypocotyls infected with *Phytophthora megasperma* var. *sojae*. Physiol Plant Path 12: 73–82 (1978).
122. Zasloff M: Magainins, a class of antimicrobial peptides from *Xenopus* skin: isolation, characterization of two active forms, and partial cDNA sequence of a precursor. Proc Natl Acad Sci USA 84: 5449–5453 (1987).

Plant Molecular Biology **19**: 123–133, 1992.
© 1992 *Kluwer Academic Publishers. Printed in Belgium.*

The search for the proteinase inhibitor-inducing factor, PIIF

Clarence A. Ryan
Institute for Biological Chemistry, Washington State University, Pullman, WA 99164-6340, USA

Key words: Proteinase inhibitor genes, induced defense response oligouronides, methyl jasmanase, systemin

Introduction

Research into the structure and function of proteinase Inhibitors I and II proteins in organs and tissues of several solanaceous plants began with the discovery and crystallization of Inhibitor I from Russet Burbank potato tubers in 1962 [1]. This was later followed by the isolation and characterization of Inhibitor II from potato tubers [2]. Inhibitor I and II are serine proteinase inhibitors that have evolved as members of two nonhomologous gene families. Inhibitor I proteins have a molecular mass of about 8000 kDa, whereas Inhibitor II proteins have a M_r of about 12 000 kDa. Inhibitor I is a potent inhibitor of chymotrypsin, while Inhibitor II is a 'double-headed' inhibitor, having evolved by gene duplicated-elongation events from a smaller ancestral gene, and it possesses two reactive sites that are specific for trypsin and chymotrypsin respectively.

In potatoes, the synthesis of the two inhibitors was originally found to be related to tuber sprouting and development but the two inhibitors were later found in leaves of both potato and tomato plants [3, 4, 5] as transient proteins whose presence could not be consistently related to any developmental process. By 1972, I became convinced that environmental factors were somehow involved in the regulation of synthesis of the inhibitor proteins in plant leaves. I was also convinced at that time, as were a few other scientists, that proteinase inhibitor proteins were part of the defensive chemicals of plants. As a consequence, the possible role of insect damage in inducing proteinase inhibitor accumulation in leaves of tomato and potato plants was tested. Dr Terry Green, a postdoctoral fellow in my lab, and I allowed Colorado potato beetles (collected from potato plants growing in a friend's yard) to attack some young tomato and potato plants in the greenhouse. The plants responded by accumulating large quantities of proteinase inhibitors, not only in the damaged leaves, but also in distal, undamaged leaves [6]. We found that any severe mechanical injury of the leaves induced the synthesis and accumulation of the two proteinase inhibitor proteins throughout the plants [6]. We called the putative systemic signal the proteinase inhibitor-inducing factor, PIIF [7]. Subsequent research has shown that the regulation of the distal synthesis of proteinase inhibitor proteins in plants in response to insect damage is widespread in nature and wound-inducible proteinase inhibitors have now been demonstrated in Fabaceae [8], Curcubitaceae [9] and *Populus* [10] and will likely be found in other plant families as well. Roles for proteinase inhibitors in plant defense against insect pests have recently received strong support through gene transfer technology. Plants from several genera have been constitutively transformed with proteinase inhibitor genes with resulting increased resistances against lepidopteran insects [11, 12]. The inhibitors are potent inactivators of proteolytic enzymes, but they also trigger physiological feedback mechanisms that cause overproduction and secretion of digestive proteases and a decrease in appetite [13]. The presence of the inhibitors in the diets of animals can therefore result in severe protein malnutrition and eventual starvation.

Five different classes of signalling molecules have now been identified and studied in some detail as candidate wound signals for the induction of the synthesis of proteinase inhibitors in plants; (i) oligouronide fragments of the plant cell walls [14, 15]; (ii) chitin and chitosan fragments from fungal cell walls [16]; (iii) a polypeptide of 18 amino acids in length from tomato leaves [17]; (iv) methyl jasmonate and its free acid, jasmonic acid [18] and a well known plant hormone, abscisic acid [19]. The polypeptide and volatile methyl jasmonate, found within the past two years to be inducers of proteinase inhibitor genes, have opened entirely new approaches for studying signal transduction leading to the activation of plant defense genes. In the following paragraphs recent research concerning the plant derived inducers will be summarized. Chitosan, being of fungal origin, is not to be considered as a plant-derived wound signal and is not discussed here in the context of the search for PIIF. The plant hormone, auxin, has recently been reported to inhibit or suppress the synthesis of Inhibitors I and II in leaves [20]. Auxin is not a candidate for being endogenous PIIF but may be involved in regulating the production of wound signals.

Regulation of wound-inducible Inhibitor I and II synthesis

Inhibitor I and II genes are coordinately expressed in leaves of wounded tomato plants [21]. Following a single severe wounding with a hemostat across the main vein of the lowest terminal leaflet of a small (2 leaf) tomato plant, messenger RNAs coding for the two inhibitors appear within 3–4 hours in both the wounded leaf and the upper, unwounded leaf as well. The inhibitor proteins can be detected in leaves shortly thereafter [21]. They are compartmented in the central vacuoles of the leaf cells [22, 23] where they have half lives of weeks or longer. The mRNA levels increase for about 8 h and then slowly decrease for several hours. The half lives for both mRNAs were calculated to be about 12 h. Successive wounds (one wound each hour for 4 h) caused about a doubling in mRNA and protein levels. The maximum cumulative levels of Inhibitor I and II mRNA produced by multiple wounds of the Bonny Best variety approach 1% of the total poly(A) mRNA population [21]. Thus, the mechanism allows for the production and storage of large quantities of inhibitor proteins at a fairly low cost to the plants.

Oligouronides as plant defense signals

By 1978, our search for PIIF had revealed that partially purified soluble pectinaceous fragments, rich in uronic acid residues, induced proteinase inhibitor synthesis when supplied to young tomato plants through their cut stems [24]. The activity was found to reside in the α-1,4-galacturonic acid oligomers that comprise the backbone of pectin [24]. Oligomers with an average degree of polymerization (DP) of about 20 uronide units [25] could be isolated from a defined fraction from tomato leaves. At this same time the Albersheim [26] and the West laboratories [27] discovered that oligouronides could activate antifungal antibiotic (phytoalexin) synthesis. The active fragments in these latter systems were thought to be produced by pectin-degrading enzymes from attacking fungi. Thus, in our system we concluded that wounding somehow mixed endogenous pectin-degrading enzymes with the walls of the wounded cells to produce oligouronide fragments that were both local and distal signals – a conclusion that later, as described below, had to be significantly modified.

Structure-function relationships of oligogalacturonides that signal proteinase inhibitor genes have shown that oligo- and polygalacturonic acids of DP = 2 through DP = 20 are active inducers of proteinase inhibitor synthesis in tomato leaves [25, 28, 29]. In addition to di- and trigalacturonic acids which are products of polygalacturonase (PGase) action on polygalacturonic acid (PGA), unsaturated di- and trigalacturonic products of pectin lyase on polygalacturonic acid were also shown to be active inducers of proteinase inhibitor in tomato leaves. In many other defensive responses only polygalacturonides with DPs above 9 are inducers (elicitors) [14, 15]. However, in at least one system, peroxidase induction in castor beans [30], oligogalacturonides of a DP of ca. 6 are active. Thus only proteinase inhibitor synthesis is known to be activated by di- and trigalacturonic acids. Structures of the oligouronides called PIIF, and its enzyme-derived oligomers are shown in Fig. 1.

Originally, small oligogalacturonides were strong candidates as systemic signals and the structure-activity relationships of di- and trigalacturonides and their derivatives was investigated [28]. Reduction of the double bonds at C4-C5 of unsaturated di- and trigalacturonides (products of pectic lyase or PGase on PGA) did not destroy inducing activities. However, reduction of the reducing end (C1) of both saturated and unsaturated di- and triuronides totally destroyed the activities. In a study of larger uronides, polyguluronic acid (PGU) (DP = 23), in which the hydroxyl group at C-4 is in the down position instead of up as in PGA, was found to be active as an inducer, but less active than PGA [29]. Polymannuronic acid (PMU), in which the C4 hydroxyl is up compared to PGA, exhibited very little activity [29]. The cumulative evidence indicated that the proteinase inhibitor-inducing activities of uronides required an intact hemiacetal

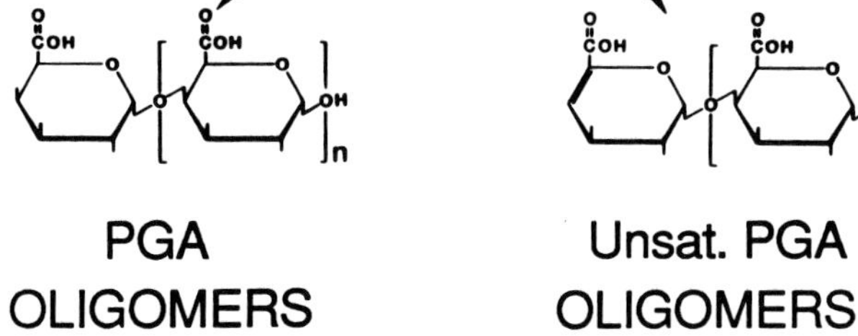

Fig. 1. Polygalacturonic acid and its enzymic products resulting from polygalacturonase ane pectin lyase degradation.

ring and a free carboxyl at C6. Activity is diminished, but not eliminated, when carbon C3 has a hydrogen or hydroxyl group in either the up or down position or when present in a double bond with C5.

Because of the widespread involvement in nature of protein phosphorylation in regulating both enzymes and gene action, the effects of defined oligouronides on the *in vitro* phosphorylation of tomato leaf plasma membrane proteins by [γ^{32}P] ATP or [γ^{35}S] ATP were investigated. *In vitro* phosphorylation of a protein of M_r 34 000 (pp34) was found to be strongly enhanced in the presence of PGA of DP = 20 in the membrane [31]. Structure-activity relationships of various polyuronides revealed that only oligogalacturonides with DPs above about 13 were active in enhancing the specific phosphorylation of pp34 [29]. Smaller galacturonides were totally inactive. PGU, which was weakly active in inducing proteinase inhibitor synthesis, was also weakly active in enhancing the *in vitro* phosphorylation of pp34 in purified plasma membranes [29]. PMA was nearly inactive in the *in vitro* assay. The differential activities of PGA, PMA and PGU in the *in vitro* phosphorylation assays were therefore similar to their activities in inducing proteinase inhibitors in leaves of young excised tomato plants [29].

Ca^{2+}-uronide complexes have been reported to form large aggregated 'egg-box' structures with PGA and PGU, but not with PMA [32]. In these structures, the uronides are the 'boxes', complexed with Ca^{2+} as the 'eggs'. Egg-box structures do not appear to form between oligouronides with DPs below about 10 uronide units and Ca^{2+} [33]. This implies that the active species of polyuronides that induce proteinase inhibitors *in vivo* and that induce protein phosphorylation *in vitro* may be large Ca^{2+}-uronide aggregates [18] The mechanism by which the smaller galacturonides (that do not form egg boxes) activate proteinase inhibitor genes in tomato leaves and peroxidase synthesis in castor bean seedlings is not clear at this time. Receptors for small β-glucans have recently been identified and characterized from soybean plasma membranes [34, 35] and probably initiate the first steps of an oligosaccharide-activated signal transduction pathway leading to localized inducible phytoalexin synthesis in response to fungal attacks. Specific receptors for oligouronides of different sizes may be involved with oligouronide signalling that, in some cases, can initiate the phosphorylation of specific plasma membrane proteins as described above.

Baydoun and Fry [36] have reported that radioactively labelled oligogalacturonides of DP = 6 or above were not transported throughout tomato plants when placed on wounds on leaves. This provided important evidence that the oligomers may not be systemic signals. Additionally, PGases and pectin lyases have so far not been reported in tomato leaves that might generate oligogalacturonide signals in response to wounding. On the other hand, reports of pectin-degrading enzymes originating from pathogens (that could produce oligouronide signals) are commonplace [14]. However, in several reports in which oligouronides produced local defensive responses against pathogens, such as phytoalexin induction, lignin induction and HPRG induction [14, 15, 37], none of these responses were not systemic in the same manner as proteinase inhibitor induction. Thus, although oligouronides are produced at pathogen sites, they are apparently not acting as systemic signals.

By 1987, the cumulative evidence suggested that the wound-induced systemic signal that initiated proteinase inhibitor synthesis in distal leaves was still to be found. We initiated a program to seek the presence of a new signalling molecule in tomato leaf extracts that would satisfy the criteria for a distal wound signal. It must be generated rapidly at wound sites; it must travel readily through the plant; and it must be a very powerful inducer of proteinase inhibitor synthesis.

The isolation of systemin, a polypeptide inducer of proteinase inhibitor genes

Using several thousand tomato plants for assay, and over 30 kg of tomato leaves for isolation, a reproducible purification scheme was developed that yielded 1 μg of a pure polypeptide having powerful inducing activity for both Inhibitor I and II genes in tomato leaves. The complete sequence of the polypeptide, called systemin, is shown in Fig. 2. The polypeptide is active at fmole/small tomato plant when supplied through their cut stems [17]. The induction is similar to wounding or to the induction of proteinase inhibitor with oligouronides in inducing both Inhibitor I and II in leaves, but the polypeptide is 10 000 times more potent than oligouronides. Unlike the oligosaccharide inducers, ^{14}C-labelled polypeptide can move readily throughout the plant with nearly the same rate as sucrose [17] or the wound inducible signal [38] making it a strong candidate for a systemic wound signal. To date, systemin is the only polypeptide hormone-like molecule that has been found in plants and has presented novel opportunities to investigate signalling pathways for plant gene expression.

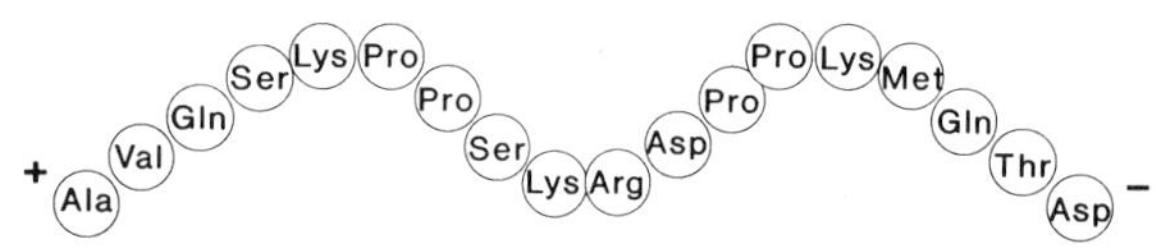

Fig. 2. Systemin, the polypeptide inducer of proteinase inhibitor synthesis tomato and tomato plants.

Methyl jasmonate and jasmonic acid as plant defense signals

Methyl jasmonate (Fig. 3) is a volatile liquid with a pleasant odor that has been used for decades in perfumes. Methyl jasmonate and/or jasmonic acid has been identified in at least 9 plant families [39]. When applied directly to plants at rel-

Fig. 3. The structures of methyl jasmonate (left) and jasmonic acid (right).

atively high concentrations, methyl jasmonate has been shown to effect a number of physiological responses in plants [40]. We recently found that volatile molecules of methyl jasmonate from a source placed near young tomato plant activated the synthesis and accumulation of Inhibitor I and II in leaves of the plants to levels much higher than could be induced by wounding [18]. Free jasmonic acid (which is not volatile) could duplicate the effects of methyl jasmonate when applied directly on leaves of tomato plants [18]. Wound-inducible proteinase inhibitors in potato, tobacco and alfalfa were also found to be strongly induced by airborne methyl jasmonate or by the application of jasmonic acid to the leaf surfaces [18]. Thus, three of the seven known proteinase inhibitor families present in plants (Inhibitor I, Inhibitor II and the Bowman-Birk) have wound-inducible family members that are also induced by methyl jasmonate vapors. More recently, we have shown that the mRNAs coding for Inhibitor I, Inhibitor II, and the alfalfa Bowman-Birk inhibitor are strongly induced by methyl jasmonate [41], suggesting that this induction, like wounding, is regulated at the transcriptional level. This conclusion was further supported by the demonstration that tobacco plants that were successfully transformed with an Inhibitor II-CAT chimeric gene and shown to possess a wound-inducible CAT expression [41], exhibit strong CAT expression when incubated in the presence of methyl jasmonate vapors [41]. This directly demonstrated that the 5′ region (950 bp) of the Inhibitor II gene used in the chimeric gene contained the response elements for both wounding and methyl jasmonate.

Artemisia tridentata (sagebrush), when placed in chambers with tomato plants, was found to cause the activation of synthesis of Inhibitor I and II proteins in the tomato leaves. The active volatile component in sagebrush leaves was isolated and identified as methyl jasmonate [18]. These experiments conclusively demonstrated that a species from one family of plants can signal defense gene activation in a species of another plant family by means of a volatile chemical. Other plants with wound-inducible proteinase inhibitors are likely to respond in a similar way to the presence of plants emitting volatile methyl jasmonate. This suggested that volatile chemicals should be seriously considered by plant ecologists as a means of interplant communication. Since some insects are known to produce large amounts of methyl jasmonate, the possibility that volatile compounds originating from animals may have roles in regulating plant genes should also be considered in defining some plant-animal interactions. If volatile chemicals are found to be integral parts of transduction pathways in plants, then such chemicals might be logical signals for interplant or animal-plant communication systems.

Abscisic acid and electrical potentials as wound signals

Recently, both abscisic acid and changes in electrical potentials have been reported to be involved in the wound-inducibility of the proteinase Inhibitor II gene in tomato and potato leaves [19, 42–44].

Spraying ABA on potato plants or on abscisic acid-deficient mutant tomato (sit) and potato (droopy) plants, induced Inhibitor II mRNA synthesis in leaves similar to wounding. However, spraying the hormone on normal tomato plants did not induce Inhibitor II mRNA [19, 42] or protein (E.E. Farmer and C.A. Ryan, unpublished data). The effects of ABA as an inducer of Inhibitor II mRNA in leaves of normal potato plants, but not in normal tomato plants is not understood, since wounding is effective in both. One possibility is that ABA, a senescence regulator, may be activating some components of the signal transduction system in potato and not tomato leaves and may not have a direct regulating role. For example, ABA might activate, synthesize, or release lipases in potato membranes, but not in tomato membranes, leading to the release of linolenic acid and the subsequent generation of jasmonic acid. The elucidation of the role of ABA in activating the proteinase inhibitor gene may yet provide some important aspects of the system that are still obscure.

The induction of proteinase inhibitors in tomato leaves by pectic fragments is sensitive to a range of chemicals that affect proton transport [43, 44], indicating that a change in protein or ion transport may be a component of the signalling by oligouronides. These oligosaccharide fragments cause rapid and reversible fluctuations in cell membrane potential [44]. Whether a relationship exists between the oligosaccharide effects on membrane potentials and ion transport with signal transduction remains to be determined.

Analysis of the structures of the wound-inducible proteinase inhibitor genes

Cis- *and* trans-*acting elements*

Wound-inducible proteinase inhibitor genes have been isolated and characterized from tomato and potato (Inhibitor I and II genes) and alfalfa (Bowman-Birk) genes [45–51]. The structure, organization, and regulation of these proteinase inhibitor gene family members have been studied to define the regulatory regions, and to characterize the factors that regulate wound inducibility.

The potato Inhibitor IIK gene [45] and the potato PIn II gene [46] have received most attention with respect to studies of *cis* and *trans* elements that are responsive to wounding. The Inhibitor IIK gene contained about 950 bp of the 5′ region [45] whereas the PIn II gene contained about 1300 bases of the 5′ region [47]. The 5′ regions of both genes can drive the wound-inducible expression of the chloramphenicol acetyl transferase (CAT) reporter gene [45, 48]. The PIn II promoter has also been shown to drive the wound-inducible expression of GUS [49].

The wound-inducible CAT activity expressed the Inhibitor IIK gene was higher when the Inhibitor IIK 3′ region was the terminator rather than the 6b and 7 terminators from the Ti plasmid [52]. Deletion analyses of the 3′ region of the gene [52] revealed that the CAT activity was apparently enhanced over the constructs with the 6b and 7 terminators because of the stability of the mRNA when the Inhibitor IIK terminator was present. The Inhibitor IIK terminator also caused a large increase in the expression of the CAT gene driven by the CaMV promoter. A 100 bp region surrounding the polyadenylation site of the Inhibitor II 3′ terminator was identified as the region essential for the highly efficient expression of the CAT gene [52].

Deletion analysis of the 5′ region of the Inhibitor IIK gene has identified a sequence of −151 to −165 that binds a nuclear protein from tomato leaves [53]. Binding was enhanced 2–5-fold when nuclear proteins were from wounded leaves. The 10 bp sequence is adjacent to a sequence from −147 to −155 that is present in several elicitor or wound-inducible genes [53]. A near repeat of the −136 to −165 sequence was found upstream within the region of −520 to −620 (C. Palm and C.A. Ryan, in preparation) that also binds nuclear factors. Thus the region from −136 to −165 and from −520 to −620 may contain a complex of *cis* elements that interact with wound-inducible or elicitor-inducible *trans*-acting factors to regulate defense genes. It will be important to determine whether the various inducers of proteinase inhibitors, including jasmonic acid, oligouronides, systemin and chitosan converge on the same or different *cis* elements of the wound-inducible genes.

Sanchez-Serrano *et al.* have analyzed the nuclear protein-binding regions of the PIn II gene that exhibits over 90% identity with the Inhibitor IIK gene [54]. A fragment of the PIn II promoter from −1297 to +33, relative to the transcription start site, was fused with either β-galacturonidase or CAT open reading frames and incorporated into the tobacco genome. Both chimeric genes were expressed in response to wounding. Deletion analyses of the PIn II promoter and assaying wound-inducible CAT gene expression [55], indicated that when a region −700 to −1300 was deleted, wound-inducible expression was greatly reduced. Deletion of −500 to −700 completely eliminated wound-inducible activity. Fusion with the enhancer of the CaMV promoter with fragments of the gene smaller than −514 bp exhibited no wound inducibility. The region from −700 to −1300 bp of PIn II appears to contain an enhancer distal to the −950 bp analyzed in PI IIK gene.

In gel shift assays, a prominent protein DNA interaction was identified that involved two nuclear proteins with a 5′ DNA fragment −1282 to −1093 bp from the transcription start site [64]. The binding was found in extracts from both tuber and leaf nuclei. Studies of nuclear protein binding with several DNA fragments that spanned the rest of the 5' region gave varying results depending upon the non-specific competitor DNA. The region of the PIn II gene from the transcription start site to −620 bp contains two regions that exhibit strong homology and that bind nuclear proteins. Deletion of the region that contains these sequences completely eliminates wound inducibility. It is clear that both the Inhibitor IIK gene and the PIn II gene will require further detailed analyses to establish the regions of the genes that are involved in wound induction or elicitor induction.

Chromatin structure

The genomic DNA of eukaryotic organisms is complexed with histone and non-histone proteins, forming the nuclear chromatin structure. The possible correlation between chromatin structure and gene expression has been extensively investigated in animals, but not in plants [56]. A detailed investigation of the chromatin structure of the tomato Inhibitor I gene in nuclei of wounded and unwounded tomato plants was initiated to help understand how wound-generated signal molecules can positively interact with the DNA to regulate the expression of the gene. A single wound-inducible tomato Inhibitor I gene, present in the Castlemart II variety of tomato, was determined to be responsible for all of the Inhibitor I synthesis in wounded leaves (A. Conconi and C.A. Ryan, submitted). Nuclei isolated from unwounded and wounded tomato leaves were incubated with DNase I or micrococcal nuclease and the DNA degradation pattern of total bulk chromatin, total ribosomal genes, and the single tomato Inhibitor I gene were determined. Striking differences were found in the chromatin structure between inactive ribosomal genes and the proteinase Inhibitor I gene. Inactive genes were found to be in closed structures, protected from DNase I, whereas the structure of the tomato Inhibitor I gene was found to be in an open, pre-active structure (A. Conconi and C.A. Ryan, submitted). The open chromatin structure of the gene, found before transcription, seems to imply a need for the gene to have a poised structure in order to be functional.

The regulatory regions of eukaryotic genes have been shown to be highly sensitive to DNase I. Often these hypersensitive sites correspond to enhancer elements, silencers, gene regulatory sequences, origins of DNA replications, etc. [57]. We are presently investigating these features in the flanking regions of the tomato Inhibitor I gene. These experiments will provide detailed information concerning the positioning of the tomato Inhibitor I gene with respect to nucleosome and chromatin structures and the relationships of these structures with the interaction of regulatory proteins with specific *cis* elements of the gene.

A working model for wound-inducible signal transduction pathways

In the model presented in Fig. 4 (E.E. Farmer and C.A. Ryan, in review) we proposed that wounding initiates a lipid-based signal transduction cascade by causing the activation or synthesis of lipases that, in turn, release linolenic acid from plant membranes. The released linolenic acid would rapidly be converted to jasmonic acid that would interact with receptors to activate proteinase inhibitor genes.

Jasmonic acid and its methyl ester have been shown to be synthesized in plants from C18 unsaturated fatty acids (ie. linolenic acid) through the action of lipoxygenase (LOX) followed by the action of a dehydrase (cyclization) and β-oxidation [58]. The structure of methyl jasmonate and its precursors are similar to the structure of some of the prostaglandins in animals that are very powerful signalling molecules [59]. The biological activities of methyl jasmonate and jas-

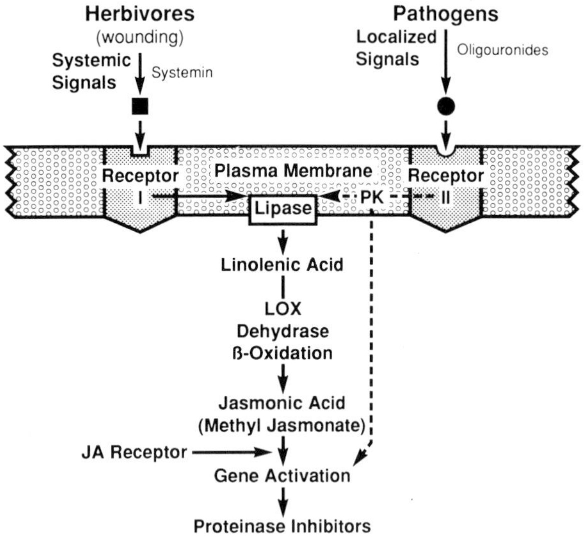

Fig. 4. A proposed model for the signal transduction pathways that regulate systemic and localized insect-induced and pathogen-induced expression of proteinase inhibitor genes in plants.

monic acid, as well as the responses of arachidonic acid [60] and traumatin [61] in plants, have often suggested the intriguing possibility that lipid based signalling systems may be important in regulating plant gene expression, and in particular, for regulating genes involved in stress responses [36, 62].

Our observation that methyl jasmonate is a powerful signalling molecule for the synthesis of proteinase inhibitors, taken together with the proposed route of methyl jasmonate biosynthesis [58], with research of several laboratories on the effects of wounding on plant membrane biosynthesis and degradation [62–66], and with research on wound-inducible lipoxygenase activity in plants [67–70], has provided the basis of our present hypothesis that the wound induction of proteinase inhibitors involves a lipid-based signal transduction system. This in turn implies that many signalling pathways involving a variety of lipid-derived signals may regulate other plant genes as well.

In this model, on the left, the polypeptide, systemin [17], is released by wounding and is transported to cells throughout the plant where it interacts with receptors on the leaf cell plasma membranes. The occupied receptors are proposed to initiate a signalling cascade by an undefined mechanism that activates or causes synthesis of lipases that release free fatty acids from plant membranes. No data as yet exists to support a role for protein kinase activity in this cascade, but this possibility is under investigation. The resulting cascade may also somehow activate an increase in the levels of LOX. Several lines of evidence have already suggested that LOX activity may be involved with stress responses, including wounding by insects [66], other mechanical damage [67], and fungal elicitors [68]. LOX enzymes, perhaps a specific enzyme, are prime candidates as key components in the wound-inducible synthesis of methyl jasmonate, or other similar compounds, that may regulate the expression of proteinase inhibitor genes in response to wounding. The net result of the interaction of systemin with plant leaf cells would be to release free linolenic acid into the pathway and facilitate its conversion to jasmonic acid.

On the upper right of the model, oligosaccharide signals are proposed to react with receptors which may activate protein kinases as part of the signalling pathway for localized responses. We have shown that protein phosphorylation can be activated in response to oligouronides [18] and protein phosphorylation has been shown to be associated with elicitor-induced phytoalexin synthesis [69, 70]. Whether oligosaccharide signalling involves the same pathway activated by the polypeptide is still not clear. Using a specific inhibitor of LOX, we have recently shown that proteinase inhibitor synthesis in response to wounding and the polypeptide both are inhibited, while the response to oligouronides is not (E.E. Farmer and C.A. Ryan, in preparation).

The mechanism by which jasmonic acid induces proteinase inhibitors in plant leaves may involve a direct intracellular interaction with a soluble receptor or *cis*-acting element. However, other mechanisms of action must be considered. Recently an odorant receptor was identified in animal nasal tissue and isolated and characterized [71]. The receptor is located on the sites of sensory olafactory neurons. These receptors on surface membranes are thought to activate G proteins and to generate action potentials along the axons. Although plants do not have axons, this mechanism for jasmonate action, or even systemin action, is attractive. It is of interest that *cis*-jasmonate was a moderate affinity ligand with the odorant binding protein of nasal tissue [71].

The search for the proteinase inhibitor inducing factor(s) and the signal transduction pathway(s) that regulate the expression of proteinase inhibitor genes continues. While the model presented in Fig. 4 is useful in designing new experimental approaches to understand the wound-induction of proteinase inhibitor genes, its purpose is, as models in the past, to accommodate the constant modification and change that will surely occur. The research has been an interesting and rewarding challenge for me and for my students, postdoctoral fellows and technicians, who have contributed so significantly to the research. It has now been 30 years since Dr Kent Balls and I excitedly stood and watched as potato tuber extracts totally inhibited the activity of chymotrypsin [1]. The incremental knowledge we have gained since then concerning the structure, evolution and function of these proteinase inhibitor proteins, and the signalling systems that regulate their synthesis, have constantly nurtured the interest, excitement and enjoyment of science of those in my laboratory. Our recent findings of the involvement of systemin and jasmonic acid in the signalling system have deepened this enthusiasm. Like curious detectives, we have uncovered many clues toward an understanding of the mystery of the signals and the pathways leading to proteinase inhibitor synthesis in response to wound-generated signals, but the plot only deepens. Perhaps within the coming years we will be able to understand how proteinase inhibitors are systemically synthesized when an insect chews on a plant. So simply stated, but so interestingly complex.

Acknowledgement

Supported in part by Project 1791 from the College of Agriculture and Home Economics, Washington State University and grants from the National Science Foundation and the McKnight Foundation.

References

1. Ryan CA, Balls AK: An inhibitor of chymotrypsin from *Solanum tuberosum* and its behavior toward trypsin. Proc Natl Acad Sci USA 48: 1839–1844 (1962).
2. Bryant J, Green T, Gurusaddaiah T, Ryan CA: Proteinase inhibitor II from potatoes: Isolation and characterization of the isoinhibitor subunits. Biochemistry 15: 3418–3423 (1976).
3. Ryan CA, Huisman OC: Chymotrypsin inhibitor I from potatoes: A transient component in leaves of young potato plants. Nature 214: 1047 (1967).
4. Ryan CA: An inducible protein in tomato and potato leaflets. Plant Physiol 43: 1880 (1968).
5. Plunkett G, Senear DF, Zuroske G, Ryan CA: Proteinase inhibitors I and II from leaves of wounded tomato plants: purification and properties. Arch Biochem Biophys 213: 463–672 (1982).
6. Green TR, Ryan CA: Wound-induced proteinase inhibitor in plant leaves: A possible defense mechanism against insects. Science 175: 776–777 (1972).
7. Ryan CA: Assay and biochemical properties of the proteinase inhibitor inducing factor, a wound hormone. Plant Physiol 54: 328–332 (1971).
8. Brown WE, Takio K, Titani K, Ryan CA: Wound-induced trypsin inhibitor in alfalfa leaves: Identity as a member of the Bowman-Birk inhibitor family. Biochemistry 24: 2105–2108 (1985).
9. Roby D, Toppan A, Esquerre-Tugaye MT: Cell surface in plant micro-organism interactions. VIII. Increased proteinase inhibitor activity in melon plants in response to infection by *Colletotrichum lagenarium* or to treatment with an elicitor fraction from this fungus. Physiol Mol Pl Path 30: 6453–460 (1987).
10. Bradshaw HDF, Hollick JB, Parsons TJ, Clarke HRG, Gordon MP: Systemically wound-responsive genes in poplar trees encode proteins similar to sweet potato sporamins and legume Kunitz trypsin inhibitors. Plant Mol Biol 14: 51–59 (1989).
11. Hilder VA, Gatehouse AMR, Sheerman SE, Barker RF, Boulter D: A novel mechanism of insect resistance engineered into tobacco. Nature 330: 160–163 (1987).
12. Johnson R, Narvaez J, An G, Ryan CA: Expression of proteinase inhibitors I and II in transgenic tobacco plants: Effects on natural defense against *Manduca sexta* larvae. Proc Natl Acad Sci USA 89: 9871–9875 (1989).
13. Ryan CA: Proteinase inhibitors in plants: Genes for improving defenses against insects and pathogens. Annu Rev Phytopath 28: 425–449 (1990).
14. Darvill AG, Albersheim P: Phytoalexins and their elici-

tors: A defense against insects and pathogens. Annu Rev Plant Physiol 35: 243–275 (1984).
15. Ryan CA: Oligosaccharide signalling in plants. Annu Rev Cell Biol 3: 295–317 (1987).
16. Walker-Simmons M, Hadwiger L, Ryan CA: Chitosans and pectic polysaccharides both induce the accumulation of the antifungal phytoalexin pisatin in pea pods and antinutrient proteinase inhibitors in tomato leaves. Biochem Biophys Res Comm 110: 194–199 (1983).
17. Pearce G, Strydom D, Johnson S, Ryan CA: A polypeptide from tomato leaves activates the expression of proteinase inhibitor genes. Science 253: 895–897 (1991).
18. Farmer EE, Ryan CA: Interplant communication: Airborne methyl jasmonate induces the synthesis of proteinase inhibitor genes in plant leaves. Proc Natl Acad Sci USA 87: 7713–7716 (1990).
19. Pena-Cortez H, Sanchez-Serrano JJ, Mertens R, Willmitzer L: Abscisic acid is involved in the wound-induced expression of the proteinase inhibitor II gene in potato and tomato. Proc Natl Acad Sci USA 86: 9851–9855 (1989).
20. Thornburg RW, Li X: Auxin levels decline in tobacco foliage following wounding. Plant Physiol 93: 500–504 (1990).
21. Graham JS, Hall G, Pearce G, Ryan CA: Regulation of synthesis of proteinase inhibitors I and II mRNAs in leaves of wounded tomato plants. Planta 169: 399–405 (1986).
22. Shumway K, Cheng V, Ryan CA: Evidence for the presence of proteinase inhibitor I in plant cell vacuolar protein bodies. Planta 129: 161–165 (1976).
23. Walker-Simmons M, Ryan CA: Immunological identification of proteinase inhibitors I and II in isolated tomato leaf vacuoles. Plant Physiol 60: 61–63 (1977).
24. Bishop P, Makus KJ, Pearce G, Ryan CA: Proteinase inhibitor inducing factor activity in tomato leaves resides in oligosaccharides enzymically released from cell walls. Proc Natl Acad Sci 78: 3536–3640 (1981).
25. Bishop P, Pearce G, Bryant JE, Ryan CA: Isolation and characterization of the proteinase inhibitor inducing factor from tomato leaves: Identity and activity of poly- and oligogalacturonide fragments. J Biol Chem 259: 13172–13177 (1984).
26. Hahn MG, Darvill AG, Albersheim P: Host-pathogen interactions. XIX. The endogenous elicitor, a fragment of a plant cell wall polysaccharide that elicits phytoalexin accumulation in soybeans. Plant Physiol 68: 1161–1169 (1984).
27. Bruce RJ, West CA: Elicitation of casbene synthetase activity in castor bean. The role of pectic fragments of the plant cell wall in elicitation by a fungal endopolygalacturonase. Plant Physiol 69: 1181–1188 (1982).
28. Moloshok T, Pearce G, Ryan CA: Oligouronide signalling of proteinase inhibitor genes in plants: Structure-activity relationships of di- and trigalacturonic acids and their derivatives: Arch. Biochem. Biophys. in press.
29. Farmer EE, Moloshok TD, Saxton MJ, Ryan CA: Oligosaccharide signaling in plants: Specificity of oligouronide-enhanced plasma membrane protein phosphorylation. J Biol Chem 266: 3140–3145 (1991).
30. Bruce RJ, West CA: Elicitation of lignin biosynthesis and isoperoxidase activity by pectic fragments in suspension cultures of castor bean. Plant Physiol 89: 889–897 (1989).
31. Farmer EE, Pearce G, Ryan CA: *In vitro* phosphorylation of plant plasma membrane proteins in response to the proteinase inhibitor inducing factor (PIIF). Proc Natl Acad Sci USA 86: 1539–1542 (1989).
32. Kohn R: Ion binding in polyuronates, alginate and pectin. Pure Appl Chem 42: 371–397 (1985).
33. Grant GT, Morris BR, Rees DA, Smith PJC, Thorn D: Biological interactions between polysaccharides and divalent cations: The eggbox model. FEBS Lett 32: 195–198 (1973).
34. Cosio EG, Frey T, Ebel J: Solubilization and characteristics of the binding sites for fungal β-glucans from soybean cell membranes. FEBS Lett 264: 235–238 (1990).
35. Cheong J-J, Hahn MG: A specific, high-affinity binding site for the hepta-β-glucoside elicitor exists in soybean membranes. Plant Cell 3: 137–147 (1991).
36. Baydoun EA-H, Fry SC: The immobility of pectic substances in injured tomato leaves and its bearing on the identity of the wound hormone. Planta 165: 269–276 (1985).
37. Bowles DJ: Defense related proteins in higher plants. Annu Rev Biochem 59: 873–907 (1990).
38. Nelson CE, Walker-Simmons M, Makus D, Wuroske G, Graham J, Ryan CA: Regulation of synthesis and accumulation of proteinase inhibitors in leaves of wounded tomato plants: In:Hedin PA (ed), Plant Resistance to Insects, pp. 103–122. American Chemical Society, Washington, DC (1983).
39. Anderson JM: Membrane-derived fatty acids as precursors to second messengers. In: Boss WF, Morre KJ (eds) Second Messengers in Plant Growth and Development, pp. 181–212. Alan Liss, New York. (1989).
40. Parthier B: Jasmonates: Hormonal regulators of stress factors in leaf senescence? J Plant Growth Regul 9: 57–63 (1990).
41. Farmer EE, Pearce G, Ryan CA: Regulation of expression of proteinase inhibitor genes by methyl jasmonate and jasmonic acid. Plant Physiol, in press (1992).
42. Peña-Cortés H, Willmitzer L, Sánchez-Serrano J: Abscisic acid mediates wound-induction but not developmental expression of the proteinase inhibitor II gene family. Plant Cell 3: 963–972 (1991).
43. Doherty HM, Bowles DJ: The role of pH and ion transport in oligosaccharide-induced proteinase inhibitor accumulation in tomato plants. Plant Cell Envir 13: 851–855 (1990).
44. Thain JF, Doherty HM, Bowles KJ, Wildon DC: Oligosaccharides that induce proteinase inhibitor activity in

tomato plants cause depolarization of tomato leaf cells. Plant Cell Envir 13: 569–574 (1990).

45. Thornburg RW, Cleveland TE, Ryan CA: Wound-inducible expression of potato inhibitor II gene in transgenic tobacco plants. Proc Natl Acad Sci USA 84: 744–748 (1987).
46. Keil M, Sánchez-Serrano J, Schell J, Willmitzer L: Primary structure of a proteinase inhibitor II gene from potato. Nucl Acids Res 14: 5641–5650 (1986).
47. Sánchez-Serrano J, Keil M, O'Connor A, Schell J, Willmitzer L: Wound-induced expression of a potato proteinase inhibitor II gene in transgenic tobacco plants. EMBO J 6: 303–306 (1987).
48. Peña-Cortés H, Sánchez-Serrano J, Rocha-Sosa M, Willmitzer L: Systemic induction of proteinase inhibitor-II gene expression in potato plants by wounding. Planta 174: 84–89 (1988).
49. Lee JS, Brown WE, Pearce G, Dreher TW, Graham JS, Ahern KG, Pearson GD, Ryan CA: Molecular characterization and phylogenetic studies with a wound-inducible proteinase inhibitor gene in Lycopersicum species. Proc Natl Acad Sci USA 83: 7277–7281 (1986).
50. Cleveland TE, Thornburg R, Ryan CA: Molecular characterization of a wound-inducible proteinase inhibitor gene from potato and the processing of the mRNA and protein. Plant Mol Biol 8: 199–207 (1987).
51. Wingate VPM, Ryan CA: Uniquely regulated proteinase Inhibitor I gene in a wild tomato species. Plant Physiol 97: 496–501 (1991).
52. An G, Mitra A, Choi HK, Costa MA, An K, Thornburg RW, Ryan CA: Functional analysis of the 3′ control region of the potato wound-inducible proteinase inhibitor II gene. Plant Cell 1: 115–122 (1989).
53. Palm CJ, Costa MA, An G, Ryan CA: Wound-inducible nuclear protein binds fragments that regulate a proteinase inhibitor II gene from potato. Proc Natl Acad Sci USA 87: 603–607 (1990).
54. Sánchez-Serrano J, Peña-Cortés H, Willmitzer L, Prat S: Identification of potato nuclear binding to the distal promoter region of the proteinase inhibitor II gene. Proc Natl Acad Sci USA 87: 7205–7209 (1990).
55. Reeves R: In: Adolph DW (ed) Chromosomes and Chromatin, vol 1, pp. 109–131. CRL Press (1988).
56. Spiker S: Plant chromatin structure. Annu Rev Plant Physiol 36: 235–253 (1985).
57. Elgin SCR: The formation and function of DNase I hypersensitive sites in the process of gene activation. J Biol Chem 263: 19259–19262 (1988).
58. Vick BA, Zimmerman DC: The biosynthesis of jasmonic acid: A physiological role for plant lipoxygenase. Biochem Biophys Res Comm 111: 470–477 (1983).
59. Samuelsson B, Goldyne M, Granstrom E, Hamberg M, Hammarstrom S, Malmsten: Prostaglandins and thromboxanes. Annu Rev Biochem 47: 997–1029 (1978).
60. Bostock RM, Kuc JA, Laine RA: Eicosapentaenoic and arachidonic acids from *Phytophthora infestans* elicit fungitoxic sesquiterpenes in the potato. Science 212: 67–69 (1981).
61. Zimmerman DC, Coudron CA: Identification of traumatin, a wound hormone, as 12-oxyOtrans-10-dodecenoic acid. Plant Physiol 623: 536–541 (1979).
62. Hildebrand DF, Hamilton-Kemp T, Legg CS, Bookjans G: Plant lipoxygenases: Occurrence, properties and possible functions. Curr Topics Plant Biochem Physiol 7: 201–219 (1988).
63. Galliard T: Degradation of acyl lipids: Hydrolytic and oxidative enzymes. Biochem Plants 4: 85–116 (1980).
64. Theologis A, Laites GG: Wound-induced membrane lipid breakdown in potato tuber. Plant Physiol 68: 53–58 (1981).
65. Hasson EP, Laites GG: Purification and characterization of an A type phospholipase from potato and its effect on potato mitochondria. Plant Physiol 57: 148–152 (1976).
66. Hildebrand DF, Rodriguez JG, Brown GC, Luu KT, Volden CS: Peroxidative responses of leaves in two soybean genotypes injured by two spotted spider mites (Acari: Tetranychidae). J Econ Ent 79: 1459–1465 (1986).
67. Galliard T: Lipolytic and lipoxygenase enzymes in plants and their action in wounded tissues. In: Kahl G (ed) Biochemistry of Wounded Plant Tissues, pp. 155–201. Walter de Gruyter, Berlin (1978).
68. Fournier J, Pelesier B, Esquerre-Tugaye: Induction of lipoxygenase activity in cultured tobacco (*Nicotiana tabacum*) cells by elicitors of ethylene from *Phytophthora parasitica* var *nicotianae*. C.R. Acad Sci Paris 303: 651–656 (1986).
69. Dietrich A, Mayer JE, Hahlbrock K: Fungal elicitor triggers rapid, transient, and specific protein phosphorylation in parsley cell suspension cultures. J Biol Chem 265: 6360–6368 (1990).
70. Felix G, Frosskopf G, Regenass M, Boller T: Protein phosphorylation is involved in signal transduction during elicitation of the plant defense response. Proc Natl Acad Sci USA in press (1991).
71. Pevsner J, Hou V, Snowman AM, Snyder SH: Odorant-binding protein: Characterization of ligand binding. J Biol Chem 265: 6118–6125 (1990).

Plant Molecular Biology **19**: 135–147, 1992.
© 1992 *Kluwer Academic Publishers. Printed in Belgium.*

Molecular basis of disease susceptibility in the Texas cytoplasm of maize

Charles S. Levings, III [1] and James N. Siedow [2]
[1] *Department of Genetics, Box 7614, North Carolina State University, Raleigh, NC 27695–7614, USA;*
[2] *Department of Botany, Duke University, Durham, NC 27706, USA*

Key words: URF13, cytoplasmic male sterility, T-toxin, corn leaf blight, pore-forming protein, T-*urf13*, pathotoxin

The Texas, or T, cytoplasm (*cms-T*) of maize (*Zea mays* L.) carries a cytoplasmically inherited male sterility. This cytoplasmic male sterility (CMS) suppresses the production of viable pollen grains and is inherited in a non-Mendelian manner. *cms-T* was first described in the line Golden June in Texas [32, 69]. The discovery of the Texas cytoplasm was important to maize geneticists and breeders because they wished to use CMS in maize hybrid production. Hybrid maize, first double-cross hybrids and later single-cross, was used to exploit hybrid vigor to achieve higher yields and greater uniformity. Many applied and basic studies have been devoted to *cms-T* to understand CMS and its relationship with mitochondrial and nuclear genes.

Male sterility is characterized by failure of anther exertion and pollen abortion in *cms-T*. Although female fertility is unaffected by the Texas male-sterile cytoplasm, comparisons of plants carrying T and normal cytoplasm reveal minor differences in several other morphological traits [15].

Two other major CMS types are well known in maize, *cms-C* and *cms-S* [46]. The various CMS types are readily distinguished by specific nuclear genes, designated restorers of fertility (*Rf*), that suppress the male-sterile effect of the cytoplasm permitting functional pollen production. Acting jointly, two genes, *Rf1* and *Rf2*, restore pollen fertility to *cms-T* but not to *cms-C* or *cms-S*. Different restorer genes are required to restore pollen fertility to C or S cytoplasms. The three male-sterile cytoplasms and normal maize can also be distinguished by mitochondrial characteristics. Mitochondrial DNA (mtDNA) restriction fragment length polymorphisms, variations in mitochondrial RNA (mtRNA), and differences in mitochondrial translational products provide additional means of differentiating the cytoplasms [16, 47, 60].

During the 1950s and 1960s, *cms-T* was used extensively to avoid hand or mechanical emasculation in the production of hybrid maize. In 1969 and 1970, the Southern corn leaf blight struck the South and Corn Belt regions of the United States, severely blighting maize carrying the Texas cytoplasm, which was more than 85% of the U.S. maize acreage [68, 75]. The blight was caused by the fungal pathogen *Bipolaris maydis* race T (formerly *Helminthosporium maydis* race T). Maize plants carrying normal, C, or S cytoplasms are generally only mildly affected by *B. maydis* race T. Because it was immediately evident that the Texas cytoplasm was directly responsible for the disease epidemics, large-scale usage of *cms-T* for hybrid seed production was abandoned. *Phyllosticta maydis*, another fungal pathogen, is also uniquely virulent to *cms-T* maize. Because this latter pathogen is restricted to the cooler northern regions of the United States, it was less serious than Southern corn leaf blight. The 1969 and 1970 blight epidemics demonstrated the perils of genetic vulnerability when a single, uniform cytoplasm is widely employed. We have prepared a review dealing with the molecular basis of disease susceptibility in the Texas cytoplasm of maize.

B. maydis race T is usually not a serious pathogen on maize unless it contains the T cytoplasm. Normal maize and maize with the male-sterile cytoplasms S (*cms-S*) and C (*cms-C*) support only limited colonization by the fungal pathogen. Generally, colonization is restricted to boat-shaped lesions on the leaves that remain small and isolated. In contrast, *B. maydis* race T can rapidly and completely colonize *cms-T* maize. Lesions enlarge rapidly, coalesce, and spread throughout the plant, causing extensive damage and sometimes plant death. Evidence indicates that susceptibility of *cms-T* maize to *B. maydis* race T is due to mitochondrial sensitivity to a host-specific toxin, BmT toxin, produced by the pathogen. Accordingly, mitochondria of disease-resistant maize types are insensitive to BmT toxin. Susceptibility to *P. maydis* has a common basis with *B. maydis* race T because it produces a structurally similar toxin to which *cms-T* mitochondria are also sensitive.

An unusual mitochondrial gene, designated T-*urf13*, is thought to be responsible for the CMS and disease susceptibility traits carried by *cms-T* maize. A 13 kDa polypeptide (URF13) that is a component of the inner mitochondrial membrane is encoded by T-*urf13* [12, 14]. The makeup of T-*urf13* is odd, reflecting its complex origin [12]. The chimeric T-*urf13* sequence appears to have arisen by multiple rearrangements involving intramolecular and intermolecular recombinations. Other plant mitochondrial genes associated with CMS also have a chimeric structure, e.g. *pcf* [79]. The coding region of T-*urf13* contains 115 codons; the first 88 codons have similarity with the 3′-flanking region of the *rrn26* gene, the next eight codons are of unknown origin, and the final 18 codons have similarity to the coding region of the *rrn26* gene. T-*urf13* contains a large 5′-flanking region of 5 kb that is similar to the 5′-flanking region of the *atp6* gene. Because of this similarity, T-*urf13* and *atp6* are thought to have similar promoters. Besides the recombinant sequences found in T-*urf13*, fully functional *atp6* and *rrn26* genes are located elsewhere in the *cms-T* mitochondrial genome. An open reading frame, designated *orf221*, is located 77 nucleotides downstream from T-*urf13*. *orf221* is transcribed and, if translated, could encode a protein of 25 kDa (221 amino acids). Although it is uncertain whether *orf221* is translated, *orf221* is common to plant mitochondrial genomes.

Even though T-*urf13* is a protein-encoding gene, its coding region consists mostly of sequences from the coding and flanking regions of *rrn26*, a structural RNA gene. Even more fortuitous is that rearrangements have placed the 345 nt (nucleotide) open reading frame immediately downstream from a region containing mitochondrial promoter activity.

T-*urf13* is uniquely found in the Texas cytoplasm; it is not encountered in other maize cytoplasms, either male-fertile or male-sterile, or in other plant species. In fact, T-*urf13* is not essential for normal mitochondrial activity. Although T-*urf13* has a complex origin, this is not unusual among plant mitochondrial genomes. MtDNA rearrangements are recognized as important factors in altering genome organization and in causing gene mutations [12, 25, 47, 51, 54–55, 60, 67, 72–73]. The structure of the *cms-T* mitochondrial genome differs in organization and size from the normal maize mitochondrial genome (*cms-T* 540 kb; normal 570 kb) because of rearrangements [19, 21].

T-*urf13* transcripts are detected in *cms-T* mitochondria from coleoptiles, ear shoots, leaves, roots, and tassels, suggesting the gene is constitutively expressed. The 13 kDa protein encoded by T-*urf13* has been confirmed by an antiserum prepared against a chemically synthesized oligopeptide derived from the predicted amino acid sequence of the gene [14, 77]. The URF13 protein is observed in *cms-T* mitochondria from all plant organs when examined by Western immunoblot analyses. Furthermore, a labeled 13 kDa protein can be immunoprecipitated with the URF13-specific antiserum from *in organello* translation products of *cms-T* mitochondria [14]. Before the discovery of T-*urf13*, Leaver and collaborators demonstrated with an *in organello* protein-synthesizing system that a unique 13 kDa protein is synthesized in *cms-T* mitochondria but not in normal or other male-sterile maize [23, 24]. Together, these findings clearly indicate that T-*urf13* and its encoded polypeptide are associated with only the *cms-T* kind of CMS.

T-*urf13* is implicated with CMS by the effect of restorer genes on T-*urf13* expression. Jointly, the dominant nuclear genes *Rf1* and *Rf2* restore pollen fertility to *cms-T* maize. *Rf1* alters the transcription profile of T-*urf13* and reduces the abundance of URF13 protein by about 80 % [14, 38, 39]. In contrast, the recessive genotype *rf2 rf2* does not affect the transcription profile or the abundance of URF13. It is uncertain how *Rf1* alters T-*urf13* expression. There is no evidence that *Rf1* affects the expression of other mitochondrial genes besides T-*urf13* or that it has other essential activities. It is unclear how *Rf2* contributes to pollen restoration except that the expression of T-*urf13* is apparently not affected by *Rf2*. *Rf2* may exert its effect on specific cells of the anther, the site of male sterility. The *Rf2* allele is common among maize inbred lines, but *Rf1* is rare.

The best genetic evidence supporting an association between the T-*urf13* gene and the CMS and disease susceptibility phenotypes has come from studies of *cms-T* revertants. Revertants, which are male-fertile and disease-resistant, are often obtained from *cms-T* cell culture experiments, even though spontaneous mutations of *cms-T* plants to male fertility and disease resistance are not observed. Toxin-resistant calli can be selected from calli of *cms-T* maize grown on a medium containing BmT toxin [27, 76]. *cms-T* calli are normally sensitive to toxin and soon die on toxin-containing media. When toxin-resistant calli are regenerated, many whole plants are obtained that are male-fertile and toxin- and disease-resistant; some regenerant plants from toxin-resistant calli remain male-sterile and disease-susceptible. Surprisingly, similar results were obtained even when the experiments were carried out without toxin selection [8, 9]. The fact that revertants can be obtained without toxin selection suggests that the T-*urf13* gene confers a selective disadvantage to cells in culture, which is independent of toxin sensitivity [66]. These findings also suggest that somaclonal variation is responsible for reversion events. Only male-fertile and disease-resistant phenotypes are obtained among true-breeding revertants. Male-sterile and disease-resistant or male-fertile and disease-susceptible plants are sometimes observed; however, these are apparently chimeras that are unstable and lost in later generations. Genetic analysis has established that reversion is caused by changes in cytoplasmically inherited genes instead of nuclear genes. Restriction enzyme mapping and nucleotide sequencing have been used to characterize changes in T-*urf13* among several revertants. Most often these studies showed that T-*urf13* is absent from the revertants and that the deletional event is probably due to homologous recombination [20, 70]. An exceptional revertant, designated T4, contains a frameshift mutation that results in a premature stop codon and a shortened URF13 polypeptide [78]. These studies show that reversion is correlated with either a deletion or mutation of T-*urf13*. Collectively, revertant analyses suggest that CMS and disease susceptibility are inseparable, and both are due to T-*urf13*.

URF13 and fungal pathogen susceptibility

After the appearance of the Southern corn leaf blight in the late 1960s, it was recognized that maize plants carrying *cms-T* cytoplasm are specifically susceptible to the fungal pathogens *B. maydis* race T and *P. maydis*. Subsequent studies established that *B. maydis* race T and *P. maydis* each produce a set of pathotoxins known as BmT and Pm toxins, respectively (collectively referred to as T-toxins), that specifically affect *cms-T* maize plants [53]. The purification and chemical characterization of T-toxins were carried out by Daly and co-workers who also established the specific nature of their interaction with *cms-T* maize [10, 42, 43]. BmT toxin consists of a group of related compounds composed of linear alkane chains 35 to 45 carbons long that contain repeated series of β-oxydioxo groups regularly spaced along the methylene backbone [42]. Pm toxin differs from BmT toxin in that the alkane chain is only 33 to 35 carbon atoms in length, and the repeated functional groups are oxy-oxo structures [10]. The toxicity of both sets of T-toxins is retained when all the carbonyl moieties are reduced to hydroxyls by the addition of sodium borohydride, although the concentration dependence of toxicity is decreased by a factor of roughly ten [34, 74].

Initial insight into the site of action of the BmT and Pm toxins was provided by the discovery that T-toxin specifically affects the functioning of mitochondria isolated from *cms-T* maize but not mitochondria isolated from other maize cytoplasms or any other plant species [58]. Reported effects of T-toxin on *cms-T* mitochondria include the inhibition of malate-stimulated electron transfer [3, 65], uncoupling of oxidative phosphorylation [2, 3, 34], promotion of mitochondrial swelling [2, 41], stimulation of NADH- or succinate-mediated state 4 respiration [2, 65], and leakage of small molecules and ions (NAD^+ and Ca^{2+}) from the matrix space [33, 56]. These effects on isolated mitochondria are observed at toxin concentrations in the range of 20 to 100 nM [2, 3, 33, 41, 56, 65]. Collectively, these results indicate that the interaction of BmT or Pm toxin with *cms-T* mitochondria leads to a rapid permeabilization of the inner mitochondrial membrane. The loss of molecules such as NAD^+ (M_r 663), but not proteins, from the mitochondrial matrix after T-toxin treatment suggests the size of the resulting membrane pore could be as large as 1.5 nm in diameter [57]. Besides T-toxins, methomyl {*S*-methyl-*N*-[(methylcarbamoyl)-oxy]thioacetimidate}, the active component of a systemic insecticide produced by Du Pont and a compound that bears no obvious structural resemblance to BmT or Pm toxins, mimics all the biological effects of T-toxins but at greater than 10^4-fold higher concentrations (1–10 mM) than are required of the pathotoxins [40, 41]. The lower activity of methomyl may be related more directly to its water solubility than to any specific structural feature. Methomyl analogues with alkyl side chains on the carbamate moiety 12 to 14 carbon atoms longer than the methyl group found in methomyl are able to affect *cms-T* mitochondria at molar concentrations comparable to those required of BmT and Pm toxins [1]. Finally, preincubation with dicyclohexylcarbodiimide (DCCD) protects isolated *cms-T* mitochondria from the effects of added T-toxin and methomyl [4, 35]. DCCD is a protein-modifying reagent that forms a stable covalent adduct with acidic amino acid side chains located in hydrophobic domains [59].

Early studies established a correlation between the presence of T-*urf13* and the susceptibility of *cms-T* maize toward *B. maydis* race T and *P. maydis* [12, 14]. Definitive proof of the relationship between the T-*urf13* gene product, URF13, and sensitivity to T-toxin was provided by Dewey *et al.* [13] who cloned T-*urf13* into an inducible expression vector and used it to express URF13 in *Escherichia coli*. Analogous to its location on the inner membrane in *cms-T* mitochondria [14], URF13 expressed in *E. coli* is exclusively associated with the plasma membrane fraction [13]. *E. coli* cells expressing URF13 are sensitive to added T-toxin or methomyl at the same concentration ranges that affect *cms-T* mitochondria. The effects on *E. coli* include inhibition of glucose-stimulated whole-cell respiration, induction of swelling in isolated spheroplasts, and the complete loss of viability of treated cells [13]. *E. coli* cells that contain the plasmid vector without the T-*urf13* insert are insensitive to either T-toxin or methomyl. The capacity of URF13 to render *E. coli* sensitive to T-toxin clearly establishes a direct connection between pathotoxin sensitivity and the presence of the T-*urf13* gene product.

Although the results outlined above link T-toxin sensitivity with the presence of URF13, they do not establish that T-toxin promotes pore formation in the plasma membrane of URF13-containing *E. coli*. Similarities do exist between the responses of *E. coli* expressing URF13 and *cms-T* mitochondria to added T-toxin. *E. coli* cells also become insensitive to T-toxin when treated with DCCD, although 10- to 50-fold lower concentrations of DCCD are required to protect *cms-T* mitochondria [7]. T-toxin stimulation of spheroplast swelling is analogous to the reported swelling of *cms-T* mitochondria after addition of T-toxin [2, 41], but the nature of the swelling effect in *cms-T* mitochondria is not well understood and is controversial [3]. The observed inhibition of whole-cell respiration could be explained by pore formation, which leads to a loss of intermediate species associated with glycolysis and the TCA cycle, as well as any unbound intracellular electrolytes. Preliminary studies using conductivity as a measure of extracellular ion concentration showed that addition of T-toxin to a concentrated suspension of *E. coli* cells expressing URF13 induces a rapid increase in the conductivity of the external medium. A more direct and quantitative evaluation of this effect is provided by the observation of a massive and immediate efflux of radiolabeled ^{86}Rb accumulated in *E. coli* cells expressing URF13 after the addition of T-toxin or methomyl [5, 7]. The rate at which ^{86}Rb is lost from cells after toxin addition (<1 min) is consistent with the formation of hydrophilic pores within the plasma membrane; added uncouplers, which break down the proton motive gradient driving Rb uptake but do not permeabilize the membrane to Rb, also promote a decrease in cellular Rb, but do so with a half-life of tens of minutes. The amount of accumulated ^{86}Rb in control cells lacking URF13 is not affected by addition of T-toxin or methomyl. These results provide direct evidence that the effects of T-toxin in *E. coli* cells expressing URF13 are analogous to those seen in *cms-T* mitochondria and result from the formation of hydrophilic pores within URF13-containing membranes.

Additional support for the relationship between URF13 expression and T-toxin sensitivity was provided by Huang *et al.* [36] who expressed T-*urf13* in baker's yeast. They constructed a chimeric gene composed of T-*urf13* fused at its amino terminus to the mitochondrial targeting sequence of the ATPase subunit 9 from *Neurospora crassa*. When this construct is transformed into yeast, it produces a protein that is localized to the mitochondrial membrane fraction and is processed to a size similar to that of standard URF13. Growth of yeast cells expressing the chimeric URF13 construct is inhibited by added T-toxin or methomyl, and NADH-supported state 4 respiration in mitochondria isolated from these cells is stimulated by these compounds [36]. Yeast transformed with full-length T-*urf13* lacking the mitochondrial transit peptide is insensitive to T-toxin or methomyl, even though immunoblotting analysis shows URF13 is expressed in these cells [36]. Analogous to the results obtained with bacteria, the studies with this heterologous eukaryotic system confirm the relationship between URF13 and T-toxin sensitivity. They also suggest that in eukaryotes URF13 needs to be localized in mitochondria to produce a toxin-sensitive phenotype. Another study has also confirmed the capacity of T-*urf13* to confer toxin sensitivity to yeast [28].

Because URF13 is capable of permeabilizing biological membranes of *E. coli*, yeast, and maize when it interacts with T-toxin, no specific association of URF13 with another mitochondrial protein is probably needed to obtain this response. With the exception of the H^+-translocating ATP synthase complex, which is an F_OF_1-type ATPase in mitochondria and *E. coli* [63], and complex II (succinate:UQ oxidoreductase), the protein complexes associated with the electron transfer chain in *E. coli* plasma membranes differ considerably from complexes I, III, and IV of the mitochondrial inner membrane [37]. An association between URF13 and the ATP synthase complex has been reported based upon a co-precipitation of URF13 and the DCCD-binding proteolipid component of F_O by a polyclonal antibody directed against URF13 [77]. A similar experimental approach has provided evidence of an association between URF13 and subunit I of cytochrome *c* oxidase [48]. Care needs to be taken in interpreting such results; fractionation of the individual electron transfer complexes from *cms-T* mitochondria [64] indicated that some URF13 is associated with each complex, presumably the result of nonspecific interactions (S. Ferguson-Miller, W. E. Peiffer, and K. L. Korth, personal communication).

The simplest model for explaining how URF13 renders biological membranes sensitive to T-toxin involves a direct interaction between URF13 and toxin. Binding studies using reduced, [^{3}H]-labeled T-toxin from either *B. maydis* race T or *P. maydis* and isolated maize mitochondria showed no significant difference in the level of T-toxin binding to mitochondria isolated from normal or *cms-T* maize [26]. Later binding studies using tritium-labeled Pm toxin and *E. coli* cells expressing URF13 indicated a significant level of specific toxin binding that does not occur in control cells not expressing URF13 [6]. The specific toxin binding is saturable, with maximum binding falling in the range 300 to 450 pmol of toxin per mg of *E. coli* protein and half-maximal binding (= apparent K_d) appearing at 40 to 75 nM free toxin. Regression analysis indicates that toxin binding is not hyperbolic but shows positive cooperativity with Hill coefficients in the range of 1.4 to 1.8. Pm toxin binding to URF13 is reversible and can be competed off with unlabeled methomyl, indicating that the two compounds either have identical binding sites on URF13 or separate binding sites with some spatial overlap. Binding studies using isolated maize mitochondria show, in agreement with Frantzen *et al.* [26], that an appreciable level of nonspecific T-toxin binding occurs in normal mitochondria. The silicon oil centrifugation technique employed by Braun *et al.* [6] is able to distinguish an additional, specific component of toxin binding associated with isolated *cms-T* mitochondria. The level of specific toxin binding to *cms-T* mitochondria, 20 pmol of toxin per mg mitochondrial protein, is well below that observed in *E. coli*, and regression analysis indicates no apparent cooperativity to the binding. Mitochondria from *cms-T*-restored maize show slightly lower levels of toxin binding, 15 pmol of toxin bound per mg mitochondrial protein. This lower level is qualitatively consistent with the reduced amount of URF13 protein detected in T-restored mitochondria [14, 23]. The amount of URF13 protein in T-restored mitochondria, however, is about 80% less than in mitochondria from the T-nonrestored line. The reason for this apparent discrepancy between URF13 protein levels and the amount of specific T-toxin binding is unknown.

The sensitivity of *E. coli* expressing URF13 to T-toxin [13], coupled with the observation that the response in bacteria is analogous to that of *cms-T* mitochondria [5, 13], provides a useful heterologous system for carrying out directed mutagenesis studies to characterize either specific amino acid residues or regions of the URF13 polypeptide that are important for interaction with T-toxin and/or induction of membrane permeabilization. Initially, three toxin-insensitive mutations of URF13 were characterized [5, 13]. A systematic deletion analysis indicated that removal of the carboxyl-terminal 32 amino acids, producing a truncated URF13 of 83 amino acids instead of the 115 found in the standard URF13, has no effect on sensitivity to T-toxin or methomyl [5]. Deletion of the next amino acid (Leu-83), to give an 82 amino acid product, however, results in a complete loss of toxin sensitivity. This result is consistent with the earlier report of a toxin-insensitive mutation (T-4) that spontaneously appeared in *cms-T* maize cell tissue culture [78]. Molecular analysis of the T-4 mutant showed the T-*urf13* reading frame has been shortened to 74 codons by a frameshift mutation. Based upon the mutational analysis carried out in *E. coli* [5], an URF13 protein containing only 74 amino acids is not expected to be toxin sensitive. Maize plants regenerated from the T-4 cell line are toxin-insensitive and male-fertile [76, 78].

A second toxin-insensitive mutant of URF13 was derived from changes at a specific amino acid residue, Asp-39 [5]. Changing this residue to valine, histidine, or glutamate produces toxin-insensitive *E. coli* and, in addition, leads to a marked decrease in the binding of [^{14}C]-DCCD to the modified URF13, compared to the standard URF13. Other mutational analyses indicate that DCCD binds to two acidic residues (Asp-39 and Asp-12) on URF13, but no decrease in toxin sensitivity is observed after DCCD treatment when only Asp-12 is modified. These results indicate that Asp-39 is required to obtain a functional interaction between URF13 and T-toxin and that Asp-39 is the site at which DCCD acts to attenuate the effects of T-toxin on URF13. A third toxin-insensitive mutation of URF13 has been described that contains an internal deletion of amino acids 2 through 11 at the protein's amino terminus [13]. More recently, over 100 additional mutations of URF13 have been generated by random mutagenesis and screening for T-toxin insensitivity, but these mutations have yet to be fully characterized.

Mutations that affect the sensitivity of URF13 to T-toxin may not all have the same mode of action. Even in the simplest scheme, a response to T-toxin requires that URF13 has the capacity to bind T-toxin and to undergo the structural changes needed to generate a pore within the membrane after toxin binding. Loss of either function results in a toxin-insensitive species. To distinguish between these alternatives, studies of toxin binding to URF13 using the three characterized toxin-insensitive mutations were carried out [6]. *E. coli* expressing two of the three mutations, the alteration from Asp-39 to Val-39 at position 39 and the 82-amino acid carboxyl-terminal deletion, bind low levels of labeled toxin, <50 pmol of toxin per mg of *E. coli* protein, compared to cells expressing unmodified URF13, >275 pmol of toxin per mg of *E. coli* protein. Cells expressing the internal deletion missing amino acid residues 2 through 11 bind labeled T-toxin at levels comparable to normal URF13-containing cells, >250 pmol of toxin per mg of *E. coli* protein, but they show a significantly lower apparent level of binding cooperativity, Hill coefficients = 1.0 to 1.2 [6]. These results underscore the importance of a T-toxin:URF13 complex to pore formation and further indicate that the amino-terminal region of URF13 contributes, in some manner, to pore formation after toxin binding.

Because URF13 is an integral membrane protein [13] that can generate a hydrophilic pore across biological membranes after interacting with T-toxin in eukaryotic [40] and prokaryotic [5] systems, the structural organization of URF13 within the membrane should have a relationship to its function. We have proposed a model to account for the structure of URF13 within the membrane (Fig. 1) [45]. Both experimental results and theoretical considerations were incorporated into the development of this model. Hy-

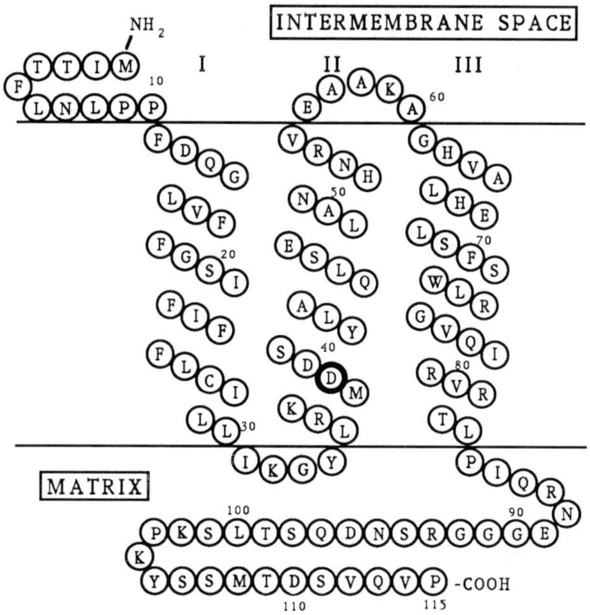

Fig. 1. Proposed model of URF13 disposition within the membrane, indicating the location of the three putative membrane-spanning helical domains. Asp(D)-39 (site of DCCD reaction) is shaded.

dropathy analysis shows an unbroken stretch of hydrophobic amino acids near the amino-terminus between residues 13 and 31 [14] that is sufficient to produce a single membrane-spanning α-helix [18]. It is also known that Asp-12 forms a stable adduct with [^{14}C]-DCCD, localizing this amino acid to a hydrophobic region of the membrane [5] and making residues 11 through 31 reasonable candidates for forming a 20 residue, membrane-spanning α-helix (Fig. 1, helix I). Following this hydrophobic membrane-spanning region, two potential membrane-spanning, amphipathic α-helices can be generated in URF13 between residues 32 and 83, the last residue required for toxin sensitivity (Fig. 1). The two putative helical sections consist of residues 35 to 55 (helix II) and residues 61 to 83 (helix III), respectively. The amphipathic nature of these helices is more apparent in helical wheel plots where the respective hydrophilic and hydrophobic faces of the two helices are delineated (Fig. 2). Calculation of the hydrophobic moments through these two regions (Fig. 3) indicates that helix II has a

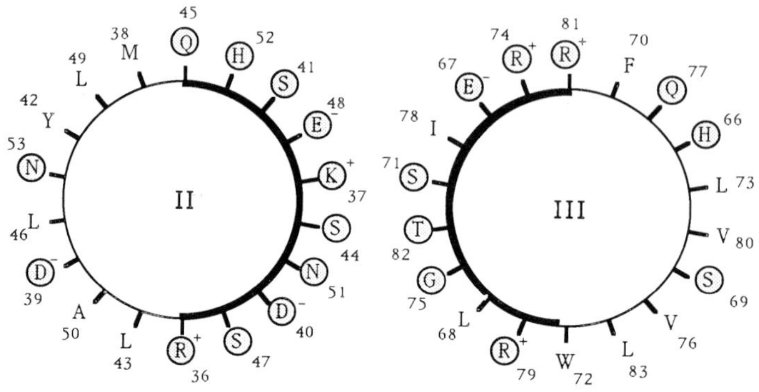

Fig. 2. Helical wheel representation of potentially amphipathic helices II and III of URF13. The residues of each helical segment are plotted as a wheel in a two-dimensional projection of the helix. The hydrophilic and hydrophobic faces are indicated, respectively, as bold and thin arcs. Charged residues are denoted by the appropriate sign.

consistently high hydrophobic moment throughout the helical span, whereas that of helix III is lower, but within the range of values reported for other postulated amphipathic helices [17]. Within helix II, Asp-39, which is essential for T-toxin binding [6], is unusual in that it is a charged amino acid located on the hydrophobic face of the helix (Fig. 2). Nevertheless, this is consistent with Asp-39 forming a stable adduct after reaction of URF13 with DCCD [5]. The failure of residues beyond Leu-83 to contribute to toxin sensitivity is consistent with their localization outside of the three membrane-spanning regions (Fig. 1). The appearance of Leu-83 at the boundary between the hydrophobic and hydrophilic phases of the

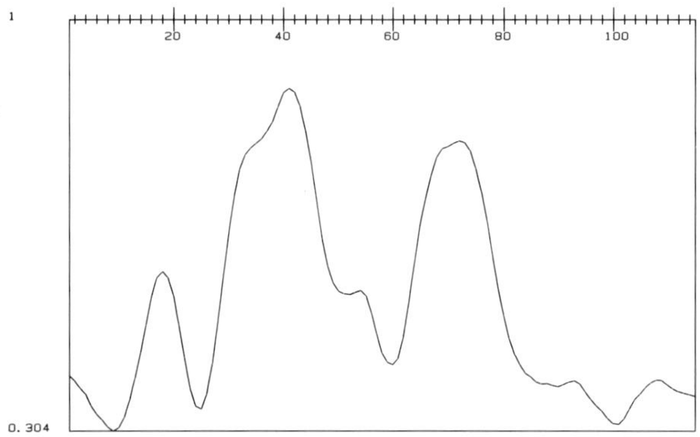

Fig. 3. Hydrophobic moment profile of URF13. The hydrophobic moment was calculated at 100 °C according to Eisenberg [17] using the consensus hydropathy scale of Eisenberg and a sliding window of 19 amino acids. Numbers on the abscissa refer to the URF13 primary sequence.

membrane could explain the sharp delineation observed between sensitivity and insensitivity to T-toxin at the residue 83/82 interface [5]. A truncated URF13 that contains only 82 amino acids may not have a sufficient number of residues in helix III to traverse the membrane in a functionally competent manner.

Recent attempts to verify the correctness of this model of URF13 have involved the use of an URF13 fusion protein expressed in *E. coli* with a 14 amino acid antigenic tag at the amino terminus [44]. Cells expressing this fusion protein show the standard responses to added T-toxin or methomyl. Protease digestion studies were used to determine the accessibility of the amino and carboxyl termini of the URF13 fusion protein in *E. coli* membrane preparations having either right-side-out or inside-out orientations. Consistent with the three-membrane-spanning model, the amino and carboxyl ends of the URF13 fusion protein are located on opposite sides of the membrane. These studies also revealed that URF13 has two topological orientations in the plasma membrane of *E. coli*, with roughly 60% of the carboxyl termini oriented toward the cytoplasmic surface and the remaining 40% facing the periplasmic space [44]. This dual orientation is not a property unique to the fusion protein; the carboxyl terminus of standard URF13 expressed in *E. coli* shows a similar mixed orientation. Without an amino-terminal-specific tag on URF13 in *cms-T* mitochondria, the relative topological positions of the amino and carboxyl ends cannot be determined; however, protease studies with right-side-out and inside-out inner membrane preparations indicate that URF13 has a single orientation in mitochondria, with the carboxyl terminus localized on the matrix surface of the membrane. If URF13 in *cms-T* mitochondria also contains three membrane-spanning regions, this places the amino terminus on the outer surface of the inner membrane. We do not know in *E. coli* whether both orientations of URF13 are capable of conferring toxin sensitivity or, analogous to the topology found in *cms-T* mitochondria, only those molecules with their carboxyl termini facing the cytoplasmic surface confer toxin sensitivity.

The current paradigm for the protein structural features needed to form a hydrophilic pore in a biological membrane postulates a series of membrane-spanning, amphipathic α-helices associated in a cylindrical fashion perpendicular to the plane of the membrane [29, 31, 62]. Helices are oriented in the membrane such that the polar faces of the helices orient inward, lining the resulting water-filled channel, whereas their hydrophobic faces point outward, interacting with the hydrocarbon phase of the lipid bilayer [62]. Because NAD^+ can pass through pores generated by the interaction of T-toxin with URF13 [56], a channel as large as 1.5 nm in diameter is suggested. With a pore diameter of only 0.8 nm, six helices are needed to form the channel lining [11]. The model outlined in Fig. 1 contains only two membrane-spanning, amphipathic helical regions per URF13 molecule. A pore 0.8–1.2 nm in diameter requires the association of URF13 monomers into at least trimeric or tetrameric structures in which a 6–8 helical pore would result from the combination of three to four sets of helices II and III interacting in some manner (Fig. 4). The possibility that URF13 exists as an oligomeric structure within the membrane is suggested by the occasional appearance of apparent dimeric species on SDS gels [30]. In addition, the cooperativity of binding of Pm toxin to URF13 in *E. coli* is consistent with an oligomeric URF13 complex [6]. It is tempting to associate a Hill coefficient of 1.4 to 1.8 with the presence of two binding sites on an URF13 dimer, but the binding of oxygen to hemoglobin, which clearly involves four sites, gives a Hill coefficient of only 2.7 [61]. The Hill coefficient for toxin binding to URF13 could result from two binding sites with a moderate degree of cooperativity or from three or more binding sites with a relatively weak cooperativity [71].

Covalent cross-linking studies have directly confirmed that a portion of the URF13 in *cms-T*

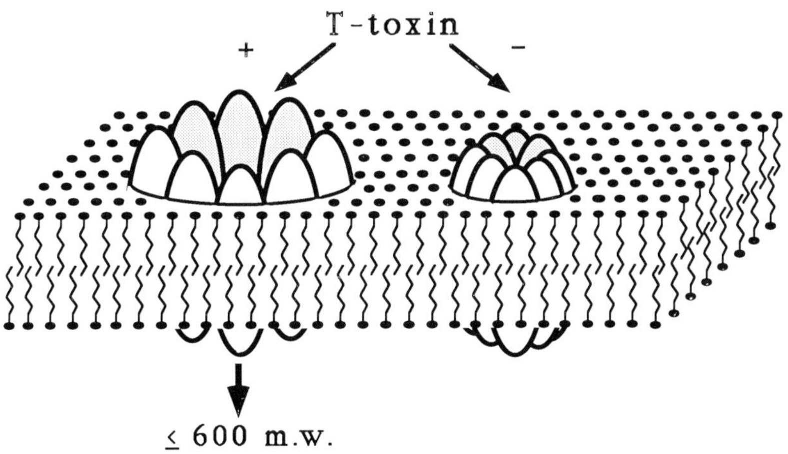

Fig. 4. Diagrammatic representation of the proposed structure formed by the interactions between helices II and III of an oligomeric URF13 complex in a lipid bilayer in the absence (right) and presence (left) of T-toxin. The structure of an URF13 tetramer is presented in the figure with each cylinder corresponding to one of the two proposed amphipathic membrane-spanning α-helices on URF13 (cf. Fig. 2).

mitochondria and *E. coli* exists in the membrane in an oligomeric state [44]. Dimers and trimers predominate, although higher-order structures are also observed. Addition of T-toxin or methomyl does not change the pattern of oligomers seen after cross-linking, indicating the changes in URF13 structure associated with pore formation are more subtle than the simple association of URF13 monomers (or even dimers) into a higher-order structure. Cross-linking studies carried out in conjunction with proteolysis experiments have also established that within a given multimer all monomers have the same orientation within the membrane [44].

Epilogue

The experimental advances provided by the molecular characterization of URF13 have clearly established the molecular basis of the susceptibility of *cms-T* maize to the fungal pathogens, *B. maydis* race T and *P. maydis*. The pathotoxins produced by these fungi inhibit *cms-T* maize mitochondrial function by their capacity to permeabilize the inner mitochondrial membrane after interaction with the T-*urf13* gene product, URF13 (Fig. 4). This, in turn, promotes the loss of metabolic integrity that results in the large-scale fungal colonization and subsequent necrotic lesions that are symptomatic of leaf blight in *cms-T* maize. Yet to be determined are the precise structural features associated with URF13 in the membrane before and after T-toxin binding that lead to membrane pore formation. The two traits, CMS and toxin sensitivity, seem to be inseparable because analysis of spontaneous revertants has shown that reversion to pollen fertility is always accompanied by simultaneous reversion to toxin insensitivity among true-breeding revertants. These results support the view that T-*urf13* is responsible for both traits and further suggest that CMS and toxin sensitivity could have a common mechanism of action. Before the discovery of T-*urf13*, it was proposed that an anther-specific substance exists that affects mitochondria in a manner like the BmT toxin of *B. maydis* race T [22]. The anther-specific substance could interact with URF13 to permeabilize the inner mitochondrial membrane in a fashion similar to the toxin-URF13 interaction. Loss of mitochondrial activity in anther cells during pollen formation would lead to pollen abortion. This explanation is interesting because it furnishes a common mechanism to account for the dual role of T-*urf13* in causing toxin sensitivity and CMS. An anther-specific compound with these properties, however, has not yet been identified, and it remains for future studies to establish the relationship between the mechanism by which URF13 is able to form membrane pores and the wholesale loss of cellular function during pollen abortion in *cms-T* maize.

Acknowledgements

Funding for these studies was provided by the Department of Energy to J.N.S. and the National Science Foundation to C.S.L.

References

1. Aranda G, Durlin P, Gauvrit C: Methomyl analogues with increased biological activity towards F_7T maize mitochondria. Phytochemistry 26: 1909–1913 (1987).
2. Bednarski MA, Izawa S, Scheffer RP: Reversible effects of toxin from *Helminthosporium maydis* race T on oxidative phosphorylation by mitochondria from maize. Plant Physiol 59: 540–545 (1977).
3. Bervillé A, Ghazi A, Charbonnier M, Bonavent J-F: Effects of methomyl and *Helminthosporium maydis* toxin on matrix volume, proton motive force, and NAD accumulation in maize (*Zea mays* L.) mitochondria. Plant Physiol 76: 508–517 (1984).
4. Bouthyette P-Y, Spitsberg V, Gregory P: Mitochondrial interaction with *Helminthosporium maydis* race T toxin: Blocking by dicyclohexylcarbodiimide. J Exp Bot 36: 511–528 (1985).
5. Braun CJ, Siedow JN, Levings CS III: The T-*urf13* gene is responsible for toxin sensitivity in maize and *E. coli*. In: Goldberg R (ed) The Molecular Basis of Plant Development, pp. 79–85. UCLA Symposia on Molecular and Cellular Biology, New Series, Vol. 92, Alan R. Liss, New York (1989).
6. Braun CJ, Siedow JN, Levings CS III: Fungal toxins bind to the URF13 protein in maize mitochondria and *Escherichia coli*. Plant Cell 2: 153–161 (1990).
7. Braun CJ, Siedow JN, Williams ME, Levings CS III:

Mutations in the maize mitochondrial T-*urf13* gene eliminate sensitivity to a fungal pathotoxin. Proc Natl Acad Sci USA 86: 4435–4439 (1989).

8. Brettell RIS, Goddard BVD, Ingram DS: Selection of Tms-cytoplasm maize tissue cultures resistant to *Drechslera maydis* T-toxin. Maydica 24: 203–213 (1979).
9. Brettell RIS, Thomas E, Ingram DS: Reversion of Texas male-sterile cytoplasm maize in culture to give fertile, T-toxin resistant plants. Theor Appl Genet 58: 55–58 (1980).
10. Danko SJ, Kono Y, Daly JM, Suzuki Y, Takeuchi S, McCrery DA: Structure and biological activity of a host-specific toxin produced by the fungal corn pathogen *Phyllosticta maydis*. Biochemistry 23: 759–766 (1984).
11. Davidson VL, Brunden KR, Cramer WA, Cohen FS: Studies on the mechanism of action of channel-forming colicins using artificial membranes. J Membrane Biol 79: 105–118 (1984).
12. Dewey RE, Levings CS III, Timothy DH: Novel recombinations in the maize mitochondrial genome produce a unique transcriptional unit in the Texas male-sterile cytoplasm. Cell 44: 439–449 (1986).
13. Dewey RE, Siedow JN, Timothy DH, Levings CS III: A 13-kilodalton maize mitochondrial protein in *E. coli* confers sensitivity to *Bipolaris maydis* toxin. Science 239: 293–295 (1988).
14. Dewey RE, Timothy DH, Levings CS III: A mitochondrial protein associated with cytoplasmic male sterility in the T cytoplasm of maize. Proc Natl Acad Sci USA 84: 5374–5378 (1987).
15. Duvick DN: Cytoplasmic pollen sterility in corn. In: Caspari EW, Thoday JM (eds) Advances in Genetics, vol 13, pp. 1–56. Academic Press, New York (1965).
16. Eckenrode VK, Levings CS III: Maize mitochondrial genes and cytoplasmic male sterility. In: Bruening GJ, Harada J, Kosuge T, Hollaender A (eds) Tailoring Genes for Crop Improvement: An Agricultural Perspective, pp. 69–84. Plenum, New York (1987).
17. Eisenberg D: Three-dimensional structure of membrane and surface proteins. Annu Rev Biochem 53: 595–623 (1984).
18. Engelman DM, Steitz TA, Goldman A: Identifying nonpolar transbilayer helices in amino acid sequences of membrane proteins. Annu Rev Biophys Chem 15: 321–353 (1986).
19. Fauron C, Havlik M: The maize mitochondrial genome of the normal type and the cytoplasmic male sterile type T have very different organization. Curr Genet 15: 149–154 (1989).
20. Fauron CM-R, Havlik M, Brettell RIS: The mitochondrial genome organization of a maize fertile cmsT revertant line is generated through recombination between two sets of repeats. Genetics 124: 423–428 (1990).
21. Fauron C, Havlik M, Lonsdale D, Nichols L: Mitochondrial genome organization of the maize cytoplasmic male sterile type T. Mol Gen Genet 216: 395–401 (1989).
22. Flavell R: A model for the mechanism of cytoplasmic male sterility in plants, with special reference to maize. Plant Sci Lett 3: 259–263 (1974).
23. Forde BG, Leaver CJ: Nuclear and cytoplasmic genes controlling synthesis of variant mitochondrial polypeptides in male-sterile maize. Proc Natl Acad Sci USA 77: 418–422 (1980).
24. Forde BG, Oliver RJC, Leaver CJ: Variation in mitochondrial translation products associated with male-sterile cytoplasms in maize. Proc Natl Acad Sci USA 75: 3841–3845 (1978).
25. Fragoso LL, Nichols SE, Levings CS III: Rearrangements in maize mitochondrial genes. Genome 31: 160–168 (1989).
26. Frantzen KA, Daly JM, Knoche HW: The binding of host-selective toxin analogs to mitochondria from normal and 'Texas' male sterile cytoplasm maize. Plant Physiol 83: 863–868 (1987).
27. Gengenbach BG, Green CE, Donovan CM: Inheritance of selected pathotoxin resistance in maize plants regenerated from cell cultures. Proc Natl Acad Sci USA 74: 5113–5117 (1977).
28. Glab N, Wise RP, Pring DR, Jacq C, Slonimski P: Expression in *Saccharomyces cerevisiae* of a gene associated with cytoplasmic male sterility from maize: Respiratory dysfunction and uncoupling of yeast mitochondria. Mol Gen Genet 223: 24–32 (1990).
29. Guy HR: A structural model of the acetylcholine receptor channel based on partition energy and helix packing calculations. Biophys J 45: 249–261 (1984).
30. Hack E, Lin C, Yang H, Horner HT: T-URF13 protein from mitochondria of Texas male-sterile maize (*Zea mays* L.). Its purification and submitochondrial localization, and immunogold labeling in anther tapetum during microsporogenesis. Plant Physiol 95: 861–870 (1991).
31. Hall JE, Vodyanoy I, Balasubramanian TM, Marshall GR: Alamethicin: A rich model for channel behavior. Biophys J 45: 233–247 (1984).
32. Harvey PH, Levings CS III, Wernsman EA: The role of extrachromosomal inheritance in plant breeding. Adv Agron 24: 1–27 (1972).
33. Holden MJ, Sze H: *Helminthosporium maydis* T toxin increased membrane permeability to Ca^{2+} in susceptible corn mitochondria. Plant Physiol 75: 235–237 (1984).
34. Holden MJ, Sze H: Dissipation of the membrane potential in susceptible corn mitochondria by the toxin of *Helminthosporium maydis*, race T, and toxin analogs. Plant Physiol 84: 670–676 (1987).
35. Holden MJ, Sze H: Effects of *Helminthosporium maydis* race T toxin on electron transport in susceptible corn mitochondria and prevention of toxin actions by dicyclohexylcarbodiimide. Plant Physiol 91: 1296–1302 (1989).
36. Huang J, Lee S-H, Lin C, Medici R, Hack E, Myers AM: Expression in yeast of the T-URF13 protein from Texas male-sterile maize mitochondria confers sensitivity to

methomyl and to Texas-cytoplasm-specific fungal toxins. EMBO J 9: 339–347 (1990).
37. Ingledew WJ, Poole RK: The respiratory chains of *Escherichia coli*. Microbiol Rev 48: 222–271 (1984).
38. Kennell JC, Pring DR: Initiation and processing of *atp6*, T-*urf13* and *ORF221* transcripts from mitochondria of T cytoplasm maize. Mol Gen Genet 216: 16–24 (1989).
39. Kennell JC, Wise RP, Pring DR: Influence of nuclear background on transcription of a maize mitochondrial region associated with Texas male sterile cytoplasm. Mol Gen Genet 210: 399–406 (1987).
40. Klein RR, Koeppe DE: Mode of methomyl and *Bipolaris maydis* (race T) toxin in uncoupling Texas male-sterile cytoplasm corn mitochondria. Plant Physiol 77: 912–916 (1985).
41. Koeppe DE, Cox JK, Malone CP: Mitochondrial heredity: A determinant in the toxic response of maize to the insecticide methomyl. Science 201: 1227–1229 (1978).
42. Kono Y, Daly JM: Characterization of the host-specific pathotoxin produced by *Helminthosporium maydis*, race T, affecting corn with Texas male sterile cytoplasm. Bioorg Chem 8: 391–397 (1979).
43. Kono Y, Suzuki Y, Takeuchi S, Knoche HW, Daly JM: Studies on the host-specific pathotoxins produced by *H. maydis*, race T and *P. maydis*: Absolute configuration of PM-toxins and HMT-toxins. Agric Biol Chem 49: 559–562 (1985).
44. Korth KL, Kaspi CI, Siedow JN, Levings CS III: URF13, a maize mitochondrial pore-forming protein, is oligomeric and has a mixed orientation in *Escherichia coli* plasma membranes. Proc Natl Acad Sci USA 88: 10865–10869 (1991).
45. Korth KL, Struck F, Kaspi CI, Siedow JN, Levings CS III: Topological orientation of the membrane protein URF13. In: Herrmann RG, Larkins BA (eds) Plant Molecular Biology, pp. 375–381. Plenum, London (1991).
46. Laughnan JR, Gabay-Laughnan S: Cytoplasmic male sterility in maize. Annu Rev Genet 17: 27–48 (1983).
47. Leaver CJ, Gray MW: Mitochondrial genome organization and expression in higher plants. Annu Rev Plant Physiol 33: 373–402 (1982).
48. Leaver CJ, Isaac PG, Small ID, Bailey-Serres J, Liddell AD, Hawkesford MJ: Mitochondrial genome diversity and cytoplasmic male sterility in higher plants. Phil Trans R Soc Lond B 319: 165–176 (1988).
49. Levings CS III: The Texas cytoplasm of maize: Cytoplasmic male sterility and disease susceptibility. Science 250: 942–947 (1990).
50. Levings CS III, Braun CJ: Insights into the Texas male-sterile cytoplasm of maize. In: Lord E, Bernier G (eds) Plant Reproduction: From Floral Induction to Pollination, vol. 1, pp. 121–127. Proceedings of the Twelfth Annual Symposium in Plant Physiology. American Society of Plant Physiologists, Rockville, MD (1989).
51. Levings CS III, Brown GG: Molecular biology of plant mitochondria. Cell 56: 171–179 (1989).
52. Levings CS III, Dewey RE: Molecular studies of cytoplasmic male sterility in maize. Phil Trans R Soc Lond B 319: 177–185 (1988).
53. Lim SM, Hooker AL: Disease determinant of Helminthosporium maydis race T. Phytopathology 62: 968–971 (1972).
54. Lonsdale DM: A review of the structure and organization of the mitochondrial genome of higher plants. Plant Mol Biol 3: 201–206 (1984).
55. Lonsdale DM: The plant mitochondrial genome. In: Marcus A (ed) The Biochemistry of Plants, A Comprehensive Treatise. Molecular Biology, vol 15, pp. 229–295. Academic Press, New York (1989).
56. Matthews DE, Gregory P, Gracen VE: *Helminthosporium maydis* race T toxin induces leakage of NAD^+ from T cytoplasm corn mitochondria. Plant Physiol 63: 1149–1153 (1979).
57. Menestrina G, Forti S, Gambale F: Interaction of tetanus toxin with lipid vesicles. Effects of pH, surface charge, and transmembrane potential on the kinetics of channel formation. Biophys J 55: 393–405 (1989).
58. Miller RJ, Koeppe DE: Southern corn leaf blight: Susceptible and resistant mitochondria. Science 173: 67–69 (1971).
59. Nałęcz MJ, Casey RP, Azzi A: Use of *N,N'*-dicyclohexylcarbodiimide to study membrane-bound enzymes. Meth Enzymol 125: 86–108 (1986).
60. Newton KJ: Plant mitochondrial genomes: Organization, expression and variation. Annu Rev Plant Physiol Plant Mol Biol 39: 503–532 (1988).
61. Ogata RT, McConnell HM: The binding of a spin-labeled triphosphate to hemoglobin. Cold Spring Harbor Symp Quant Biol 36: 325–336 (1972).
62. Ojcius DM, Young JD-E: Cytolytic pore-forming proteins and peptides: Is there a common structural motif? Trends Biochem Sci 16: 225–229 (1991).
63. Pedersen PL, Carafoli E: Ion motive ATPases. I. Ubiquity, properties, and significance to cell function. Trends Biochem Sci 12: 146–150 (1987).
64. Peiffer WE, Ingle RT, Ferguson-Miller S: Structurally unique plant cytochrome *c* oxidase isolated from wheat germ, a rich source of plant mitochondrial enzymes. Biochemistry 29: 8696–8701 (1990).
65. Peterson PA, Flavell RB, Barratt DHP: Altered mitochondrial membrane activities associated with cytoplasmically-inherited disease sensitivity in maize. Theor Appl Genet 45: 309–314 (1975).
66. Pring DR, Gengenbach BG, Wise RP: Recombination is associated with polymorphism of the mitochondrial genomes of maize and sorghum. Phil Trans R Soc Lond B 319: 187–198 (1988).
67. Pring DR, Lonsdale DM: Molecular biology of higher plant mitochondrial DNA. Int Rev Cytol 97: 1–46 (1985).
68. Pring DR, Lonsdale DM: Cytoplasmic male sterility and maternal inheritance of disease susceptibility in maize. Annu Rev Phytopathol 27: 483–502 (1989).

69. Rogers JS, Edwardson JR: The utilization of cytoplasmic male-sterile inbreds in the production of corn hybrids. Agron J 44: 8–13 (1952).
70. Rottmann WH, Brears T, Hodge TP, Lonsdale DM: A mitochondrial gene is lost via homologous recombination during reversion of CMS T maize to fertility. EMBO J 6: 1541–1546 (1987).
71. Segel IH: Multisite and allosteric enzymes. In: Segel IH (ed) Enzyme Kinetics: Behavior and Analysis of Rapid Equilibrium and Steady State Enzyme Systems, pp. 346–384. Wiley Interscience, New York (1975).
72. Small I, Suffolk R, Leaver CJ: Evolution of plant mitochondrial genomes via substoichiometric intermediates. Cell 58: 69–76 (1989).
73. Stern DB, Palmer JD: Extensive and widespread homologies between mitochondrial DNA and chloroplast DNA in plants. Proc Natl Acad Sci USA 81: 1946–1950 (1984).
74. Suzuki Y, Tegtmeier KJ, Daly JM, Knoche HW: Analogs of host-specific phytotoxin produced by *Helminthosporium maydis*, race T. II. Biological activities. Bioorg Chem 11: 313–321 (1982).
75. Ullstrup AJ: The impacts of the southern corn leaf blight epidemics of 1970–1971. Annu Rev Phytopathol 10: 37–50 (1972).
76. Umbeck PF, Gengenbach BG: Reversion of male-sterile T-cytoplasm maize to male fertility in tissue culture. Crop Sci 23: 584–588 (1983).
77. Wise RP, Fliss AE, Pring DR, Gengenbach BG: *Urf13-T* of T cytoplasm maize mitochondria encodes a 13 kD polypeptide. Plant Mol Biol 9: 121–126 (1987).
78. Wise RP, Pring DR, Gengenbach BG: Mutation to male fertility and toxin insensitivity in Texas (T)-cytoplasm maize is associated with a frameshift in a mitochondrial open reading frame. Proc Natl Acad Sci USA 84: 2858–2862 (1987).
79. Young EG, Hanson MR: A fused mitochondrial gene associated with cytoplasmic male sterility is developmentally regulated. Cell 50: 41–49 (1987).

Plant Molecular Biology **19**: 149–168, 1992.
© 1992 *Kluwer Academic Publishers. Printed in Belgium.*

The chloroplast genome

Masahiro Sugiura
Center for Gene Research, Nagoya University, Chikusa, Nagoya, Japan 464-01

Key words: Cloroplast, DNA genome gene, intron, splicing, *trans*-splicing

Introduction

Chloroplasts are intracellular organelles in plants which contain the entire machinery necessary for the process of photosynthesis. They also participate in the biosynthesis of amino acids, nucleotides, lipids and starch. Mendel's law was rediscovered at the beginning of this century, and in 1909 Baur and Correns separately published the first reports of non-Mendelian inheritance based on studies of variegation in higher plants. Some of the green-and-white variegated leaves were shown to be caused by factors inherited in a non-Mendelian manner. Further analysis of variegation in higher plants revealed that the genetic determinants for these characters were associated with chloroplasts. However, the difficulty of obtaining specific chloroplast mutations has limited the study of non-Mendelian genetics in higher plants. Our knowledge of extranuclear genetics came primarily from studies using the unicellular alga *Chlamydomonas*.

The demonstration of a unique DNA species in chloroplasts [e.g. 94] has led to intensive studies of both the structure of chloroplast DNA and its expression. These studies have been accelerated by gene cloning and DNA sequencing techniques developed in the mid-1970s. The first physical map of chloroplast DNA was constructed for maize in 1976 [3] and the first chloroplast gene was cloned in 1977 [5]. These studies and others established a new field, 'chloroplast molecular biology,' and the organization and expression of chloroplast genomes were among the most extensively studied fields in plant molecular biology. After 10 years the entire sequence of the chloroplast DNA was determined in tobacco, liverwort and then in rice. Sequences for defined regions of many other chloroplast DNAs have also been completed, but the identification and expression analysis of many chloroplast genes have mostly been done with several representative higher plants and green algae.

The purpose of this paper is to review briefly the historical background as well as our latest knowledge of the chloroplast genome, emphasizing its structural aspect, followed by a short discussion of future research. The references will attempt to cite the first one or two reports, or reviews. Other aspects of the chloroplast genome have been presented in several recent reviews [10, 40, 82, 88, 128].

Genome structure

The presence of unique, double-stranded and high-molecular-weight DNA in chloroplasts was demonstrated by the distinct and characteristic buoyant density of these molecules in CsCl gradients [18, 94]. Since 1963, CsCl gradient centrifugation has been used widely to identify and isolate chloroplast DNA from plants [95]. However, difficulties were encountered early in isolating the chloroplast DNA of higher plants because their buoyant densities were similar to those of the nuclear DNAs. Chloroplast DNA is now prepared from highly purified intact chloroplasts. The circularity of chloroplast DNA was first reported by Manning *et al.* [75], who used an electron microscope and observed circular DNA molecules with a contour length of 44.5 μm in lysates of *Euglena* chloroplasts. Circular chloroplast DNAs have since been observed in many other plants. Chloroplast DNA molecules appeared to have a uniform contour length within a given plant species. Convincing evidence for the homogeneity of chloroplast DNA molecules was provided by digestion with class II restriction endonucleases [4]. Furthermore, mapping the restriction endonuclease fragments always yielded a circular map. The first physical map of chloroplast DNA was thus determined from maize [3]. Restriction site mapping is routinely used to determine the size of chloroplast DNA, and is almost a prerequisite for further studies.

Almost all chloroplast DNAs fall into the size range of 120 to 160 kb [88]. Among chloroplast genomes for which an accurate size estimate exists, the siphonous green alga *Codium fragile* has the smallest chloroplast DNA known (85 kb) while the green alga *Chlamydomonas moewusii* has the largest (292 kb). The chloroplast genome of the giant green alga *Acetabularia* is more complex than those of other plants and its genome size appears to be 2000 kb. The population of chloroplast DNA in a plant species is generally homogeneous. However, the chloroplast genome of the brown alga *Pylaiella littoralis* has been shown to be composed of two different circular DNA molecules of 133 kb and 58 kb in size [74]. Though genetically homogeneous, chloroplast DNA often consists of two groups of molecules differing only in the relative orientation of the single-copy regions [87]. A small proportion of the molecules exists in dimer, trimer and tetramer forms [20].

One of the outstanding features of the chloroplast DNAs found in most plants is the presence of a large inverted repeat (IR) which ranges from 6 to 76 kb in length [88]. Most of the size variation among land plant chloroplast DNAs can be accounted for by changes in the length of the IR. The segments of the IR are separated by one large and one small single-copy region (LSC and SSC, respectively). Pea, broad bean, alfalfa and pine chloroplast DNAs are exceptions to this pattern and lack IRs [e.g. 63]. It has been suggested that the IR was present in the common ancestor of land plants and one segment of the IR was lost in some legumes and conifers during evolution [88]. However, our recent analysis indicates that the loss of the IR is partial at least in black pine as its genome retains a residual IR (unpublished). The chloroplast DNA from *Euglena gracilis* contains three tandem repeats, each of which contains an rRNA gene cluster. Thus, chloroplast DNAs can be classified into three groups: chloroplast DNAs lacking IRs (group I), chloroplast DNAs containing IRs (group II) and chloroplast DNAs with tandem repeats (group III).

The entire nucleotide sequence of the chloroplast DNA is now established for tobacco (155 844 bp) [103], liverwort (121 024 bp) [85] and rice (134 525 bp) [53]. The gene order present in tobacco (Fig. 1) is most representative of land plants, probably reflecting the ancestral gene order among higher plants. Most of the chloroplast DNA from maize, pea and *Euglena* have already been sequenced, and the determination of complete chloroplast DNA sequences from several other plants including *Arabidopsis* and pine is in progress. It was originally believed that the gene organization of chloroplast DNAs was relatively uniform from species to species. More recent analyses of chloroplast genomes from a variety of land plants and algae has revealed that this is not always the case. New genes not present in vascular plant chloroplast DNAs have been found one after another in chloroplast genomes from *Euglena*, marine algae and the cyanelles of *Cyanophora*. This gives impetus to diversify the range of plant species in which chloroplast genome analysis is undertaken. The accumulation of chloroplast DNA sequence data facilitates further analysis of chloroplast origin and evolution.

Gene structure

The average chloroplast genome contains about 120 kb of unique sequence, which is enough to encode 120 genes if one assumes that an average gene contains about 1 kb. The number of protein coding genes seems to be about 100 in addition to rRNA and tRNA genes [24].

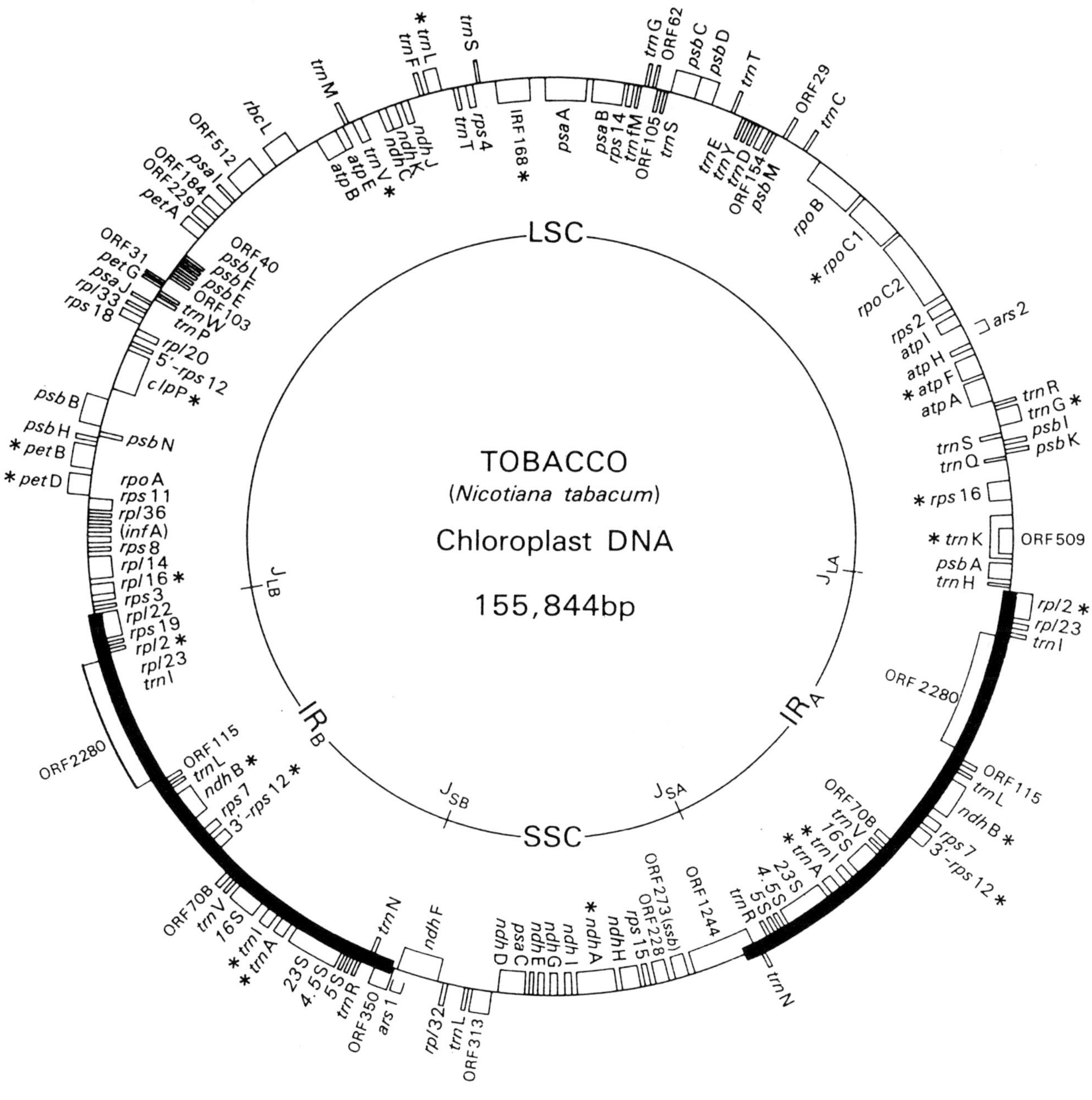

Fig. 1. Gene map of the tobacco chloroplast genome. Genes shown inside the circle are transcribed clockwise, and genes on the outside are transcribed counter-clockwise. Asterisks denote split genes. Major ORFs are included. IRF, intron-containing reading frame; IR, inverted repeat; LSC, large single-copy region; SSC, small single-copy region; J, junctions between IR and LSC/SSC. From the Research Grant Progress Report (1989) with minor revisions.

Initially, there were essentially three methods used to locate genes in the chloroplast genome. The first method was that of standard genetic analysis. Chloroplast genes coding for ribosomal proteins in *Chlamydomonas* and fraction I proteins in tobacco were analyzed by this method [95]. Analysis of the plastome mutants of *Antirrhinum* and *Pelargonium* suggested that some components of the thylakoid membrane are encoded in the chloroplast genome [46]. The second method was RNA-DNA hybridization experiments which demonstrated the presence of rRNA and tRNA genes in the chloroplast genome [4]. Third, the study of proteins synthesized by isolated chloroplasts was a powerful technique used to detect chloroplast-encoded proteins [26]. The analysis of ribosome-deficient chloroplast mutants also provided information on proteins synthesized within the chloroplast [46]. Later, the application of gene cloning and DNA sequencing technologies gave us the primary structure of chloroplast genes and predicted the presence of new chloroplast genes.

Identification of chloroplast genes has been pursued extensively in maize, wheat, spinach, pea, tobacco, *Euglena* and *Chlamydomonas*. Chloroplast DNAs are now known to contain all the chloroplast rRNA genes (3–5 genes), 30–31 tRNA genes and about 100 genes encoding proteins, which fits the number of genes estimated from the size of the chloroplast DNA. The chloroplast genes that have been sequenced (including putative genes) are presented in Table 1. Gene nomenclature follows the proposal of Hallick [47]. The progress of chloroplast DNA analysis has been rapid and exceeds that of chloroplast protein analysis. At present there are many open reading frames (ORFs), which are potential polypeptide genes in the chloroplast genome. Homology searches in protein databases have predicted that some of the ORFs are protein genes. However, the final identification of chloroplast genes encoding polypeptides requires the analysis of the translation products. *In vitro* transcription-translation of specific DNA fragments followed by immunoassay and the determination of partial amino acid sequences of isolated chloroplast proteins are the two principal procedures used for this purpose.

Genetic system genes

Ribosomal RNA genes

Chloroplasts contain a 70S class of ribosomes which are distinct from the 80S ribosomes found in the cytoplasm. A 23S, a 5S and a 4.5S rRNA are all associated with the 50S subunit. The 4.5S rRNA has been found in higher-plant chloroplasts [e.g. 12] and it is homologous to the 3′ end of the 23S rRNA of prokaryotes. Maize chloroplast rRNA genes (*rDNA*s) were the first chloroplast genes cloned [5]. Sequencing of maize and tobacco chloroplast *rDNA*s revealed a gene order of 16S-23S-4.5S-5S and an interspersion of tRNA genes within this cluster [e.g. 97, 112]. In *Chlamydomonas reinhardtii* the *rDNA* cluster consists of 16S, 7S, 3S, 23S and 5S in this order [93]. The presence of two small *rDNA*s (3S and 7S) is unique in this alga. The *Chlamydomonas* 23S *rDNA* has one intron and was the first split gene found in the chloroplast genome. Moreover, the 23S *rDNA* from *C. eugametos* contains six introns and three short internal transcribed spacers that are post-transcriptionally excised to yield four rRNA species [115]. In *Euglena* (strain Z), there are three copies of the *rDNA* cluster arranged tandemly and an extra copy of the 16S *rDNA* [e.g. 38].

The chloroplast *rDNA*s described above are all arranged as in *Escherichia coli* (16S-23S-5S). However, the *rDNA* cluster of *Chlorella ellipsoidea* is split into two back-to-back operons; operon 1 comprising 16S rRNA, $tRNA^{Ile}$(GAU) and operon 2 comprising $tRNA^{Ala}$(UGC), 23S rRNA, 5S rRNA [121]. In *P. littoralis* chloroplasts, the large DNA contains two sets of *rDNA* in reverse orientation while the small DNA has a 16S rRNA pseudogene and a split 23S rRNA sequence separated by at least 4 kb [76]. The sequences of individual chloroplast *rDNA* are highly homologous to each other and to those from eubacteria, but *rDNA* operons differ significantly when land plants and algae are compared. They vary in the presence or absence of introns, additional small *rDNA* species and in *rDNA* organization. This has led to speculation that the origin and evolutionary history of chloroplast genomes are diverse.

Transfer RNA genes

Saturation hybridization of a total chloroplast tRNA fraction to chloroplast DNA indicated the presence of 20–40 tRNA genes on the chloroplast genome [e.g. 45]. Chloroplast genomes are thus

Table 1. Chloroplast genes.

Genes	Products	Remarks
Genes for the genetic system		
23S rDNA	23S rRNA	
16S rDNA	16S rRNA	
7S rDNA	7S rRNA	in *Chlamydomonas*
5S rDNA	5S rRNA	
4.5S rDNA	4.5S rRNA	in land plants
3S rDNA	3S rRNA	in *Chlamydomonas*
trnA-UGC	Ala-tRNA (UGC)	
trnR-ACG	Arg-tRNA (ACG)	
trnR-UCU	Arg-tRNA (UCU)	
trnR-CCG	Arg-tRNA (CCG)	in liverwort
trnN-GUU	Asn-tRNA (GUU)	
trnD-GUC	Asp-tRNA (GUC)	
trnC-GCA	Cys-tRNA (GCA)	
trnQ-UUG	Gln-tRNA (UUG)	
trnE-UUC	Glu-tRNA (UUC)	
trnG-GCC	Gly-tRNA (GCC)	
trnG-UCC	Gly-tRNA (UCC)	
trnH-GUG	His-tRNA (GUG)	
trnI-GAU	Ile-tRNA (GAU)	
trnI-CAU	Ile-tRNA (CAU)	
trnL-UAA	Leu-tRNA (UAA)	
trnL-CAA	Leu-tRNA (CAA)	
trnL-UAG	Leu-tRNA (UAG)	
trnK-UUU	Lys-tRNA (UUU)	
trnfM-CAU	fMet-tRNA (CAU)	
trnM-CAU	Met-tRNA (CAU)	
trnF-GAA	Phe-tRNA (GAA)	
trnP-UGG	Pro-tRNA (UGG)	
trnS-GGA	Ser-tRNA (GGA)	
trnS-UGA	Ser-tRNA (UGA)	
trnS-GCU	Ser-tRNA (GCU)	
trnT-GGU	Thr-tRNA (GGU)	
trnT-UGU	Thr-tRNA (UGU)	
trnW-CCA	Trp-tRNA (CCA)	
trnY-GUA	Tyr-tRNA (GUA)	
trnV-GAC	Val-tRNA (GAC)	
trnV-UAC	Val-tRNA (UAC)	
tscA	small RNA	in *Chlamydomonas*
rps2	30S r-protein CS2	
rps3	CS3	
rps4	CS4	
rps7	CS7	2 pieces in *Chlamydomonas*
rps8	CS8	
rps9	CS9	in *Cryptomonas*
rps10	CS10	in *Cryptomonas*
rps11	CS11	
rps12	CS12	
rps14	CS14	
rps15	CS15	
rps16	CS16	
rps18	CS18	
rps19	CS19	
rpl2	50S r-protein CL2	
rpl5	CL5	in *Euglena*
rpl13	CL13	in *Cryptomonas*
rpl14	CL14	
rpl16	CL16	
rpl20	CL20	
rpl21	CL21	in liverwort
rpl22	CL22	not in legumes
rpl23	CL23	pseudogene in spinach
rpl32	CL32	
rpl33	CL33	
rpl36	CL36	
rpoA	RNA polymerase subunit α	
rpoB	subunit β	
rpoC1	subunit β′	
rpoC2	subunit β′′	
tufA	elongation factor Tu	in algae
infA	initiation factor 1	pseudogene in tobacco
clpP	ATP-dependent protease, proteolytic subunit	
Genes for the photosynthetic system		
rbcL	Rubisco large subunit	
rbcS	small subunit	in red and brown algae
psaA	PSI P700 apoprotein A1	
psaB	P700 apoprotein A2	
psaC	9 kDa protein	
psaI	I-protein	
psaJ	J-protein	
psbA	PSII D1-protein	
psbB	47 kDa protein	
psbC	43 kDa protein	
psbD	D2-protein	
psbE	cytochrome b559 (8 kDa)	
psbF	cytochrome b559 (4 kDa)	
psbH	10 kDa phosphoprotein	
psbI	I-protein	
psbK	K-protein	
psbL	L-protein	
psbM	M-protein	
psbN	N-protein	
petA	*b/f* complex cytochrome *f*	
petB	cytochrome *b6*	
petD	subunit IV	
petG	subunit V	
atpA	H^+-ATPase subunit CF_1 α	
atpB	subunit CF_1 β	
atpE	subunit CF_1 ε	
atpF	subunit CF_0 I	
atpH	subunit CF_0 III	
atpI	subunit CF_0 IV	
ndhA	NADH dehydrogenase ND1	
ndhB	ND2	
ndhC	ND3	
ndhD	ND4	
ndhE	ND4L	
ndhF	ND5	
ndhG	ND6	
ndhH	49 kDa protein	
ndhI (*frxB*)	18 kDa protein	
ndhJ		ORF158–169 in LSC
ndhK (*psbG*)	27 kDa protein	
frxC	31 kDa protein	

believed to encode all the tRNA species used in chloroplast protein synthesis, although plant mitochondria take up some of their tRNAs from the cytoplasm. The tRNA gene (*trn*) maps were constructed by hybridizing purified individual tRNAs to chloroplast DNA fragments [e.g. 23] followed by the sequencing of tRNA genes [e.g. 35, 86]. All the tobacco chloroplast DNA fragments that hybridized to total chloroplast tRNAs have been sequenced, and thirty different tRNA genes were found [116]. A subsequent search of the complete sequence of tobacco chloroplast DNA yielded no new tRNA genes. Hence these 30 tRNA genes are probably all of the tRNA genes encoded in the chloroplast genome of tobacco and perhaps of other higher plants. The liverwort genome contains an additional $tRNA^{Arg}$(CCG) sequence.

No chloroplast genes have been found which code for a tRNA with a 3′-CCA end. All the tRNAs deduced from the DNA sequences can form the clover leaf structure, and none exhibits an abnormal form. The presence of introns in chloroplast tRNA genes was first demonstrated in maize *trnI* and *trnA* located in the long spacer separating the 16S and 23S *rDNA*s [61]. Six chloroplast tRNA genes from land plants are now known to harbor long single introns (0.5–2.5 kb). The *trnG*-UCC gene contains an intron in the D-stem region, a feature unique to chloroplasts [21]. In contrast, no split tRNA genes have been found in algal chloroplasts.

In land plants, the tRNA genes are scattered over the chloroplast genome, while in *Euglena* most of the tRNA genes are clustered [49]. The *Euglena* genome contains an additional pseudo $tRNA^{Ile}$ gene in the *16S rDNA* leader region, the first pseudogene found in the chloroplast genome [86]. In monocot chloroplast genomes at least five pseudo tRNA genes have been found. These are located near the inversion endpoints, and the involvement of tRNA genes in genome inversions during evolution has been proposed [e.g. 100].

All 61 possible codons are used in chloroplast genes which encode polypeptides. The minimum number of tRNA species required for translation of all 61 codons is 32 if normal wobble base-pairing occurs in codon-anticodon recognition. As shown in Table 1, no tRNAs which recognize codons CUU/C(Leu), CCU/C(Pro), GCU/C(Ala) or CGC/A/G(Arg) [or CGC/A in liverwort] according to normal wobble base-pairing have been found. If the 'two-out-of-three' mechanism operates in the chloroplast, four single tRNA species, $tRNA^{Pro}$(UGG), $tRNA^{Ala}$(UGC) and $tRNA^{Arg}$(ACG) species can read all four Pro, Ala and Arg codons, respectively (note that these tRNAs form only GC pairs in their first and second codon-anticodon interactions). There is a tRNA gene in which the $tRNA^{Leu}$ anticodon is UAG and if this tRNA has an unmodified U in the first position of the anticodon, it can read all four Leu codons (CUN) by 'U:N wobble.' The bean, spinach and soybean $tRNAs^{Leu}$(UAG) have unmodified Us in their anticodons ($UA^{m7}G$) [90]. Thus, 30 tRNAs are probably sufficient to read all 64 codons [103]. These hypotheses have been supported by a recent study in which these four chloroplast tRNAs were purified from bean and their decoding properties were analyzed in a tRNA-dependent wheat germ protein synthesizing system [89].

Ribosomal protein genes

Chloroplast ribosomes contain about 60 different protein components, one-third of which are thought to be encoded by chloroplast DNA according to protein synthesis studies in isolated chloroplasts [e.g. 27]. Genes encoding chloroplast ribosomal proteins have been deduced through their homology with *E. coli* ribosomal protein genes [e.g. 110]. Twenty-one different ORFs potentially coding for polypeptides homologous to *E. coli* ribosomal proteins have been found in the tobacco, liverwort and rice chloroplast genomes [111]. The tobacco and rice genomes lack *rpl21* but contain *rps16* which has an intron. The black pine genome apparently lacks *rps16* (unpublished observations). Many of their translation products have been identified in spinach, pea and tobacco through partial amino acid sequencing [114]. Several nuclear-encoded chloroplast ribosomal proteins and their genes have also been analyzed. Among them two have no similarity to any bacterial ribosomal proteins, indicating the uniqueness of chloroplast ribosomes [32, 58].

The *rpl23*, *rpl2*, *rps19*, *rpl22*, *rps3*, *rpl16*, *rpl14*, *rps8*, *infA*, *rpl36*, *rps11* and *rpoA* genes are clustered in this order in the *rpl23* operon and the arrangement corresponds to that of the homologous genes in the *E. coli* S10, *spc* and α operons [e.g. 113] (Fig. 2). This raises the possibility that the genes for ribosomal proteins of higher plant chloroplasts and *E. coli* may have evolved from a common ancestral gene set. However, *rpl22* was not found in legume chloroplast DNAs [105]. *rpl23* is split into two overlapping reading frames in spinach and several other higher plants [128], and its translation product has been identified and shown to be functional in tobacco [123]. The *rpl23* operon in *Euglena* is also similar to the gene arrangement in the *E. coli* operons [17] (Fig. 2). *Euglena rpl5* is a new gene not reported for any land plant chloroplast genomes to date, as are *rpl13*, *rps9* and *rps10* in the marine alga *Cryptomonas* [22].

An intron within a potential chloroplast ribosomal protein gene was first found in *Nicotiana debneyi rpl2*, but this intron is absent in spinach and some related dicots [128]. The higher-plant *rps16* and *rpl16* sequences also contain single introns while *Euglena rpl16*, *rpl23*, *rps2*, *rps3*, *rps8*, *rps11* and *rps19* contain multiple introns. More unusual is *rps12* in land plants, which consists of three exons and requires *trans*-splicing for expression (see the next section). The putative *rps7* gene of *C. reinhardtii* is split into two segments. The 5′ and 3′ portions of *rps7* are separately transcribed and no common transcript has been detected. A Shine-Dalgarno sequence occurs upstream of the 3′ portion, but the consensus sequences for chloroplast introns are absent. These results suggest that *trans*-splicing is probably not involved in *rps7* expression, but rather that the protein may be made in two pieces [33].

Translational factors

A sequence similar to the *E. coli* EF-Tu gene (*tufA*) has been found upstream of *rps12*/*rps7* in *Euglena* [81] and several other algae, but not in any land plant chloroplast DNAs sequenced to date. A putative gene for the initiation factor IF-1 (*infA*) was found in the *rpl23* gene cluster between *rps8* and *rpl36* in spinach [104]. Among the *infA*s so far sequenced, a tobacco *infA* homologue does not constitute an ORF but is transcribed along with neighboring genes. Recently a cDNA potentially coding for chloroplast IF-1 has been isolated, suggesting that *infA* has transferred into the nucleus in tobacco (T. Wakasugi *et al.*, unpublished). It is curious that only one of the *rpl23* gene cluster (containing 12 genes) is duplicated

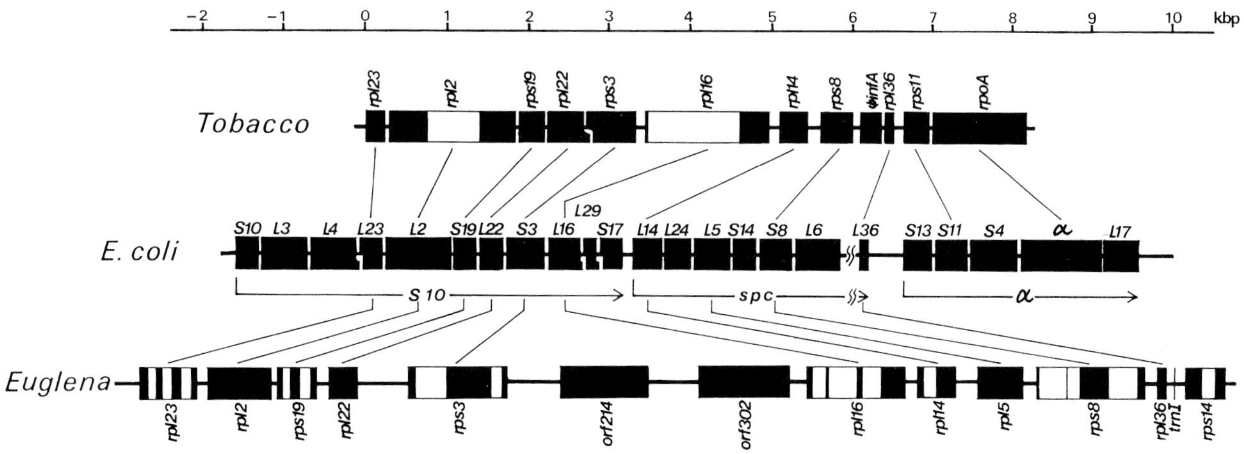

Fig. 2. Comparison of tobacco and *Euglena* chloroplast *rpl23* operons with the *E. coli* S10, *spc* and α operons. Exons are shown by filled boxes and introns by open boxes.

and transferred to the nucleus. A putative wheat chloroplast gene (*clpP*) encoding the proteolytic subunit of an ATP-dependent protease has been reported [37]. This enzyme degrades incomplete polypeptides and unassembled proteins in chloroplasts.

RNA polymerase subunit genes
It had been suggested that higher-plant chloroplast RNA polymerase is encoded in the nuclear genome [e.g. 71]. However, chloroplast DNA sequences hybridizing with the *E. coli* RNA polymerase genes were reported in *Chlamydomonas* [117]. Subsequent sequence analysis revealed that chloroplast DNA regions potentially coding for polypeptides similar to *E. coli* RNA polymerase α (*rpoA*), β (*rpoB*) and β' (*rpoC*) subunits were found in land plants [e.g. 84, 104]. An *E. coli rpoC* homologue is split into two parts, *rpoC1* and *rpoC2*, of which only *rpoC1* contains an intron [e.g. 85]. However in rice and maize, *rpoC1* is a continuous gene and *rpoC2* contains an extra sequence (380–450 bp) in the middle of it [57, 99]. No sequences similar to a bacterial *rpoD* have been found. The amino-terminal sequences of the 38 kDa, 120 kDa, 78 kDa and 180 kDa subunits of maize chloroplast RNA polymerase have recently been determined and found to correspond precisely to the sequences deduced from the maize *rpoA*, *rpoB*, *rpoC1* and *rpoC2* genes, respectively [55, 56]. These findings indicate that chloroplasts are the site of synthesis of some if not all of the chloroplast RNA polymerase subunits. We cannot rule out the possibility that a distinct species of RNA polymerase is imported from the cytoplasm (see the next section).

Photosynthetic system genes

Rubisco subunit genes
Ribulose-1,5-bisphosphate carboxylase/oxygenase (Rubisco) is the major stromal protein in chloroplasts and is composed of eight identical large subunits (LS) of 55 kDa and eight identical small subunits (SS) of 12 kDa. LS is encoded by the chloroplast DNA and SS is encoded by the nuclear DNA in higher plants and green algae. In contrast, the SS gene (*rbcS*) has been found in the chloroplast DNA from brown and red algae [e.g. 92]. The maize chloroplast gene for LS (*rbcL*) was the first chloroplast protein gene cloned and sequenced [78]. *rbcL* has become the most widely sequenced gene, enabling comparison for the determination of phylogenetic relationships among plant species [128]. The *rbcL* genes of higher plants and *Chlamydomonas* contain no introns while nine introns have been found in the *Euglena rbcL* gene [64]. In the chloroplast genomes which contain it, *rbcS* is located downstream from *rbcL* and constitutes an operon with *rbcL* as has been reported in cyanobacteria and cyanelles. No intron has been found in chloroplast *rbcS* genes while the nuclear *rbcS* genes have one to three introns.

Photosystem II genes
The thylakoid membranes have four distinct complexes [e.g. 36]: photosystems I and II (PSI, PSII), the cytochrome *b*/*f* complex and ATP synthase. The genes encoding thylakoid proteins have usually been isolated and identified through protein analysis, in which a protein component is purified and its antibody is prepared. A cloned DNA fragment is then placed in a transcription-translation system, or an isolated mRNA (or hybrid-selected mRNA) is translated in a cell-free system. The protein product is then identified using the specific antibody [e.g. 52]. After chloroplast DNA sequences became available, some genes were identified by western blotting analysis using antibodies against synthetic oligopeptides deduced from the DNA sequences [e.g. 106] and by comparing partial amino acid sequences of isolated proteins with the amino acid sequences derived from the DNA sequences [e.g. 28].

At least 12 components of PSII are encoded in the chloroplast genome. The gene for the 32 kDa protein Q_B or D1 protein (*psbA*) was the first PS gene sequenced in spinach and *Nicotiana debneyi* [126]. The 32 kDa protein binds to herbicides such as atrazine and DCMU. Therefore *psbA* is agronomically important and is another widely analyzed gene. The *psbA* genes isolated from herbicide-resistant mutants have point mutations at codon 264 of the protein which result in substitution of glycine or alanine for serine [e.g. 54]. In land plants all the PSII genes are continuous while some of the algal *psb* genes are split by one to six introns [e.g. 60]. In higher plants the *psbD* gene overlaps *psbC* by about 50 bp [e.g. 1], suggesting that chloroplasts must have a specific mechanism for producing a proper amount of each component of a given complex.

Photosystem I genes
Five components of PSI are encoded in the chloroplast genome. The genes for subunits A1 and A2 of the P700 chlorophyll *a* apoprotein (*psaA* and *psaB*) were first sequenced from maize [29]. The *psaA* and *psaB* genes in higher plants contain no introns, are situated tandemly and are about 45% homologous at the amino acid level. The predicted A1 and A2 products contain a leucine-zipper motif, which is probably involved in dimerization of these subunits [65, 118]. In *Chlamydomonas* the *psaA* gene is divided into three exons scattered around the chloroplast genome, while *psaB* is uninterrupted [68]. The three distantly separated exons of *psaA* produce a functional mRNA by *trans*-splicing (see the next section).

Cytochrome b/f *complex genes*
The cytochrome *b/f* complex consists of six components, four of which are encoded by the chloroplast genome [e.g. 51, 120]. The *petB* and *petD* genes are clustered with *psbB* and *psbH* in higher plants and they constitute a transcription unit [e.g. 52]. In higher plants both *petB* and *petD* contain single introns with short first exons (6–8 bp). In the green alga KS3/2 *petD* contains a 3.5 kb intron, the largest chloroplast intron reported to date, which has an ORF (608 codons) showing significant homology with reverse transcriptase genes [67].

ATP synthase genes
ATP synthase consists of two parts, CF_1 and CF_0. CF_1 is composed of five different subunits and CF_0 is composed of four different subunits. The genes for six subunits are present in the chloroplast genome. Genes for the β and ε subunits (*atpB* and *atpE*) were first sequenced from maize and spinach [66, 127]. The *atpB* and *atpE* genes in most higher plants overlap by 4 bp, so that the first two bases of the TGA stop codon of *atpB* and an A residue preceding it form the ATG initiation codon of *atpE*. The genes for the three CF_0 subunits (*atpI*, *atpH*, *atpF*) are clustered just before *atpA* [e.g. 8]. The deduced amino acid sequences of these six subunits show homology with their counterparts in *E. coli*.

ndh *genes*
Eleven chloroplast DNA sequences (*ndh*) whose predicted amino acid sequences resemble those of components of the respiratory-chain NADH dehydrogenase from mitochondria have been found in a variety of plants [e.g. 85, 103]. The *ndhA* and *ndhB* genes contain single introns. As most of these sequences are actively transcribed and the *ndhA* and *B* transcripts are spliced rapidly, they are likely to be the genes for components of a chloroplast NADH dehydrogenase [e.g. 77]. These observations suggest the existence of a respiratory-chain in chloroplasts, although it remains to be determined whether or not all of these transcripts are translated into functional proteins.

Gene expression

Transcription and promoters

RNA polymerase
Chloroplasts contain at least two different RNA polymerase activities, a soluble enzyme and a transcriptionally active chromosome (TAC) [40]. A soluble DNA-dependent RNA polymerase was highly purified from maize chloroplasts and its subunit composition was analyzed (see the previous section). A TAC was first isolated from *Euglena* chloroplasts [50]. RNA polymerase associated with TAC is tightly bound to chloroplast DNA and preferentially transcribes the rRNA genes.

For the precise initiation of transcription, RNA

polymerase requires additional factors. Such factors have been isolated from maize (S-factor [59], BF fraction [125]), from spinach (a σ-like polypeptide [70]) and from mustard [14]. In parallel with the characterization of RNA polymerases and their accessory factors, *in vitro* transcription systems have been developed to identify chloroplast promoters [e.g. 42, 73]. Chloroplast primary transcripts are known to harbor 5′-triphosphates which can be specifically labeled with [^{32}P]GTP and guanylyltransferase (*in vitro* capping). The *in vitro* capping assay can therefore identify the initiation site of transcription *in vivo*.

Promoter sequences

The upstream regions of many initiation sites determined by *in vitro* capping contain DNA sequences similar to the '– 10' and '– 35' *E. coli* promoter motifs. The identification and characterization of chloroplast promoters have been done using deleted and mutated genes and *in vitro* transcription systems [43, 44, 73]. These analyses have confirmed that '– 35'- and '– 10'-like sequences are required for proper transcription (Fig. 3). However, a class of chloroplast tRNA genes has been identified which do not require their 5′ upstream regions for transcription *in vitro* [41]. Relevant to this, one of the four primary transcripts for spinach *atpB* completely lacks an untranslated leader; the transcription start site is at the translation start codon [6]. The *psbA* genes in higher plants contain both prokaryotic-type '– 35' and '– 10' regions and between them a sequence motif similar to the nuclear TATA box. Mutation experiments have shown that the TATA box-like region is also critical for correct *psbA* transcription *in vitro* [25]. Thus at present the chloroplast genome is known to contain at least three structurally distinct upstream regions: regions containing the '– 35'/'– 10' promoter motifs, the TATA box-like promoter and no upstream promoter. Furthermore, they imply that there are multiple RNA polymerase species and/or multiple σ-like factors in chloroplasts. This is consistent with nuclear and chloroplast origins of chloroplast RNA polymerases (see 'Genetic system genes,' above).

Transcription from deleted or mutated genes has been studied *in vitro* as described above, but with recently developed systems for stably introducing foreign genes into chloroplast genomes it has become possible to do so *in vivo*. This method will permit testing of the conclusions derived from *in vitro* studies [e.g. 9, 13].

Transcript processing and introns

The chloroplast genome contains over 120 genes and about 50 transcription units, suggesting that

		"–35" "TATA" "–10"
Mustard	*psbA*	TTGGTTGACATGGCTATATAAGTCATGTTATACTGTTCAAT
Spinach	*psbA*	TTGGTTGACACGGGCATATAAGGCATGTTATACTGTTGAAT
	rbcL	TGGGTTGCGCCATATATATGAAAGAGTATACAATAATGATG
	atpB	TCTTGACAGTGGTATATGTTGTATATGTATATCCTAGATGT
	atpB (5)	ATTTTTGCAAAAAATTTCGACATACTTTACTATATATTATG (met)
	trnM	TTATATTGCTTATATATAATATTTGATTTATAATCAATCTA

Fig. 3. Chloroplast promoter regions identified by using deleted/mutated genes and *in vitro* transcription systems. Mustard *psbA* [25, 73] spinach *psbA* to *trnM* [43, 44] and spinach *atpB*(5) (the promoter for a fifth transcript starting at the translation initiation codon [6]) are shown.

chloroplast genes are generally cotranscribed. The detection of primary transcripts has actually shown that most of the chloroplast genes are transcribed polycistronically. Multiple transcripts are observed for most chloroplast gene clusters and these are mainly the results of multiple RNA processing of the primary transcripts. Processing of rRNA and tRNA precursors and of precursors from split genes is required to form functional RNA molecules. A couple of the chloroplast operons consist of functionally distinct genes; e.g. the *psbDC* operon contains two PSII genes and a tRNA gene [7], and the *psaA* operon has two PSI genes and a ribosomal protein gene [79]. Some chloroplast operons are known to be constitutively transcribed. These findings suggest that posttranscriptional processing of primary transcripts represents an important step in the control of chloroplast gene expression [40].

RNA cutting

Chloroplast polycistronic transcripts are generally processed into many overlapping shorter RNA species. Some of the shorter RNAs are monocistronic but others are not. Several transcript sets contain multiple 5′ ends, which result from the cutting of precursor RNAs and from multiple transcriptional initiation. Detailed analysis of polycistronic transcripts have been made for the rRNA gene cluster [e.g. 19], the tRNA gene cluster [39], the *psbB* operon [119] and the *psbDC* operon [e.g. 31]. For example, the RNA pattern of the spinach *psbB* operon (*psbB-psbH-petB-petD*) is complex and resolves into 18 major RNA species [119]. All RNA species arise from the cutting of 5.6 kb primary transcript rather than from multiple transcription initiation and termination events. Processing results ultimately in the formation of monocistronic mRNAs for each of the two PSII proteins and a dicistronic mRNA for both *pet* subunits. These mono- and dicistronic mRNAs are thought to be major translatable mRNAs. Almost all of the transcripts from the maize *psbB* operon cosediment with polysomes, suggesting that they are translated. Intercistronic cutting is not always required for translation of these RNAs [2].

Most chloroplast transcription units contain short inverted repeats at their 3′ ends, which were originally thought to function as transcription terminators. The role of such inverted repeats has been examined using an *in vitro* transcription system. It was found that these inverted repeats are ineffective as transcription terminators *in vitro* but serve as accurate and efficient RNA-processing signals [107]. The stability of RNAs containing inverted repeats at their 3′ ends is greatly enhanced.

The stability of chloroplast mRNAs and the protein interaction with their 3′ -inverted repeats have been investigated in spinach [e.g. 108], *Chlamydomonas* [e.g. 69, 109], barley [e.g. 30, 98] and mustard [83]. The observations suggest that nuclear-encoded proteins function in chloroplast mRNA maturation and differential mRNA stability, which are major control steps in chloroplast gene expression. Recently a 28 kDa RNA-binding protein which is responsible for processing the 3′ ends of chloroplast mRNA has been isolated [96] and related proteins containing RNA-binding domains have also been reported [72, 122].

Introns and splicing

Introns in chloroplast genes were first reported for the *23S rDNA* of *C. reinhardtii* [93]. Chloroplast genes which have been found to contain introns are listed in Table 2. Most genes possessing introns in higher plants contain single introns, while *Euglena* and *Chlamydomonas* polypeptide encoding genes have multiple introns [e.g. 48, 91]. Six chloroplast tRNA genes in higher plants have introns but none are known in algae. The presence of introns can be predicted by sequence homology with known genes (e.g. tRNA genes, ATP synthase genes, ribosomal protein genes) and by conserved intron boundary sequences (see below). However, experimental analyses are required to confirm the existence of introns and to determine the splice sites of pre-RNAs.

Introns found in chloroplast genes can be classified into four groups on the basis of the intron boundary sequences and possible secondary structures [16, 102]. Chloroplast group I introns

Table 2. Chloroplast genes containing introns.

Gene	Number of introns			Remarks
	higher plants	*Euglena*	*Chlamydomonas*	
23SrDNA	0	0	1	ORF, an intron in *Chlorella*
			6	in *C. eugametos*
16SrDNA	0	0	1	in *C. moewusii*
trnL-UAA	1	0	0	
trnI-GAU	1	0	0	
trnA-UGC	1	0	0	
trnV-UAC	1	0	0	
trnG-UCC	1	0	0	
trnK-UUU	1	0		ORF
rps2	0	4		
rps3	0	2	0	
rps8	0	3		
rps11	0	2	0	
rps12	3 exons	0	0	*trans*-splicing
rps14	0	1		
rps16	1			
rps19	0	2		
rpl2	1	0	0	no intron in spinach
rpl14	0	1	0	
rpl16	1	3	0	
rpl22	0	1		
rpl23	0	3		pseudogene in spinach
rpoB	0	8		
rpoC1	1	11		no intron in rice and maize
rpoC2	0	2		
tufA	no gene	3	0	
clpP	2			no intron in wheat and rice
rbcL	0	9	0	
psaA	0	3	3 exons	*trans*-splicing
psaB	0	6	0	
psbA	0	4	4	2 introns in *C. moewusii*
psbB	0	4		
psbC	0	1	0	ORF
psbE	0	2		
psbF	0	1		an-intron-within-an-intron
petB	1	1	0	no intron in *Chlorella*
petD	1		0	no intron in *Chlorella*
			1	3.5 kb intron in alga KS3/2
atpF	1	3		
ndhA	1			
ndhB	1			
IRF167–170	2			

ORF means the presence of an ORF in an intron. IRF indicates an intron-containing reading frame which is present in front of the *psaA* operon.

can be folded with a secondary structure typical of group I introns of fungal mitochondrial genes [11, 80]. The introns of *trnL*, the *23S rDNA* and *C. moewusii psbA* belong to this group. Introns of *trnI* and *trnA* in the 16S-23S spacer can be folded into a secondary structure which is similar to the postulated structure of group II introns in fungal mitochondrial genes, but their boundary sequences are different from those of chloroplast group III introns. Chloroplast group III introns have conserved boundary sequences GTGYGRY at the 5′ ends and RYCNAYY(Y)YNAY at the 3′ ends, and include introns in protein-encoding genes, *trnV*-UAC, *trnG*-UCC and *trnK*-UUU from higher plants. Their postulated secondary structures are similar to those of group II introns (therefore groups II and III are sometimes combined). It should be noted that introns in protein-encoding genes and some tRNA genes in higher plant chloroplasts show common features. This is not true of nuclear genes which are split. A fourth intron group has been described for *Euglena rpl14*, *rpl16*, *rpl23*, *rps2*, *rps3*, *rps8*, *rps11*, *rps14*, *rps19* and *tufA* [e.g. 16]. These introns are uniform in size (95–109 bp), share common features with each other and are distinct from chloroplast group I – III introns.

The tobacco gene for a ribosomal protein, CS12, is divided into one copy of *5′-rps12* and two copies of *3′-rps12*. *5′-rps12* contains exon 1 consisting of 38 codons and *3′-rps12* consists of exon 2 (78 codons), a 536 bp intron and exon 3 (7 codons). This gene structure was designated as a 'divided' gene [103]. The *5′*- and *3′-rps12* segments are separated by 28 kb and are transcribed independently. These two transcripts are spliced in *trans* to produce a mature mRNA for CS12 [e.g. 124] (Fig. 4). The 3′ -flanking sequence of exon 1 and the 5′ -flanking sequence of exon 2 fit the conserved boundary sequences of chloroplast group III introns. It is noteworthy that the tobacco *rps12* gene requires both *cis* and *trans* splicing in order to produce the mature mRNA. Liverwort *rps12* is divided into two parts; the *5′*- and *3′-rps12* segments are present in single copies and are located on opposing DNA strands [85]. The mRNA is also produced by *trans*-splicing [62].

The *Chlamydomonas psaA* gene is also divided into three parts [68]. The first exon of 30 codons is 50 kb away from the second exon (60 codons), which is itself 90 kb away from exon 3 (661 codons). All exons are flanked by the consensus intron boundary sequences. The three exons are transcribed independently as precursors, and the synthesis of mature *psaA* mRNA involves *trans* assembly of these three separate transcripts [e.g. 15]. Interestingly, exon 2 is cotranscribed with the upstream *psbD* gene, and *psaB* is uninterrupted as are *psaA* and *psaB* of higher plants (Fig. 4). At least one additional chloroplast locus (*tscA*) is required for *trans*-splicing of exons 1 and 2 and produces a small RNA of about 430 bases [34].

Chloroplast introns can be classified into three to four groups, suggesting the presence of multiple splicing pathways. Clear self-splicing of pre-RNAs from split chloroplast genes has not been demonstrated *in vitro*. The group III conserved intron boundary sequence is similar to that found in nuclear protein-encoding genes, suggesting that at least one group of chloroplast intron sequences is removed by a mechanism similar to that operating in the nucleus. Splicing of nuclear pre-mRNAs is catalyzed by protein-RNA complexes. RNA molecules are not thought to be imported into chloroplasts from the cytoplasm, which suggests that the RNA components of these complexes, if there are any, should be encoded in the chloroplast genome. Aside from *Chlamydomonas tscA* RNA, the tobacco chloroplast genome has been found to encode a small RNA species which is not tRNA or rRNA (unpublished observations). No *in vitro* splicing systems in chloroplasts are currently available. This makes it difficult to analyze individual steps in splicing and to detect factors involved in splicing in chloroplasts.

Conclusions

Thirty-eight different genes for RNA components and 74 different genes for polypeptides (including putative genes) have now been reported (see Table 1). Most of the putative genes (some *rps*/*rpl*,

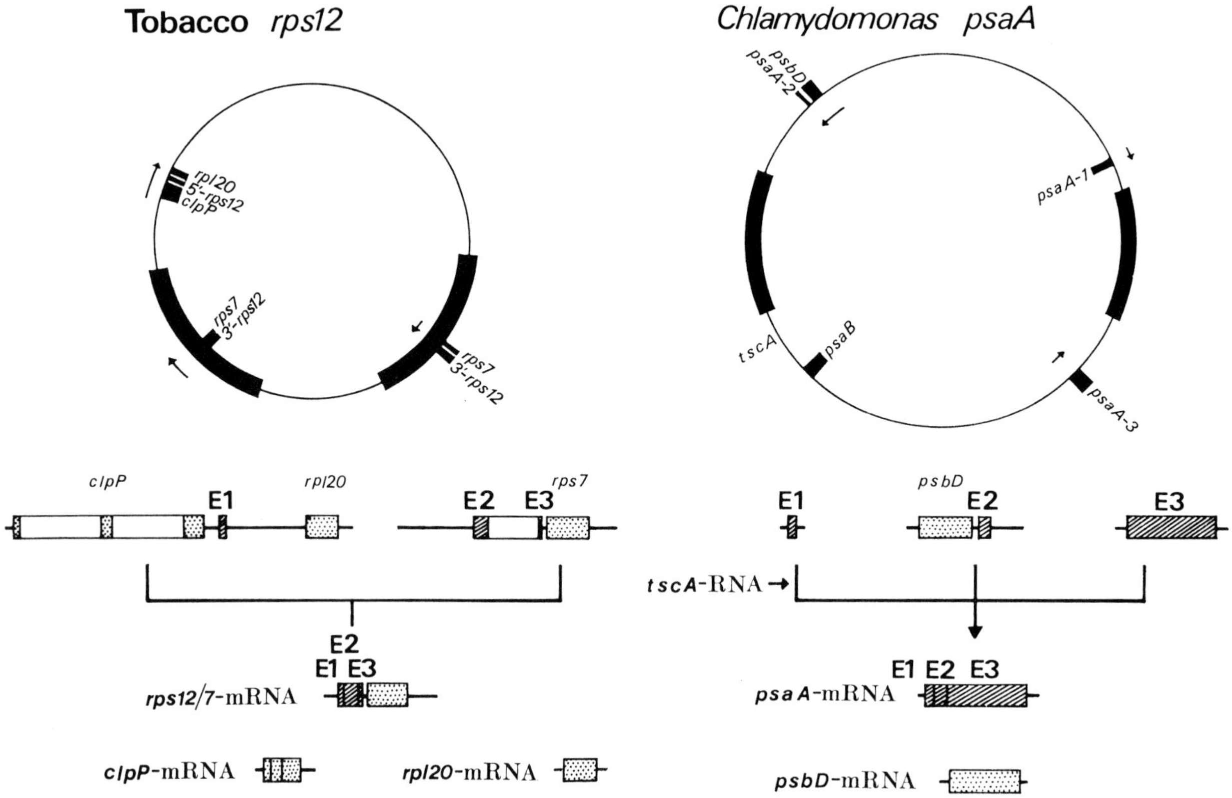

Fig. 4. Scheme for *rps12* and *psaA* mRNA maturation from separate pre-mRNAs [15, 124]. Upper circles show the location of genes and the direction of transcription. Bold lines, IR. In tobacco, *5'-rps12* contains exon 1 (E1), and *3'-rps12* has exon 2 (E2)-intron-exon 3 (E3). In *Chlamydomonas*, *psaA*-1, 2 and 3 contain exons 1, 2 and 3 (E1, E2, E3), respectively. The pathway is shown in the lower part. Slashed/dotted boxes indicate exons and open boxes, introns.

tuf, *inf*, *clp*, *ndh* and *frx*), which have been identified through homology with the corresponding genes of other organisms, are transcribed in chloroplasts, but their translation products remain to be isolated and characterized. There are still 26–34 ORFs (each over 29 codons in length) and twelve of them are conserved in size and sequence among the chloroplast genomes which have been completely sequenced [101]. Further efforts must be made to determine whether these ORFs (including putative genes) are functional chloroplast genes coding for polypeptides. Isolation of translation products followed by amino acid sequencing appears to be the best way to do so. There remain several long sequenced regions that have been assigned no genes and contain no significant ORFs. The *tscA* RNA was found in such a region from *Chlamydomonas* chloroplast DNA and further small RNA species might be encoded there.

The sequence analysis of algal chloroplast DNA and cyanelle DNA of *Cyanophora* has revealed the presence of new genes not found in land plant chloroplast DNAs as well as significant differences in genome organization in comparison both with each other and with land plants. Further analysis of chloroplast genomes from distantly related plant species as well as cyanelle and cyanobacterial genomes will provide the fundamental data needed to estimate the origins and process of chloroplast genome evolution as well as the phylogenetic relationships among plant species. Relevant to this, portions of chloroplast genomes are found in both nuclear and mitochondrial genomes (called 'promiscuous sequences'). The process of chloroplast DNA sequence transfer and the functional significance within the nucleus and the mitochondrion are interesting subjects for future study.

The molecular mechanism of chloroplast gene expression and its control are very interesting subjects. The number of RNA polymerase species in chloroplasts still remains to be answered. The structure of chloroplast genes and their modes of expression as presently understood exhibit both prokaryotic and eukaryotic features. This implies the presence of multiple RNA polymerase species. The control of chloroplast gene expression operates during several steps, transcription, post-transcription, translation and post-translation during chloroplast gene expression. It is also affected by environmental factors such as light. Recent studies suggest that the role of transcription in controlling gene expression is rather limited and that gene expression is more tightly controlled at the post-transcriptional level. This includes, RNA processing (cutting), RNA splicing and RNA stabilization. Chloroplast mRNA stability is currently being studied in detail.

The mechanism of chloroplast pre-RNA splicing is also very intriguing. For example, the tobacco chloroplast genome contains 18 distinct split genes and a comparable number of split genes have been found to date in *Euglena*. The introns found in chloroplast genes can be classified into three to four groups based on their structures. The largest group includes introns in some protein-encoding genes and in tRNA genes. Two genes, land plant *rps12* and *Chlamydomonas psaA* genes, require *trans*-splicing for expression. Some introns contain ORFs the predicted polypeptides of which show homology with maturases, reverse transcriptases and DNA endonucleases. Splicing of chloroplast pre-RNA is a most promising target for future study. For study of this process it is essential to develop *in vitro* splicing systems for chloroplasts. Recently chloroplast transformation systems have been developed in *Chlamydomonas* and tobacco and this method appears to be promising not only for *in vivo* the analysis of transcription but also for analysis of splicing pathways. At this time the method needs improvement before it can be used routinely.

The chloroplast contains a compact genetic system with circular DNAs of about 150 kb. This system has many interesting features as described in this paper. Many researchers have taken advantage of the relative simplicity of the chloroplast genetic system compared to nuclear and prokaryotic systems to produce new findings and hypotheses which are not only restricted to the plant world but apply to other organisms as well. It can be expected that further research on the chloroplast genome as a model system will continue to yield such exciting results.

References

1. Alt J, Morris J, Westhoff P, Herrmann RG: Nucleotide sequence of the clustered genes for the 44 kd chlorophyll *a* apoprotein and the '32 kd'-like protein of the photosystem II reaction center in the spinach plastid chromosome. Curr Genet 8: 597–606 (1984).
2. Barkan A: Proteins encoded by a complex chloroplast transcription unit are each translated from both monocistronic and polycistronic mRNAs. EMBO J 7: 2637–2644 (1988).
3. Bedbrook JR, Bogorad L: Endonuclease recognition sites mapped on *Zea mays* chloroplast DNA. Proc Natl Acad Sci USA 73: 4309–4319 (1976).
4. Bedbrook JR, Kolodner R: The structure of chloroplast DNA. Annu Rev Plant Physiol 30: 593–620 (1979).
5. Bedbrook JR, Kolodner R, Bogorad L: *Zea mays* chloroplast ribosomal RNA genes are part of a 22,000 base pair inverted repeat. Cell 11: 739–749 (1977).
6. Bennett DC, Rogers SA, Chen LJ, Orozco Jr EM: A primary transcript in spinach chloroplasts that completely lacks a 5′ untranslated leader region. Plant Mol Biol 15: 111–119 (1990).
7. Berends T, Gamble PE, Mullet JE: Characterization of the barley chloroplast transcription units containing psaA-psaB and psbD-psbC. Nucl Acids Res 15: 5217–5240 (1987).
8. Bird CR, Koller B, Auffret AD, Huttly AK, Howe CJ, Dyer TA, Gray JC: The wheat chloroplast gene for CF_0 subunit I of ATP synthase contains a large intron. EMBO J 4: 1381–1388 (1985).
9. Blowers AD, Ellmore GS, Klein U, Bogorad L: Transcriptional analysis of Endogenous and foreign genes in chloroplast transformants of *Chlamydomonas*. Plant Cell 2: 1059–1070 (1990).
10. Bogorad L, Vasil IK (eds): The Molecular Biology of Plastids. Academic Press, San Diego (1991).
11. Bonnard G, Michel F, Weil JH, Steinmetz A: Nucleotide sequence of the split tRNA-Leu (UAA) gene from *Vicia faba* chloroplasts: evidence for structural homologies of the chloroplast tRNA-Leu intron with the intron from the autosplicable *Tetrahymena* ribosomal RNA precursor. Mol Gen Genet 194: 330–336 (1984).

12. Bowman CM, Dyer TA: 4.5S ribonucleic acid, a novel ribosome component in the chloroplasts of flowering plants. Biochem J 183: 605–613 (1979).
13. Boynton JE, Gillham NW, Harris EH, Hosler JP, Johnson AM, Jones AR, Randolph-Anderson BL, Robertson D, Klein TM, Shark KB, Sanford JC: Chloroplast transformation in *Chlamydomonas* with high velocity microprojectills. Science 240: 1534–1538 (1988).
14. Bülow S, Link G: Sigma-like activity from mustard (*Sinapis alba* L.) chloroplasts conferring DNA-binding and transcription specificity to *E. coli* core RNA polymerase. Plant Mol Biol 10: 349–357 (1988).
15. Choquet Y, Goldschmidt-Clermont M, Girard-Bascou J, Kück U, Bennoun P, Rochaix JD: Mutant phenotypes support a *trans*-splicing mechanism for the expression of the tripartite *psaA* gene in the *C. reinhardtii* chloroplast. Cell 52: 903–913 (1988).
16. Christopher DA, Hallick RB: *Euglena gracilis* chloroplast ribosomal protein operon: a new chloroplast gene for ribosomal protein L5 and description of a novel organelle intron category designated group III. Nucl Acids Res 17: 7591–7607 (1989).
17. Christopher DA, Hallick RB: Complex RNA maturation pathway for a chloroplast ribosomal protein operon with an internal tRNA cistron. Plant Cell 2: 659–671 (1990).
18. Chun EHL, Vaugham MH, Rich A: The isolation and characterization of DNA associated with chloroplast preparations. J Mol Biol 7: 130–141 (1963).
19. Delp G, Igloi GL, Kössel H: Identification of *in vivo* processing intermediates and of splice junctions of tRNAs from maize chloroplasts by amplification with the polymerase chain reaction. Nucl Acids Res 19: 713–716 (1991).
20. Deng XW, Wing RA, Gruissem W: The chloroplast genome exists in multimeric forms. Proc Natl Acad Sci USA 86: 4156–4160 (1989).
21. Deno H and Sugiura M: Chloroplast tRNA-Gly gene contains a long intron in the D stem: nucleotide sequences of tobacco chloroplast genes for tRNA-Gly (UCC) and tRNA-Arg (UCU). Proc Natl Acad Sci USA 81: 405–408 (1984).
22. Douglas SE (1991): Unusual organization of a ribosomal protein operon in the plastid genome of *Cryptomonas* Φ: evolutionary considerations. Curr Genet 19: 289–294 (1991).
23. Driesel AJ, Crouse EJ, Gordon K, Bohnert HJ, Herrmann RG, Steinmetz A, Mubumbila M, Keller M, Burkard G, Weil JH: Fractionation and identification of spinach chloroplast transfer RNAs and mapping of their genes on the restriction map of chloroplast DNA. Gene 6: 285–306 (1979).
24. Dyer TA: The chloroplast genome: its nature and role in development. In: Baker NR, Barber J (eds) Chloroplast Biogenesis, pp. 23–69. Elsevier, Amsterdam (1984).
25. Eisermann A, Tiller K, Link G: *In vitro* transcription and DNA binding characteristics of chloroplast and etioplast extracts from mustard (*Sinapis alba*) indicate differential usage of the *psbA* promoter. EMBO J 9: 3981–3987 (1990).
26. Ellis RJ: Chloroplast proteins: synthesis, transport and assembly. Annu Rev Plant Physiol 32: 111–137 (1981).
27. Eneas-Filho J, Hartley MR, Mache R: Pea chloroplast ribosomal proteins: characterization and site of synthesis. Mol Gen Genet 184: 484–488 (1981).
28. Fish LE, Bogorad L: Identification and analysis of the maize P700 chlorophyll a proteins of PSI-A1 and PSI-A2 by high pressure liquid chromatography analysis and partial sequence determination. J Biol Chem 261: 8134–8139 (1986).
29. Fish LE, Kück U, Bogorad L: Two partially homologous adjacent light-inducible maize chloroplast genes encoding polypeptides of the P700 chlorophyll a-protein complex of photosystem I. J Biol Chem 260: 1413–1421 (1985).
30. Gamble PE, Mullet JE: Blue light regulates the accumulation of two *psbD-psbC* transcripts in barley chloroplasts. EMBO J 8: 2785–2794 (1989).
31. Gamble PE, Sexton TB, Mullet J: Light-dependent changes in *psbD* and *psbC* transcripts of barley chloroplasts: accumulation of two transcripts maintains *psbD* and *psbC* translation capability in mature chloroplasts. EMBO J 7: 1289–1297 (1988).
32. Gantt JS: Nucleotide sequences of cDNAs encoding four complete nuclear-encoded plastid ribosomal proteins. Curr Genet 14: 519–528 (1988).
33. Gillham NW, Harris EH, Randolph-Anderson BL, Boynton JE, Hauser CR, McElwain KB, Newman SM: Molecular genetics of chloroplast ribosomes in *Chlamydomonas*. In: Mache R, Stutz E, Subramanian AR (eds) The Translational Apparatus of Photosynthetic Organelles, pp. 127–144. Springer-Verlag, Berlin (1991).
34. Goldschmidt-Clermont M, Choquet Y, Girard-Bascou J, Michel F, Schirmer-Rahire M, Rochaix JD: A small chloroplast RNA may be required for *trans*-splicing in *Chlamydomonas reinhardtii*. Cell 65: 135–143 (1991).
35. Graf L, Kössel H, Stutz E: Sequencing of 16S-23S spacer in a ribosomal RNA operon of *Euglena gracilis* chloroplast DNA reveals two tRNA genes. Nature 286: 908–910 (1980).
36. Gray JC: Genetics and synthesis of chloroplast membrane proteins in Photosynthesis. In: Amesz J (ed) Photosynthesis, pp. 319–342. Elsevier, Amsterdam (1987).
37. Gray JC, Hird SM, Dyer TA: Nucleotide sequence of a wheat chloroplast gene encoding the proteolytic subunit of an ATP-dependent protease. Plant Mol Biol 15: 947–950 (1990).
38. Gray PW, Hallick RB: Physical mapping of the *Euglena gracilis* chloroplast DNA and ribosomal RNA gene region. Biochem 17: 284–289 (1978).
39. Greenberg BM, Gruissem W, Hallick RB: Accurate pro-

cessing and pseudouridylation of chloroplast transfer RNA in a chloroplast transcription system. Plant Mol Biol 3: 97–109 (1984).

40. Gruissem W: Chloroplast RNA: transcription and processing. In Marcus A (ed) Biochemistry of Plants, pp. 151–191. Academic Press, San Diego (1989).
41. Gruissem W, Elsner-Menzel C, Latshaw S, Narita JO, Schaffer MA, Zurawski G: A subpopulation of spinach chloroplast tRNA genes does not require upstream promoter elements for transcription. Nucl Acids Res 14: 7541–7556 (1986).
42. Gruissem W, Greenberg BM, Zurawski G, Prescott DM, Hallick RB: Biosynthesis of chloroplast transfer RNA in a spinach chloroplast transcription system. Cell 35: 815–828 (1983).
43. Gruissem W, Zurawski G: Identification and mutational analysis of the promoter for a spinach chloroplast transfer RNA gene. EMBO J 4: 1637–1644 (1985).
44. Gruissem W, Zurawski G: Analysis of promoter regions for the spinach chloroplast *rbc* L, *atp* B and *psb*A genes. EMBO J 4: 3375–3383 (1985).
45. Haff LA, Bogorad L: Hybridization of maize chloroplast DNA with transfer ribonucleic acids. Biochem 15: 4105–4109 (1976).
46. Hagemann R: Genetics and molecular biology of plastids of higher plants. In: Stadler Symposium vol. 11, pp. 91–116. University of Missouri, Columbia (1979).
47. Hallick RB: Proposals for the naming of chloroplast genes. II Update to the nomenclature of genes for thylakoid membrane polypeptides. Plant Mol Biol Rep 7: 266–275 (1989).
48. Hallick RB, Gingrich JC, Johanningmeier U, Passavant CW: Introns in *Euglena* and *Nicotiana* chloroplast protein genes. In: van Vloten-Doting L, Groot GSP, Hall TC (eds) Molecular Form and Function of the Plant Genome, pp, 211–231. Plenum Press, New York (1985).
49. Hallick RB, Hollingsworth MJ, Nickoloff JA: Transfer RNA genes of *Euglena gracilis* chloroplast DNA. Plant Mol Biol 3: 169–175 (1984).
50. Hallick RB, Lipper C, Richards OD, Rutter WJ: Isolation of a transcriptionally active chromosome from chloroplasts of *Euglena gracilis*. Biochem 15: 3039–3045 (1976).
51. Heinemeyer W, Alt J, Herrmann RG: Nucleotide sequence of the clustered genes for apocytochrome b6 and subunit 4 of the cytochrome *b*/*f* complex in the spinach plastid chromosome. Curr Genet 8: 543–549 (1984).
52. Herrmann RG, Westhoff P, Alt J, Tittgen J and Nelson N: Thylakoid membrane protein and their genes. In: van Vloten-Doting L, Groot GSP, Hall TC (eds) Molecular Form and Function of the Plant Genome, pp. 233–256. Plenum Press, New York (1985).
53. Hiratsuka J, Shimada H, Whittier RF, Ishibashi T, Sakamoto M, Mori M, Kondo C, Honji Y, Sun CR, Meng BY, Li Y, Kanno A, Nishizawa Y, Hirai A, Shinozaki K, Sugiura M: The complete sequence of the rice (*Oryza sativa*) chloroplast genome: intermolecular recombination between distinct tRNA genes accounts for a major plastid DNA inversion during the evolution of the cereals. Mol Gen Genet 217: 185–194 (1989).
54. Hirschberg J, McIntosh L: Molecular basis of herbicide resistance in *Amaranthus hybridus*. Science 222: 1346–1349 (1983).
55. Hu J, Bogorad L: Maize chloroplast RNA polymerase: the 180-, 120-, and 38-kilodalton polypeptides are encoded in chloroplast genes. Proc Natl Acad Sci USA 87: 1531–1535 (1990).
56. Hu J, Troxler RF, Bogorad L: Maize chloroplast RNA polymerase: the 78-kilodalton polypeptide is encoded by the plastid *rpoC1* gene. Nucl Acids Res 19: 3431–3434 (1991).
57. Igloi GL, Meinke A, Döry I, Kössel H: Nucleotide sequence of the maize chloroplast rpo $B/C_1/C_2$ operon: comparison between the derived protein primary structures from various organisms with respect to functional domains. Mol Gen Genet 221: 379–394 (1990).
58. Johnson CH, Kruft V, Subramanian AR: Identification of a plastid-specific ribosomal protein in the 30 S subunit of chloroplast ribosomes and isolation of the cDNA clone encoding its cytoplasmic precursor. J Biol Chem 265: 12790–12795 (1990).
59. Jolly SO, Bogorad L: Preferential transcription of cloned maize chloroplast DNA sequences by maize chloroplast RNA polymerase. Proc Natl Acad Sci USA 77: 822–826 (1990).
60. Karabin GD, Farley M, Hallick RB: Chloroplast gene for M_r 32 000 polypeptide of photosystem II in *Euglena gracilis* is interrupted by four introns with conserved boundary sequences. Nucl Acids Res 12: 5801–5812 (1984).
61. Koch W, Edwards K, Kössel H: Sequencing of the 16S-23S Spacer in a ribosomal RNA operon of zea mays chloroplast DNA reveals two split tRNA genes. Cell 25: 203–213 (1981).
62. Kohchi T, Umesono K, Ogura Y, Komine Y, Nakahigashi K, Komano T, Yamada Y, Ozeki H, Ohyama K: A nicked group II intron and *trans*-splicing in liverwort, *Marchantia polymorpha*, chloroplasts. Nucl Acids Res 16: 10025–10036 (1988).
63. Koller B, Delius H: *Vicia faba* chloroplast DNA has only one set of ribosomal RNA genes as shown by partial denaturation mapping and R-loop analysis. Mol Gen Genet 178: 261–269 (1980).
64. Koller B, Gingrich JC, Stiegler GL, Farley MA, Delius H, Hallick RB: Nine introns with conserved boundary sequences in the *Euglena gracilis* chloroplast ribulose-1,5-bisphosphate carboxylase gene. Cell 36: 545–553 (1984).
65. Kössel H, Döry I, Igloi G, Maier R: A leucine-zipper motif in photosystem I. Plant Mol Biol 15: 497–499 (1990).
66. Krebbers ET, Larrinua IM, McIntosh L, Bogorad L:

The maize chloroplast genes for the β and ε subunits of the photosynthetic coupling factor CF1 are fused. Nucl Acids Res 10: 4985–5002 (1982).

67. Kück U: The intron of a plastid gene from a green alga contains an open reading frame for a reverse transcriptase-like enzyme. Mol Gen Genet 218: 257–265 (1989).
68. Kück U, Choquet Y, Schneider M, Dron M, Bennoun P: Structural and transcriptional analysis of two homologous genes for the P700 chlorophyll *a*-apoproteins in *Chlamydomonas reinhardtii*: evidence for *in vivo trans*-splicing. EMBO J 6: 2185–2195 (1987).
69. Kuchka MR, Goldschmidt-Clermont M, van Dillewijn J, Rochaix JD: Mutation at the *Chlamydomonas* nuclear *NAC2* locus specifically affects stability of the chloroplast *psbD* transcript encoding polypeptide D2 of PS II. Cell 58: 869–876 (1989).
70. Lerbs S, Bräutigam E, Mache R: DNA-dependent RNA polymerase of spinach chloroplasts: characterization of α-like and σ-like polypeptides. Mol Gen Genet 211: 459–464 (1988).
71. Lerbs S, Bräutigam E, Parthier B: Polypeptides of DNA-dependent RNA polymerase of spinach chloroplast: characterization by antibody-linked polymerase assay and determination of sites of synthesis. EMBO J 4: 1661–1666 (1985).
72. Li Y, Sugiura M: Three distinct ribonucleoproteins from tobacco chloroplasts: each contains a unique amino terminal acidic domain and two ribonucleoprotein consensus motifs. EMBO J 9: 3059–3066 (1990).
73. Link G: DNA sequence requirements for the accurate transcription of a protein-coded plastid gene in a plastid *in vitro* system from mustard (*Sinapis alba* L.). EMBO J 3: 1697–1704 (1984).
74. Loiseaux-de Goër S, Markowicz Y, Dalmon J, Audren H: Physical maps of the two circular plastid DNA molecules of the brown alga *Pylaiella littoralis* (L.) Kjellm. Curr Genet 14: 155–162 (1988).
75. Manning JE, Wolstenholme DR, Ryan RS, Hunter JA, Richards OC: Circular chloroplast DNA from *Euglena gracilis*. Proc Natl Acad Sci USA 68: 1169–1173 (1971).
76. Markowicz Y, Loiseaux-de Goër S, Mache R: Presence of a 16S rRNA *pseudogene* in the bi-molecular plastid genome of the primitive brown alga *Pylaiella littoralis*. Evolutionary implication. Curr Genet 14: 599–608 (1988).
77. Matsubayashi T, Wakasugi T, Shinozaki K, Shinozaki KY, Zaita N, Hidaka T, Meng BY, Ohto C, Tanaka M, Kato A, Maruyama T, Sugiura M: Six chloroplast genes (*ndh A* to *F*) homologous to human mitochondrial genes encoding components of the respiratory-chain NADH dehydrogenase are actively expressed: determination of the splice-sites in *ndh A* and *ndh B* pre- mRNAs. Mol Gen Genet 210: 385–393 (1987).
78. McIntosh L, Poulsen C, Bogorad L: Chloroplast gene sequence for the large subunit of ribulose bisphosphate carboxylase of maize. Nature 288: 556–560 (1980).
79. Meng BY, Tanaka M, Wakasugi T, Ohme M, Shinozaki K, Sugiura M: Cotranscription of the genes encoding two P700 chlorophyll *a* apoproteins with the gene for ribosomal protein CS14: determination of the transcriptional initiation site by *in vitro* capping. Curr Genet 14: 395–400 (1988).
80. Michel F, Dujon B: Conservation of RNA secondary structure in two intron families including mitochondrial-, chloroplast- and nuclear-encoded members. EMBO J 2: 33–38 (1983).
81. Montandon PE, Stutz E: The genes for the ribosomal proteins S12 and S7 are clustered with the gene for the EF-Tu protein on the chloroplast genome of *Euglena gracilis*. Nucl Acids Res 12: 2851–2859 (1984).
82. Mullet JE: Chloroplast development and gene expression. Annu Rev Plant Physiol 39: 475–502 (1988).
83. Nickelsen J, Link G: Interaction of a 3′ RNA region of the mustard *trnK* gene with chloroplast proteins. Nucl Acids Res 17: 9637–9648 (1989).
84. Ohme M, Tanaka M, Chunwongse J, Shinozaki K, Sugiura M: A tobacco chloroplast DNA sequence possibly coding for a polypeptide similar to *E. coli* RNA polymerase β subunit. FEBS Lett 200: 87–90 (1986).
85. Ohyama K, Fukuzawa H, Kohchi T, Shirai H, Sano T, Sano S, Umesono K, Shiki Y, Takeuchi M, Chang Z, Aota S, Inokuchi H, Ozeki H: Chloroplast gene organization deduced from complete sequence of liverwort *Marchantia polymorpha* chloroplast DNA. Nature 322: 572–574 (1986).
86. Orozco EM Jr, Rushlow KE, Dodd JR, Hallick RB: *Euglena gracilis* chloroplast ribosomal RNA transcription units II. Nucleotide sequence homology between the 16S-23S ribosomal RNA spacer and the 16S ribosomal RNA leader regions. J Biol Chem 255: 10997–11003 (1980).
87. Palmer JD: Chloroplast DNA exists in two orientations. Nature 301: 92–93 (1983).
88. Palmer JD: Comparative organization of chloroplast genomes. Annu Rev Genet 19: 325–354 (1985).
89. Pfitzinger H, Weil JH, Pillay DTN, Guillemaut P: Codon recognition mechanisms in plant chloroplasts. Plant Mol Biol 14: 804–814 (1990).
90. Pillay DTN, Guillemaut G, Weil JH: Nucleotide sequences of three soybean chloroplast tRNAs-Leu and re-examination of bean chloroplast $tRNA_2$-Leu sequence. Nucl Acids Res 12: 2997–3001 (1984).
91. Plant AL, Gray JC: Introns in chloroplast protein-coding genes of land plants. Photosynth Res 16: 23–39 (1988).
92. Reith M, Cattolico RA: Inverted repeat of *Olisthodiscus luteus* chloroplast DNA contains genes for both subunits of ribulose-1,5-bisphosphate carboxylase and the 32 000-dalton Q_B protein: phylogenetic implications. Proc Natl Acad Sci USA 83: 8599–8603 (1986).
93. Rochaix JD, Malnoe P: Anatomy of the chloroplast ribosomal DNA of *Chlamydomonas reinhardtii*. Cell 15: 661–670 (1978).

94. Sager R, Ishida MR: Chloroplast DNA in *Chlamydomonas*. Proc Natl Acad Sci USA 50: 725–730 (1963).
95. Sager R, Schlanger G: Chloroplast DNA: physical and genetic studies. In: King RC (ed) Handbook of Genetics, vol. 5, pp. 371–423. Plenum Press, New York (1976).
96. Schuster G, Gruissem W: Chloroplast mRNA 3′ end processing requires a nuclear-encoded RNA-binding protein. EMBO J 10: 1493–1502 (1991).
97. Schwarz Z, Kössel H: The primary structure of 16S rDNA from *Zea mays* chloroplast is homologous to *E. coli* 10S rRNA. Nature 283: 739–742 (1980).
98. Sexton TB, Christopher DA, Mullet JE: Light-induced switch in barley *psbD-psbC* promoter utilization: a novel mechanism regulating chloroplast gene expression. EMBO J 9: 4485–4494 (1990).
99. Shimada H, Fukuta M, Ishikawa M, Sugiura M: Rice chloroplast RNA polymerase genes: the absence of an intron in *rpoC1* and the presence of an extra sequence in *rpoC2*. Mol Gen Genet 221: 395–402 (1990).
100. Shimada H, Sugiura M: Pseudogenes and short repeated sequences in the rice chloroplast genome. Curr Genet 16: 293–301 (1989).
101. Shimada H, Sugiura M: Fine structural features of the chloroplast genome: comparison of the sequenced chloroplast genomes. Nucl Acids Res 19: 983–995 (1991).
102. Shinozaki K, Deno H, Sugita M, Kuramitsu S, Sugiura M: Intron in the gene for the ribosomal proteins S16 of tobacco chloroplast and its conserved boundary sequences. Mol Gen Genet 202: 1–5 (1986).
103. Shinozaki K, Ohme M, Tanaka M, Wakasugi T, Hayashida N, Matsubayashi T, Zaita N, Chunwongse J, Obokata J, Shinozaki, KY, Ohto C, Torazawa K, Meng BY, Sugita M, Deno H, Kamogashira T, Yamada K, Kusuda J, Takaiwa F, Kato A, Tohdoh N, Shimada H, Sugiura M: The complete nucleotide sequence of the tobacco chloroplast genome: its gene organization and expression. EMBO J 5: 2043–2049 (1986).
104. Sijben-Müller G, Hallick R, Alt J, Westhoff P, Herrmann RG: Spinach plastid genes coding for initiation factor IF-1, ribosomal protein S11 and RNA polymerase α-subunit. Nucl Acids Res 14: 1029–1044 (1986).
105. Spielmann A, Roux E, von Allmen JM, Stutz E: The soybean chloroplast genome: complete sequence of the *rps*19 gene, including flanking parts containing exon 2 of *rpl* 2 (upstream), but lacking *rpl* 22 (downstream). Nucl Acids Res 16: 1199 (1988).
106. Steinmetz AA, Castroviejo M, Sayre RT, Bogorad L: Protein PSII-G: an additional component of photosystem II identified through its plastid gene in maize. J Biol Chem 261: 2485–2488 (1986).
107. Stern DB, Gruissem W: Control of plastid gene expression: 3′ inverted repeats act as mRNA processing and stabilizing elements, but do not terminate transcription. Cell 51: 1145–1157 (1987).
108. Stern DB, Jones H, Gruissem W: Function of plastid mRNA 3′ inverted repeats: RNA stabilization and gene-specific protein binding. J Biol Chem 264: 18742–18750 (1989).
109. Stern DB, Radwanski ER, Kindle KL: A 3′ stem/loop structure of the *Chlamydomonas* chloroplast *atpB* gene regulates mRNA accumulation *in vivo*. Plant Cell 3: 285–297 (1991).
110. Sugita M, Sugiura M: Putative gene of tobacco chloroplast coding for ribosomal protein similar to *E. coli* ribosomal protein S19. Nucl Acids Res 11: 1913–1918 (1983).
111. Sugiura M, Torazawa K, Wakasugi T: Chloroplast genes coding for ribosomal proteins in land plants. In: Mache R, Stutz E, Subramanian AR (eds) The Translational Apparatus of Photosynthetic Organelles, pp. 59–69. Springer-Verlag, Berlin (1991).
112. Takaiwa F, Sugiura M: The nucleotide sequence of 4.5S and 5S ribosomal RNA genes from tobacco chloroplasts. Mol Gen Genet 180: 1–4 (1980).
113. Tanaka M, Wakasugi T, Sugita M, Shinozaki K, Sugiura M: Genes for the eight ribosomal proteins are clustered on the chloroplast genome of tobacco (*Nicotiana tabacum*): similarity to the S10 and *spc* operons of *Escherichia coli*. Proc Natl Acad Sci USA 83: 6030–6034 (1986).
114. Thomas F, Massenet O, Dorne AM, Briat JF, Mache R: Expression of the *rpl23*, *rpl2* and *rps19* genes in spinach chloroplasts. Nucl Acids Res 16: 2461–2472 (1988).
115. Turmel M, Boulanger J, Schnare MN, Gray MW, Lemieux C: Six group I introns and three internal transcribed spacers in the chloroplast large subunit ribosomal RNA gene of the green alga *Chlamydomonas eugametos*. J Mol Biol 218: 293–311 (1991).
116. Wakasugi T, Ohme M, Shinozaki K, Sugiura M: Structures of tobacco chloroplast genes for tRNA-Ile (CAU), tRNA-Leu (CAA), tRNA-Cys (GCA), tRNA-Ser (UGA) and tRNA-Thr (GGU): a compilation of tRNA genes from tobacco chloroplasts. Plant Mol Biol 7: 385–392 (1986).
117. Watson JC, Surzycki SJ: Both the chloroplast and nuclear genomes of *Chlamydomonas reinhardtii* share homology with *Escherichia coli* genes for transcriptional and translational components. Curr Genet 7: 201–210 (1983).
118. Webber AN, Malkin R: Photosystem I reaction-centre proteins contain leucine zipper motifs: a proposed role in dimer formation. FEBS Lett 264: 1–4 (1990).
119. Westhoff P, Herrmann RG: Complex RNA maturation in chloroplasts: the *psb*B operon from spinach. Eur J Biochem 171: 551–564 (1988).
120. Willey DL, Auffret AD, Gray JC: Structure and topology of cytochrome *f* in pea chloroplast membranes. Cell 36: 555–562 (1984).
121. Yamada T, Shimaji M: Peculiar feature of the organization of rRNA genes of the *Chlorella* chloroplast DNA. Nucl Acids Res 14: 3827–3839 (1986).
122. Ye L, Li Y, Fukami-Kobayashi K, Go M, Konishi T,

Watanabe A, Sugiura M: Diversity of a ribonucleoprotein family in tobacco chloroplasts: two new chloroplast ribonucleoproteins and a phylogenetic tree of ten chloroplast RNA-binding domains. Nucl Acids Res 19: 6485–6490 (1991).
123. Yokoi F, Tanaka M, Wakasugi T, Sugiura M: The chloroplast gene for ribosomal protein CL23 is functional in tobacco. FEBS Lett 281: 64–66 (1991).
124. Zaita N, Torazawa K, Shinozaki K, Sugiura M: *Trans* splicing *in vivo*: joining of transcripts from the 'divided' gene for ribosomal protein S12 in the chloroplasts of tobacco. FEBS Lett 210: 153–156 (1987).
125. Zaitlin D, Hu J, Bogorad L: Binding and transcription of relaxed DNA templates by fractions of maize chloroplast extracts. Proc Natl Acad Sci USA 86: 876–880 (1989).
126. Zurawski G, Bohnert HJ, Whitfeld PR, Bottomley W: Nucleotide sequence of the gene for the M_R 32,000 thylakoid membrane protein from *Spinacia oleracea* and *Nicotiana debneyi* predicts a totally conserved primary translation product of M_R 38,950. Proc Natl Acad Sci USA 79: 7699–7703 (1982).
127. Zurawski G, Bottomley W, Whitfeld PR: Structure of the genes for the β and ε subunit of spinach chloroplast ATPase indicates a dicistronic mRNA and an overlapping translation stop/start signal. Proc Natl Acad Sci USA 79: 6260–6264 (1982).
128. Zurawski G, Clegg MT: Evolution of higher-plant chloroplast DNA-encoded genes: implications for structure-function and phylogenetic studies. Annu Rev Plant Physiol 38: 391–418 (1987).

Plant Molecular Biology **19**: 169–191, 1992.
© 1992 *Kluwer Academic Publishers. Printed in Belgium.*

The biochemistry and molecular biology of plant lipid biosynthesis

Antoni R. Slabas and Tony Fawcett
Plant Molecular Biology Group, Department of Biological Sciences, South Road, University of Durham, Durham DH1 3LE, UK

Key words: Oil seed rape, lipids, fatty acid synthesis, biotechnology, desaturases, mutants

The importance of lipids

In the areas of cell biology, biophysics and the biochemistry of signal perception and transduction major advances have come about from an appreciation of the importance of biological membranes. Such membranes not only provide a barrier to the outside of the cell but also define the limits of subcellular organelles contained within them. Subcellular organelles have been fractionated and biochemical compartmentalization of enzyme function deduced by classical work of de Duve and colleagues [16]. Initial electron microscopy observations concentrated on the lipid bilayer description of a biological membrane [18]. Subsequent studies by Singer and Nicholson [114] have led to an appreciation of the importance of the proteins in membranes resulting in the fluid mosaic model. The importance of biological membranes has been additionally highlighted with the development of the chemiosmotic hypothesis of energy generation in biological systems. The systems have as a central factor the insertion of an ATPase in a membrane, so allowing vectoral transport [66]. Apart from the lipid bilayer providing the correct environment for ATPase to function, it is now becoming increasingly apparent that many other enzymes require a lipid environment in order to function. An example of this is the bacterial glycerol-3-P acyltransferase (G-3P-AT). The homogeneous protein was found to be inactive, but activity was regained on the addition of sonically disrupted *Escherichia coli* phospholipids, pure phospholipids or mixtures of phospholipids [58]. The enzyme could also be reconstituted in single walled vesicles; chymotrypsin treatment of which caused 95% inactivation of enzymic activity, suggesting that G-3P-AT was inserted assymetrically with the active site facing outwards [24].

The importance of lipids is further exemplified by (1) the cutin coat of plants which forms a barrier against fungal attack [53], (2) sulphated lipo-oligosaccharides of *Rhizobium meliloti* which act as signals to elicit root nodule organogenesis [59] and (3) the effect of desaturation on chilling tolerance, as demonstrated very elegantly by the group of Murata [146]. From an industrial perspective, vegetable oils are a major raw resource for both food and detergent based products and have an increasingly important role in the chemical industry. Major sources are soybean, sunflower, rape and oil palm in which lipids constitute up to 40% of the dry weight of the seed. Different fatty acids may be esterified to the glycerol backbone to generate the enormous variety of natural triglycerides which have been reported [36]. It is thus clear that great potential exsists for genetically engineering plant lipids providing that there is sufficient information on the biochemical pathways involved in addition to the temporal and tissue specific expression of the genes.

It is somewhat surprising, given the central importance of lipids in plants that initially advances in understanding their biosynthesis were restricted to a few pioneering laboratories. In this review we will consider advances made over the last 10 years in the understanding of plant lipid biosynthesis and its regulation. As two excellent reviews have recently appeared we suggest that

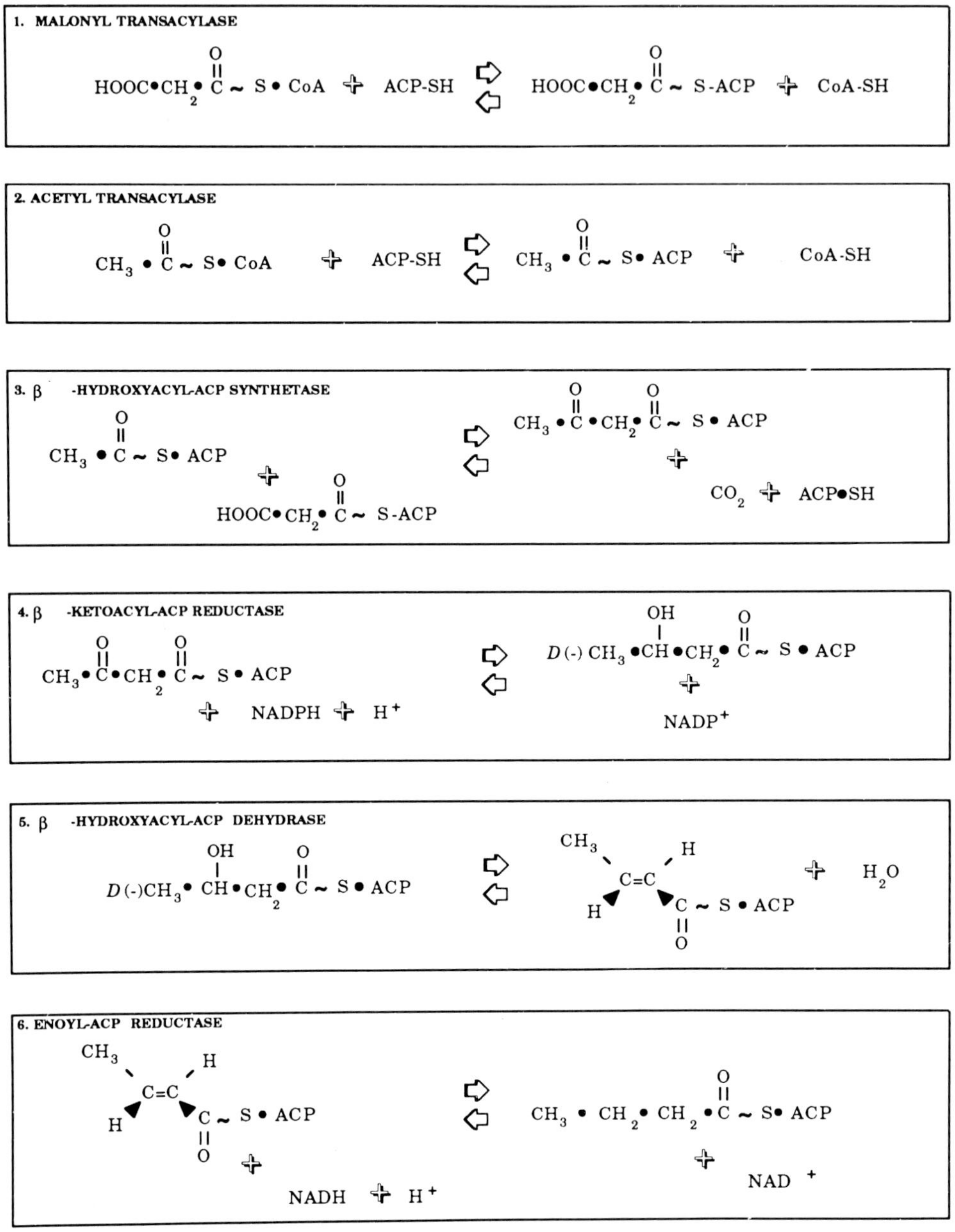

Fig. 1. Enzymic reactions involved in fatty acid synthesis.

additional information can be obtained from these and we have attempted not to duplicate their content. [127, 80].

Historic perspective

Prior to 1982, very little work had been performed on the purification of enzymes involved in plant lipid biosynthesis. It had been known for some time that the biosynthesis of fatty acids up to C_{18} in chain length occurred in chloroplasts [67]. Earlier Stumpf and Barber [134] described the conversion of [^{14}-C] acetate to palmitic and oleic acids by a 10 000 × *g* particulate fraction from avocado mesocarp. In 1975, Canvin [7] reported the biosynthesis of long chain fatty acids in the developing castor bean and Weaire and Kekwick reported that plastids from avocado mesocarp, prepared by density gradient centrifugation, could incorporate [1-^{14}C]-acetate into palmitate and oleate [148]. In the same period, Yamada and his colleagues documented that the 10 000 × ***g*** fraction of castor bean endosperm could convert [^{14}C] sucrose to palmitate, stearate and oleate. This indicated that proplastids contained all of the enzymes required to convert sucrose to acetyl-CoA and for the synthesis of fatty acids. It is curious that despite indications that lipid biosynthesis could be studied advantageously in lipid-rich tissues, such as avocado mesocarp and developing oil seeds, this point was overlooked by many laboratories in subsequent years primarily because other systems were more convenient to work with.

Little attention was initially paid to purifying the proteins involved in lipid biosynthesis with one notable exception – that of acyl carrier protein (ACP). The core reactions of fatty acid biosynthesis had been described (see Fig. 1) and it was realized that, like bacterial systems, fatty acid synthesis in crude plant extracts was stimulated by the addition of ACP. ACP is a low-molecular-weight molecule (ca. 10 kDa), first isolated from *E. coli* [86], to which the intermediates of fatty acid biosynthesis are attached via the terminal sulphydryl of the 4′ phosphopantetheine group, which is in turn attached to the protein through a phosphodiester bond to a serine residue; ACP closely resembles coenzyme A in structure (Fig. 2).

ACP has two other claims to fame – it was one of the first proteins to be synthesized chemically [87] and it is one of the most abundant proteins in *E. coli*, ca. 5 × 10^4 molecules per cell. It is however rarely seen on two-dimensional gels due to its acidic nature and low binding capacity for Coomassie blue [16, 143].

ACP was purified from avocado mesocarp and spinach leaf by Simoni *et al.* [113]. It was assumed at that time that plant fatty acid synthetase was just like that of *E. coli* i.e. composed of several separate catalytic polypeptides. However despite the fact that the component *E. coli* system enzymes had been resolved from one another and several of them purified to homogeneity, no such attempts had been made with systems. Perhaps

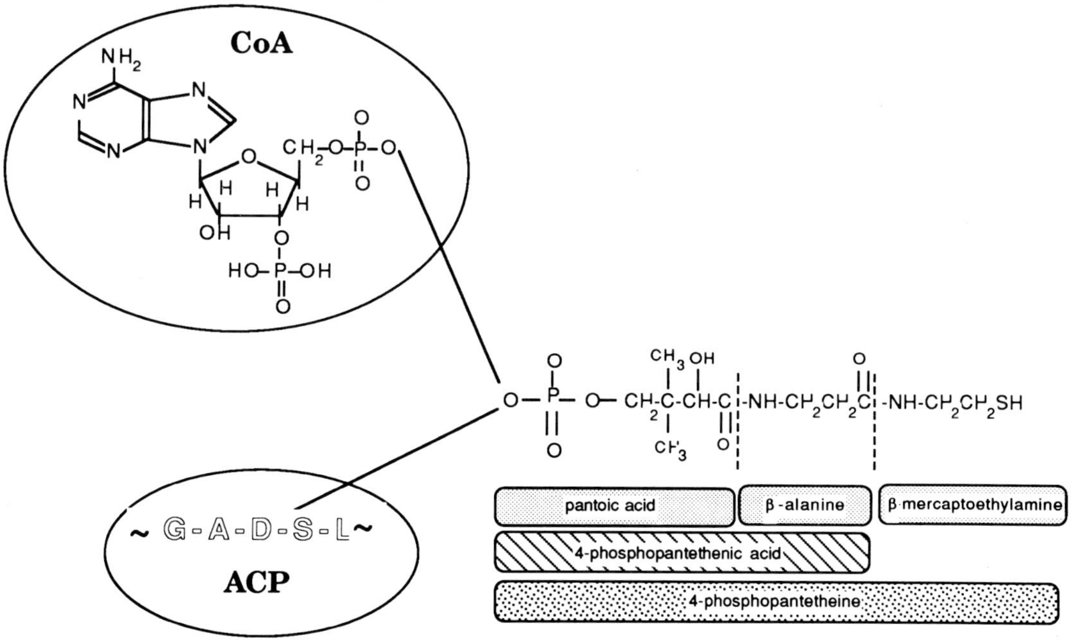

Fig. 2. Structure of coenzyme A and acyl carrier protein. Each has a 4-phosphopantetheine moiety.

because of the attitude 'it would be the same as *E. coli* anyway'. Knowledge of the compartmentalization of lipid biosynthesis had forwarded a model (Fig. 3) in which acetate entered the plastid, was initially converted to acetyl-CoA and then, by acetyl-CoA carboxylase, to malonyl-CoA. Both acetyl- and malonyl-CoA were used to synthesize fatty acids up to C_{18} in chain length as their ACP derivatives. Desaturation then occurred at $C_{18:0}$ by a soluble stearoyl-ACP desaturase. Fatty acid moieties could then either be incorporated into the complex plastid lipids MGDG and DGDG or be exported to the cytoplasm, following hydrolysis to their free fatty acids by acyl-ACP thioesterase. Subsequent desaturation, elongation reactions and synthesis of triglyceride occurs outside of the plastid, predominantly on microsomes. Triacyl glycerols are subsequently stored in discrete oil bodies. Although the broad brush strokes had been painted little was known about the details. This has been the subject of increasingly intense research over the past few years.

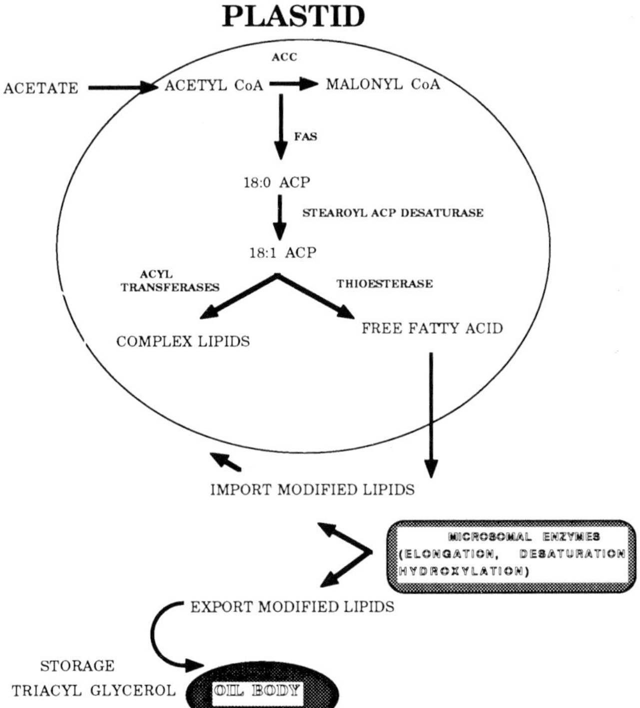

Fig. 3. Compartmentalization of fatty acid synthesizing and modifying enzymes. Acetate enters the plastid, is converted to acetyl-CoA and by acetyl-CoA carboxylase (ACC) to malonyl-CoA. These two substrates are used to synthesize fatty acids up to C_{18} in chain length, as their ACP derivatives, by fatty acid synthetase (FAS). Desaturation can then occur, by the action of soluble stearoyl-ACP desaturase, before the lipids are either incorporated into complex plastid lipids by acyltransferases, or hydrolysed to their free fatty acids by acyl-ACP thioesterase. Following export from the plastid, further elongation, desaturation and hydroxylation may occur (predominantly on microsomes) before the lipid is either returned to the plastid or incorporated in storage triacylglycerols.

The quest for the soluble enzymes and proteins of fatty acid synthetase in plants

There are basically two types of fatty acid synthetase. Type I represented by yeast and mammalian systems have all of their component activities on one or two polypeptide chains. For yeast the functional enzyme is $\alpha_6\beta_6$ and for animals α_2. The second type, type II, is represented by *E. coli* in which the component activities are present on separate polypeptide chains. Whilst the proteins have been isolated and the cDNA cloned from both yeast and animal sources [99, 100], little attention has been paid to the plant enzymes.

In 1982, four independent laboratories reported the separation of component biological activities of plant fatty acid synthetase (FAS) from avocado [8], barley [39], parsley suspension culture [98], safflower [107] and spinach chloroplast [108]. The multi-polypeptide nature of the synthetase was starting to be established but this was only the onset of the quest for these soluble enzymes. To aid readers a schematic representation of plant FAS is given in Fig. 4 showing the relationship of the various components.

ACP

Initial emphasis on characterization of fatty acid synthesis was placed on ACP. This was partly historic, being based on the work of Simoni *et al.* [113], but also because of a number of well based scientific reasons, namely: (1) it was a small mol-

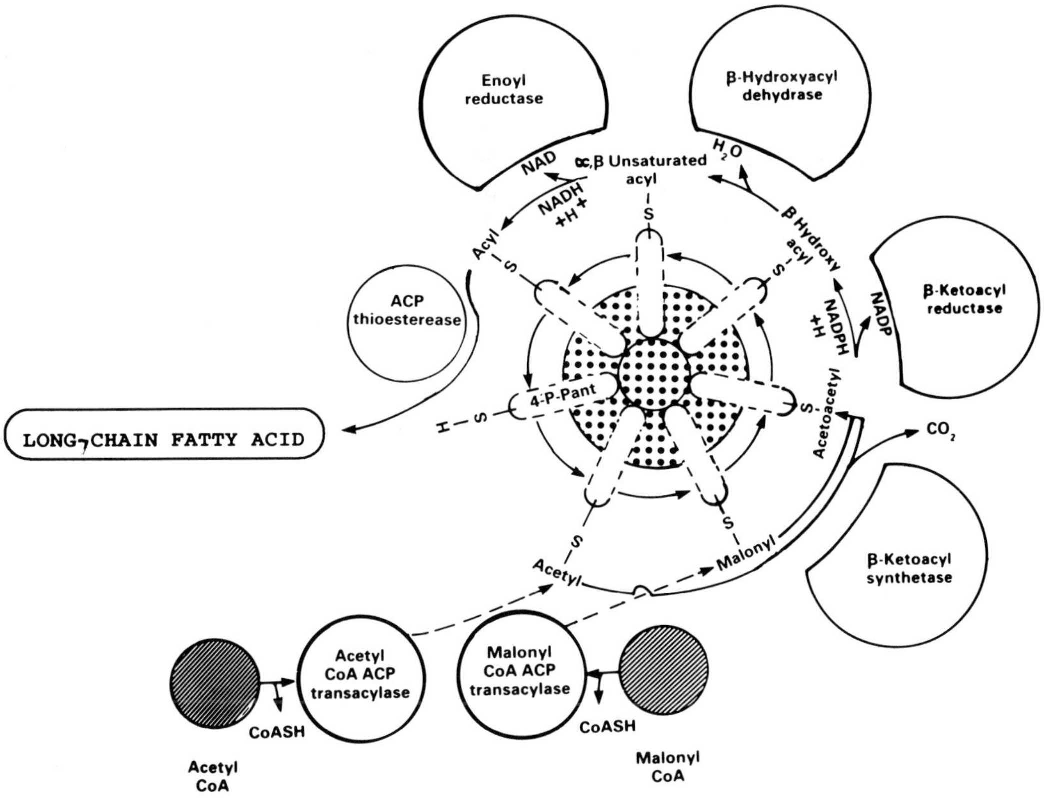

Fig. 4. Schematic representation of plant fatty acid synthetase.

ecule and hence should yield information relatively easily, both at the protein and cDNA level; (2) it appeared to be enzymically relatively stable (the *E. coli* and spinach enzymes were both prepared using acid precipitation and heat denaturation as part of their purification); and (3) it was known to be of central importance, not only as a component of FAS, but also as a substrate for complex lipid biosynthesis. In comparison to other components of plant FAS, ACP would be expected to be present in more than stoichometric quantity due to its many roles. The purification of ACP from spinach leaf [55] and barley leaf [39] was accomplished and the amino acid sequence of both determined. Two independent lines of evidence, from protein purification and western blotting, started to reveal the complexity of ACP types in plants. In barley leaf there are at least three isoenzymes as determined by protein purification. Immunological studies performed by Ohlrogge [78] pointed to two distinct molecular weight forms of ACP in leaf and one in seed. This is somewhat difficult to understand as it would have been predicted that seed would have two forms of ACP, a core FAS component and one involved in storage triglyceride biosynthesis, and that leaf would only have one. The position still remains to be resolved as experiments using ACP as substrates for acyltransferases and acylthioesterases have thrown little light on the matter [27]. The level of ACP activity in developing rape seed [118] was investigated to (1) determine if the activity of ACP was correlated with the deposition of storage lipid (and hence potentially be encoded by a temporally and tissue regulated gene) and (2) select a stage at which to purify the protein from a seed source. ACP activity appeared just prior to the onset of storage lipid biosynthesis and thus the gene was potentially a candidate for both tissue and temporal specific regulation. Studies on ACP levels in developing soybean seed had yielded the same results [78]).

Additional measurements have also been made for several of the enzymes of rape seed FAS for the same reason.

Purification of rape seed ACP provided its own challenges, unlike the leaf counterpart, it was not freely soluble but required detergent solubilization. Unfortunately, it was not as stable as *E. coli* ACP being neither heat- or acid-stable and it rapidly lost biological activity following chromatographic separation. Purification to apparent homogeneity was achieved by using *E. coli* acyl ACP synthetase to specifically radiolabel ACP, in a crude extract, with [^{14}C]palmitic acid. This step introduced a tag with which to follow ACP and convert it from a non-hydrophobic species into a hydrophobic one, and was used as the basis for purification (Fig. 5). Amino acid sequence comparison of ACP from various sources has revealed several areas of high sequence homology (see Fig. 6). Of special note is the region surrounding the phosphopantethylated serine. ACP has not been purified from seed and leaf of the same plant species; the amino acid sequence of leaf forms seem to start AK, only one seed form has been purified to date and this starts AAK which could indicate the existence of a different processing enzyme in seed and leaves.

Ten independent ACP cDNA clones were obtained by screening a size fractionated rape embryo λgt10 cDNA library with redundant oligonucleotides synthesized against the conserved region of ACP. Sequence analysis of these cDNAs demonstrated that the protein (which was nuclear-encoded) had a 51 amino acid N-terminal extension. Amino acid heterogeneity in the mature coding sequences revealed the presence of at least 3 ACP species, which verified heterogeneity seen in sequencing the mature protein [95]. The

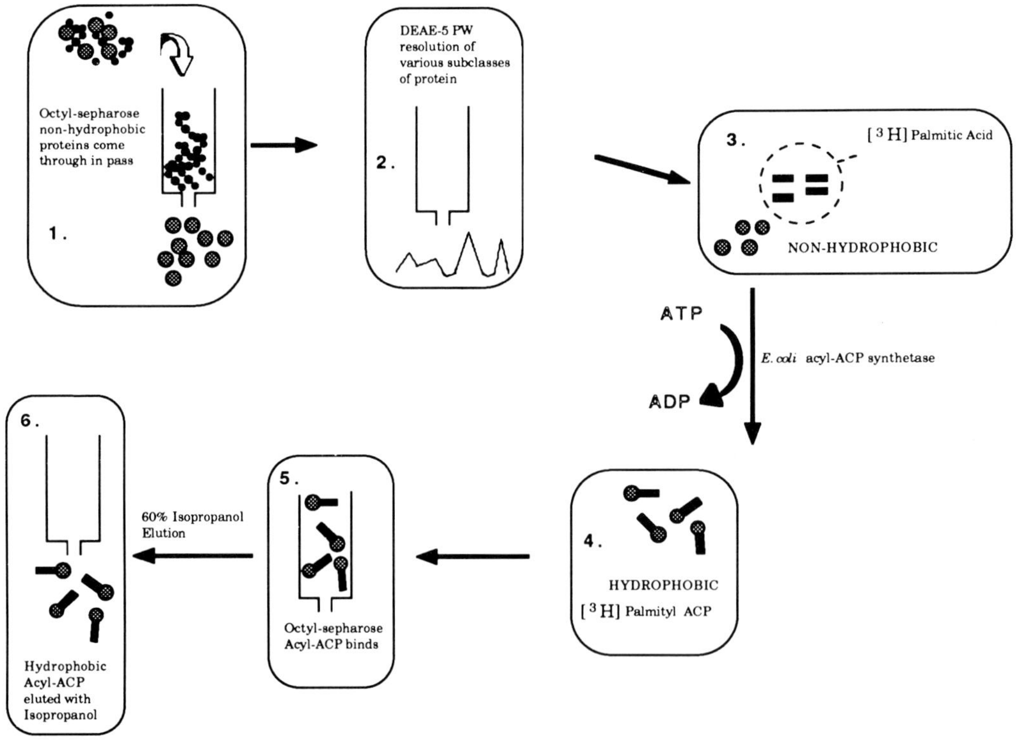

Fig. 5. Purification scheme of rape seed ACP. 1. Hydrophobic proteins (●) are separated from non-hydrophobic proteins by octyl-sepharose chromatography. 2. Ion exchange separation of ACP (followed by biological activity). 4. Quantitative labelling of ACP (a non-hydrophobic species) with radio-labelled palmitic acid using *E. coli* acyl-ACP synthetase. 5. Binding of hydrophobic acyl-ACP to octyl column. 6. Elution by organic solvent.

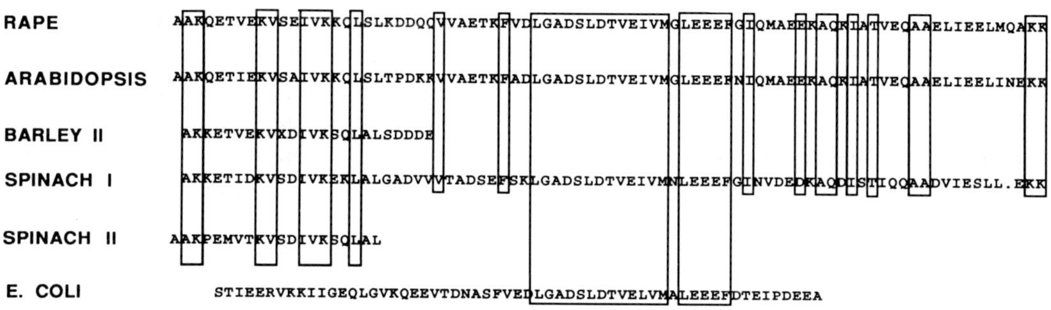

Fig. 6. Sequence homologies of various ACPs. The rape, *Arabidopsis*, spinach I and *E. coli* sequences are translated from their nucleotides, whilst the spinach II and barley II data were obtained by direct amino acid sequencing of the proteins.

ACPs could be broadly classified into two classes on the basis of heterogeneity in the transit peptide [112].

Whilst investigating potential similarity of *Brassica napus* seed and leaf ACP a rape seed cDNA clone was used as a probe on a northern blot containing both embryo and leaf poly(A)$^+$ RNA [95]. The probe hybridized to the embryo mRNA but, unexpectedly, did not hybridize to the leaf mRNA. The use of totally degenerate oligonucleotide probe to the conserved amino acid motif 'LEEEF' resulted in strong hybridization to both seed and leaf derived poly(A)$^+$ RNA. Although there is an apparent strong conservation of amino acid sequence between the leaf and seed forms of ACP from different species, this is not reflected at the nucleotide level. A spinach leaf ACP cDNA has been isolated [96]; when used as a probe for a northern blot, low hybridization to the embryo mRNA resulted, further supporting a concept of different codon usage between leaf and seed isoforms.

NMR studies have been performed on *E. coli* ACP to determine the three dimensional structure [40, 49]. Such investigations have become possible with plant ACP obtained from *E. coli* expression systems [28, 50]. Genomic clones for ACP have been reported from *Arabidopsis* [57, 84] and *B. napus* [112]. The genes cloned from *B. napus* are known to be expressed, because of absolute sequence identity to two cDNAs isolated from a rape embryo library. They contain three introns which are in the same positions in the two genes. Intron 1 occurs in the transit peptide, intron 2 occurs between amino acids 1 and 2 of the mature protein and intron 3 is situated in the middle of the conserved phosphopantetheine binding region. The *Arabidopsis* gene appears to have the same pattern of introns based on protein sequence alignments [84]. Oil seed rape contains approximately 35 copies of seed expressed ACP gene per haploid genome [112].

We will not go into further complexities of the compartmentalization/organellar location of ACP at this stage but will mention three significant observations in the literature: (1) following a demonstration of ACP in *Neurospora* mitochondria [5], ACP has also been immunologically localized to plant mitochondria [11], suggesting a wider metabolic role than that reported to date. Recent experiments from Walkers' lab indicate it is a component of the NADH-ubiquinone oxidoreductase from bovine heart mitochondria [93]; (2) ACP has been found in multiple copies in the erythromycin synthesizing gene of *Saccharopolyspora erythraea* [13] and hence has an important role in the synthesis of macrocylic compounds; (3) sequence similarity between the NodF gene product of *Rhizobium leguminosarum* and ACP has been reported [102] indicating a role of fatty acids in host recognition in the nodulation process. Recent evidence [105] shows that the β-keto-acyl [ACP] reductase from *B. napus* has sequence similarity to the NodG gene product, supporting the contention that some Nod genes encode components of fatty acid synthase.

Condensation reactions and the nature of the first step in plant fatty acid synthetase: the role of acetyl CoA:ACP transacylase

The initial condensation reaction in both bacterial and plant fatty acid synthesis has been held to be the condensation between acetyl-ACP and malonyl-ACP with the formation of acetoacetyl-ACP. Acetyl-CoA:ACP transacylase is considered to be the rate-limiting step in fatty acid synthetase [133]. Following this initial condensation, the same enzyme, β-ketoacyl-ACP synthase (KAS), elongates the acyl-ACP by further addition of C2 units from malonyl-ACP until $C_{18:0}$ ACP is synthesized. Shimakata and Stumpf [109] clearly resolved two KAS isoforms in spinach leaf extracts and carried out reconstitution experiments with purified spinach leaf FAS components. Using hexanoyl-ACP or palmitoyl-ACP as the primers and [^{14}C]-malonyl-CoA, KAS I was shown to be involved in synthesizing fatty acids up to C_{16}, and KAS II to be responsible for the C_{16} to C_{18} conversion. Further, KAS I was shown to have little activity when palmitoyl-ACP was the primer and KAS II had a substrate specificity for chain lengths C_{12} or greater. The reconstituted system, with either KAS I or KAS II, could not extend stearoyl-ACP, indicating that other condensation enzymes must be involved in this process. Condensation reactions are of central importance in plant metabolism, they are involved in elongation to long-chain fatty acids [149], and represent the key reaction of chalcone synthase [29] and resveratrol synthase [97] amongst others. β-ketoacyl synthase I has been shown to have high amino acid sequence homology to the *nodE* gene product [41].

KAS I and KAS II, from spinach leaf are sensitive to covalent modification and inactivation by the antibiotic cerulenin; KAS I being much more sensitive than KAS II. Purification of KAS I to homogeneity was first achieved from rape seed [63], the key to the procedure being the development of a rapid *in vitro* complementation assay in which fractions were assayed for their ability to restore fatty acid synthesis to a cerulenin-inhibited *E. coli* extract. A similar assay had been used to purify bacterial KAS I [91]. Cerulenin labelling has also been used to purify the condensing enzyme from barley chloroplasts; here it was essential to radiolabel the chloroplasts whilst they were intact, with [^{3}H]-cerulenin [111]. Barley KAS I is composed of α and β subunits; using oligonucleotides derived from amino acid sequence data for the pure β subunit, cDNA was amplified using the PCR reaction. This approach yielded a 311 bp cDNA sequence which was used as a probe to isolate the full-length cDNA clone. The deduced primary structure of the β subunit and the *fabB*-encoded β-ketoacyl-ACP synthase from *E. coli* share 49% similarity, including 35% identity [114], but the barley protein has less similarity to other plant condensing type enzymes such as chalcone synthetase.

The use of an antibiotic proved vital in the isolation of KAS I and in the separation of KAS I from KAS II activity. Antibiotics have also played a key role in the identification of a new condensing enzyme, KAS III, involved in the initial reaction of fatty acid synthesis in both bacteria and plants.

Bacterial fatty acid synthesis is strongly inhibited by the action of cerulenin. However, Jackowski and Rock [44] demonstrated, in *E. coli*, that acyl-ACP formation *in vivo* was not blocked by this antibiotic and short-chain (4–8-carbon) acyl-ACPs accumulated in cerulenin-treated cells, indicating the presence of a short-chain condensing enzyme. Following this, a cerulenin-insensitive short-chain condensing enzyme was demonstrated in spinach leaves [45] and in other plant species [147].

Earlier experiments [74, 75] showed that the antibiotic thiolactomycin selectively inhibits type II, dissociable, fatty acid synthases and that in the *E. coli* system, acetyl-CoA:ACP transacylase and KAS were the thiolactomycin-sensitive enzymes. Lowe and Rhodes [61] have purified acetyl CoA:ACP transacylase from *E. coli* and found it not to be thiolactomycin-sensitive; this difference may be explained by an observation of Jaworski *et al.* [46], which showed that inhibition of KAS III is actually caused by an oxidative breakdown product of thiolactomycin.

Whilst measuring the activity of a number of lipid synthesis enzymes in plant material, Stumpf and Shimakata [135] concluded that (1) acetyl-CoA:ACP transacylase was probably rate limiting and (2) elevation of its level in reconstitution experiments gave rise to the ability to synthesize medium-chain fatty acids. If there is indeed an alternative KAS, KAS III, which utilizes acetyl-CoA and not acetyl ACP, what then is the function of acetyl CoA:ACP transacylase and does such an activity really exist? The results concerning a potential role for acetyl-CoA:ACP transacylase in the synthesis of medium-chain fatty acids should be treated with some caution as rape seed, which does not make medium-chain fatty acids, has a high *in vitro* level of this activity [135]. Lowe and Rhodes [61] recently reinvestigated acetyl-CoA:ACP transacylase activity from *E. coli* and found that it only represented a minor component in the original purification. Some insight into what could be occurring comes from consideration of another bacterial enzyme, acyl-ACP synthetase, which we have previously mentioned in the isolation of seed ACP. The requirement for this biological activity in *E. coli* was questioned, as acyl-ACP are the synthetic product of fatty acid synthetase, it is now clear that it is a partial reaction of another enzyme involved in phosphatidyl ethanolamine metabolism [12]. There are strong indications that acetyl-CoA:ACP transacylase could be a partial activity of KAS III, but only purification of KAS III to homogeneity will resolve this question. What then is the function of acetyl-CoA:ACP transacylase activity? Is it really required for lipid synthesis in plants? It may well be that plants are plastic in their metabolism having alternative metabolic routes [89]. In *E. coli* KAS III (*fabH*) has recently been cloned (C.O. Rock, unpublished) and experiments are underway to construct a strain in which this gene is totally inactive to test if its activity is essential (C.O. Rock, pers. comm.). Figure 7 shows the relationship between the three different KAS activities in the synthesis of fatty acids.

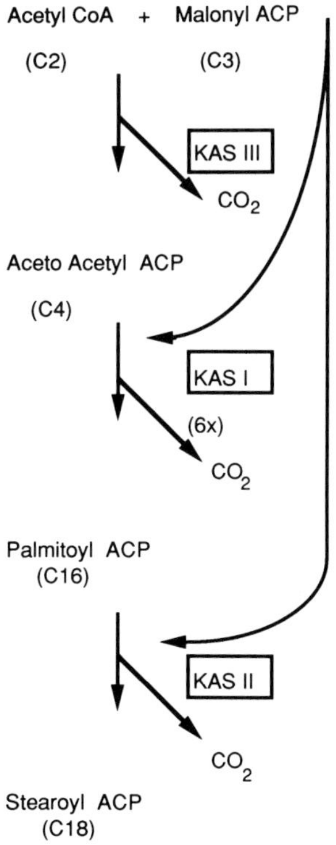

Fig. 7. Role of β-ketoacyl [ACP] synthase isoforms in the synthesis of saturated fatty acids.

Malonyl-CoA:ACP transacylase and β-hydroxyacyl dehydratase

Malonyl-CoA:ACP transacylase has been purified to homogeneity from avocado (Hilt, 1984, cited [133]) and a number of other plant sources have been used to produce partially purified preparations: barley [38]; leek [60], spinach [129] and soybean [26]. In the case of soybean and leek, evidence for the presence of isoforms was found, but these have no ascribed functional significance. A gene for malonyl-CoA:ACP transacylase has recently been cloned from *E. coli* by complementation of the *fabD* mutant [30; A.R. Stuitje, pers. comm.), so it may be possible to clone the plant gene by complementation of the *E. coli* mutant (cf. the complementation of an *E. coli* enolase-deficient mutant with a maize cDNA [56]). The dehydratase has been purified to homogeneity from spinach leaves [110] and partially purified from developing safflower seeds

[107], but since this work there has been little activity in this area.

The reductive steps of fatty acid synthesis

Two reductive steps are required for fatty acid biosynthesis. The enzymes for both steps have been purified, using acyl-CoA substrate analogues, but it has subsequently been shown that ACP is by far the preferred substrate. The first reductive step is catalysed by β-ketoacyl-ACP reductase. This enzyme has been purified to homogeneity from spinach [110], avocado [105] and rape [103]. Amino acid sequence data has been obtained for the avocado enzyme which is NADPH-specific. The N-terminus shows homology to Cyt f of *Marchantia* and there is extensive internal sequence homology to the NodG gene product [105]. The rape enzyme, like the avocado enzyme, is both cold-labile and dilution-sensitive [106].

Recently, a cDNA clone has been obtained for the rape seed enzyme and this was used to clone a full-length cDNA from *Arabidopsis*. Extensive nucleotide sequence homology has been found between the rape and *Arabidopsis* cDNAs. The amino acid sequence of the β-ketoacyl-ACP reductase is highly homologous to that of the NodG gene product (Fig. 8).

The NodG gene product was previously of unknown biochemical function and it can now be concluded that it is probably a β-ketoacyl-ACP reductase. The structure of the sulphated lipo-oligosaccharide involved in *R. meliloti* host specificity for nodulation has recently been described [140]. From the structure (Fig. 9) it can be seen that both ACP and β-ketoreductase could have an important role in the synthesis of the lipid moiety. Significant homology is also observed between NodE and condensing enzyme [41], adding evidence to the hypothesis that some Nod genes encode for components of fatty acid synthase (A. Downie, pers. comm.). Amino acid

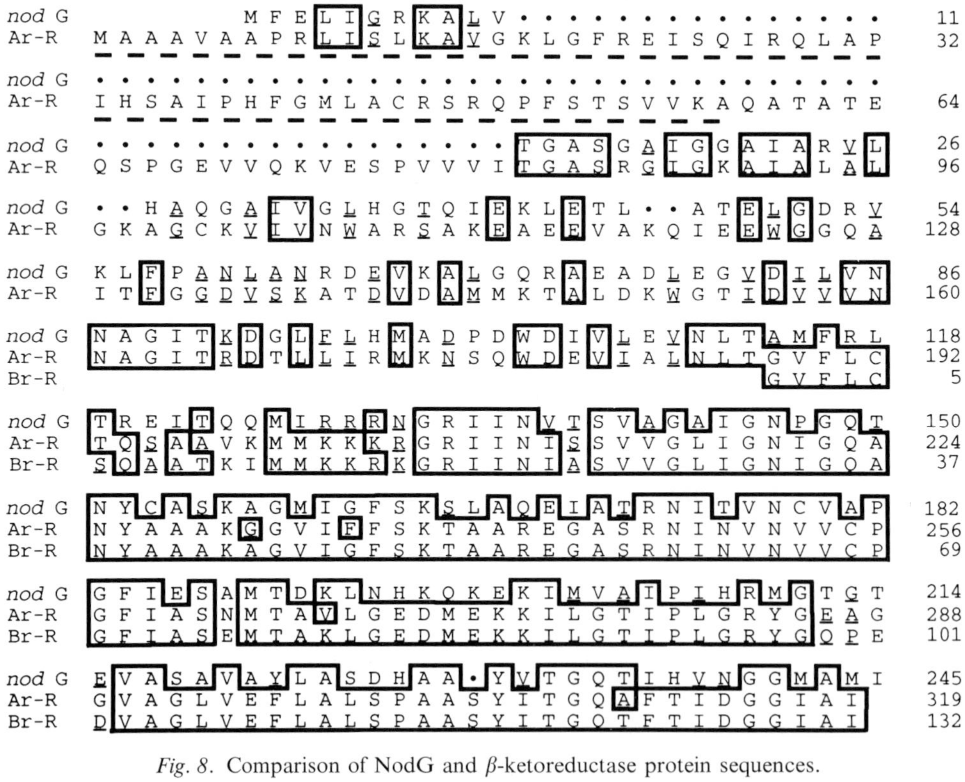

Fig. 8. Comparison of NodG and β-ketoreductase protein sequences.

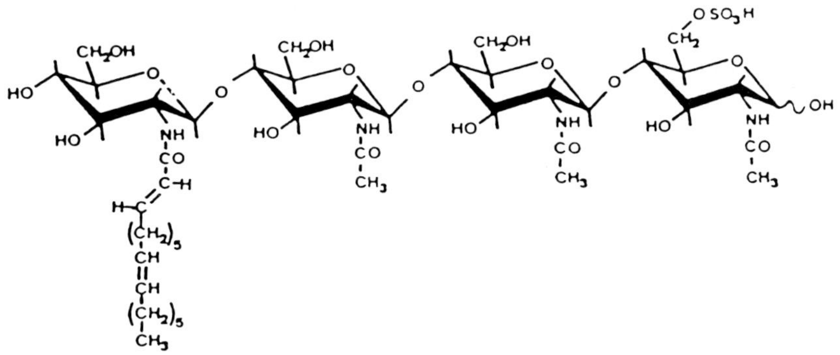

Fig. 9. Sulphated lipo-oligosaccharide involved in *Rhizobium meliloti* host specificity for nodulation.

sequence homology derived from genes involved in lipid synthesis have thus revealed the probable function of certain nodulation genes. What is somewhat unclear is why the endogenous bacterial fatty acid synthesis components are unable to perform these reactions. This is probably due to absolute substrate specificity of the individual reactions involved in the synthesis of the signalling molecule.

Two types of enoyl-ACP reductases have been reported for plant material. The NADH-specific enzyme has been purified from avocado [8], spinach [110] and rape [117]. In rape seed, this enzyme is induced prior to the deposition of lipid and the biological activity and protein appearance shows the same type of kinetics as ACP. It is therefore also a potential target for a seed-specific and temporally regulated gene. Initial experimentation demonstrated that the enzyme is tetrameric. Based on the presence of two separate bands on SDS-PAGE it was concluded that the enzyme was $\alpha_2\beta_2$ [117]. Subsequent separation of the α and β components by reversed-phase chromatography and N-terminal amino acid sequencing showed that the two polypeptides differed by a six amino acid amino terminal extension [14]. Analysis by western blotting on freshly homogenized rape seed indicated that the enzyme actually has an α_4 structure.

The lower β band was an isolational artefact due to an endogenous protease which cleaved between a serine and lysine residue [120]. Recent experimentation in our laboratory with higher-resolution gels has demonstrated that the situation in rape is not so simple. In crude extracts of both rape leaf and rape seed two bands are present. These have an approximately equal intensity. The newly identified polypeptide runs above α and we have termed it α'. The nature of this new species still remains to be elucidated and is the subject of current investigations. It could be either a membrane-bound form or, alternatively, a long acyl chain-specific NADPH form which has been reported from two separate laboratories. The latter enzyme requires long-chain substrates for detection and could have been lost in the original purifications which used the short-chain crotonyl (C_4) substrate.

Extensive amino acid sequence data is available on the rape seed enzyme [121]. This enzyme has a strong affinity for ACP [120], and covalent modification of an arginine residue in enoyl reductase using phenylglyoxal destroyed this interaction [15]. The cDNA has been cloned from a rape embryo library and its complete sequence determined showing a transit peptide of 73 amino acids [48]. This has enabled the peptides obtained by amino acid sequencing to be ordered [121].

Two independent genomic clones from *Arabidopsis* have been isolated, by the group of Stuitje, and fully sequenced (A.R. Stuitje, pers. comm.). *Arabidopsis* cDNA clones have also been isolated by the group at Durham using a rape enoyl reductase cDNA probe provided by Stuitje (in preparation). Northern analysis on leaf, seed and flower RNA indicates that the rape gene is constitutively expressed, whereas protein extracts from leaf and flower reveal relatively little immunological cross-reaction on western blots [48]. This is somewhat different from the situation with ACP where the embryo cDNA isolated failed to hybridize to leaf mRNA. If a seed specific enoyl reductase gene does exist then it may require a more sophisticated probe for its detection, possibly derived from the transit peptide sequence or the 5′ untranslated region.

Import and export of lipids in the plastid

Early experimentation demonstrated that acetate could enter plastids and be incorporated into lipids [132]. However, is acetate the true substrate which is imported into plastids? Recent results from the laboratory of Thomas have indicated that acetyl groups could enter chloroplasts as acetylcarnitine. This is similar to the mitochondrial systems [64], where a specific carnitine translocator exists. In feeding experiments on isolated chloroplasts L-[1-^{14}C acetyl] carnitine gave a five-fold greater incorporation of radioactivity into fatty acids than [1-^{14}C] acetate [64].

Following the synthesis of long-chain acyl-ACPs they are either hydrolysed to free fatty acids, by acyl-ACP thioesterase, and are exported from the plastid – possibly by a carnitine route. Alternatively, they are incorporated into the plastid lipids by acyl transferases [92]. Independently the acyl-ACP thioesterase has been purified to homogeneity from rape seed [34] and to near homogeneity from squash [42]. In rape, enoyl-ACP reductase copurified with acyl-ACP thioesterase through several successive steps indicating a close affinity between the two enzymes (A. Hellyer and A.R. Slabas, unpublished observation). Interest in the thioesterase has arisen for two reasons: (1) the balance of thioesterase and acyltransferase has been hypothesized to be a regulatory point for the prokaryotic versus eukaryotic pathway [92] and (2) as a possible mechanism, by analogy with animal systems, for the synthesis of medium-chain fatty acid biosynthesis in plants.

Nature of the chain termination mechanism and the synthesis of medium-chain fatty acids.

Several mechanisms have been proposed for the synthesis of medium-chain fatty acids in plants. They include: specific thioesterases; specific acyltransferases; high levels of acetyl-CoA:ACP transacylase; separate compartmental synthetases; controlled β-oxidation and specific β-keto acyl synthetase [133]. There is no definitive evidence for the mechanism in plants, although extensive studies have been performed in crude systems. In rats and mallard duck the synthesis of medium-chain fatty acids is brought about by the premature chain termination of fatty acid synthesis by interaction with a specific medium-chain thioesterase-thioesterase II, alternatively called medium-chain hydrolase [126, 86]. The protein has been purified from both duck and rat and the cDNA cloned [94]. Transformation of rat fibroblast 3T3 cells with the gene resulted in the synthesis of medium-chain fatty acids. It should be borne in mind, however, that this is not a universal mechanism of medium-chain fatty acid synthesis in animals. Goat fatty acid synthetase is capable of synthesizing medium-chain fatty acids merely by supplementation with microsomes and therefore has no apparent requirement for thioesterase II [52]. Recently a medium-chain specific acyl-ACP thioesterase from California bay laurel has been isolated [83]. Introduction of this gene into the *fadD* mutant of *E. coli* has resulted in the production of high quantities of medium-chain fatty acids (T. Volker, pers. comm.). It will be interesting to see in the future if the gene encoding this enzyme will be capable of altering the lipid profile of plants.

Regulation of the rate and extent of lipid synthesis

In animal systems it is well documented that the rate of fatty acid biosynthesis is limited by acetyl-CoA carboxylase (ACC). This enzyme has a subunit M_r of 250 000 and has recently been cloned both from chicken and rat [20]. ACC is highly regulated, in response to hormonal balance, by phosphorylation. The sequence of the phosphorylation sites has been determined [81]. Many of the interactions of animal FAS with other components, when investigated *in vitro*, have required rate limitation on the availability of malonyl-CoA [51]. What then is the situation in plants both with respect to the subunit size and regulation of ACC? There are reports in the literature of varying subunit M_r of the basic polypeptide of ACC (summarized in [33]) and this has led to the belief that there is possibly more than one form of ACC. Perhaps note should have been taken on the first reports of the purification of ACC as a high-molecular-weight polypeptide subunit from parsley, wheat germ [21] and rape seed [115]. In these instances there was clearly one high-molecular-weight species. We have used antibodies raised against wheat germ ACC to probe leaf material and found similar high M_r (220 kDa plus) cross-reacting proteins in leaf material [122]. Also direct purification of rape leaf ACC has demonstrated that it is a high M_r species of 220 kDa [116]. This draws us to the conclusion that in rape and wheat a single high-M_r form is present in both leaf and seed. Other biotin-containing peptides, which are carboxylases, have been found in animals [9] and plants [150], but these probably have other functions. The possibility cannot, however, be ruled out that a different cytoplasmic form of ACC exists which has a lower molecular weight. Indeed, Nikolau has purified a 50 kDa protein from embryogenic carrot cells and analysis of the cDNA clone isolated has indicated that this is possibly a cytoplasmic ACC [73]. Proof of this will come from immunogold localization studies. Attempts have been made to look for phosphorylation and the regulation of plant ACC by citrate, but there is no evidence that such regulation exists. Is such regulation however required in plants? ATP, generated by photophosphorylation, is one of the key substrates of ACC which is apparently located within plastids [133]. Any reduction in the pool size of ATP would rapidly reduce the rate of the carboxylase and hence slow down lipid synthesis. Perhaps no other control at this point is required. If a separate cytoplasmic form of ACC exists it will probably be subject to a different regulatory mechanism.

Measurement of the major acyl-ACP intermediates of fatty acid biosynthesis in light- and dark-grown material has shown an increase of acetyl-ACP in the dark. This is consistent with the light/dark control of the rate of fatty acid biosynthesis being by ACC [85].

Two further points are worthy of mention here: (1) a biotin-binding sequence from tomato has been identified [37] and it contains the amino acid sequence A-M-K-M, a highly conserved motif in carboxylase sequences; and (2) acetyl-CoA carboxylase seems to be the site of action of differential herbicide tolerance between monocots and dicots [82]. The latter point and its potential regulatory role has led to a number of groups concentrating on the cloning of ACC. In rape seed the level of ACC activity has been measured during seed maturation [141] like other enzymes of lipid synthesis the activity rises prior to the onset of lipid deposition and rises continually through lipid synthesis. However, by contrast, whilst the level of other enzymes of lipid metabolism remains high [119] that of ACC is rapidly reduced once maximal lipid accumulation has been achieved [141]. ACC could be acting as a central regulator, effectively shutting down lipid synthesis even though an active FAS is present, by removing the substrate malonyl-CoA.

Stumpf and Shimakata [135] have measured the rate of the component enzymes in plant FAS in various plant extracts and have come to the conclusion that probably acetyl-CoA:ACP transacylase was rate-limiting. With the current confusion over the role of acetyl-CoA:ACP transacylase in plants this whole question requires re-evalution. It is important to remember that these are *in vitro* results and may not reflect the *in vivo* situation which will be dependent on subcellular location of substrate and enzyme. It is also not clear what controls the extent of lipid synthesis (it could be ACC as alluded to above) and little information is available concerning the partitioning between lipid synthesis and the synthesis of other major storage products: starch and protein. Alterations to the total quantity of lipid deposited and the ratio of the storage products in seeds is a potential agronomic target for the future.

Desaturases in higher plants

Desaturation has a marked effect on the physical properties of lipids. Indeed, several plants specifically change their degree of unsaturation in response to cold acclimatization so that the membranes can retain their degree of fluidity [62, 69]. Alterations in the chilling sensitivity of plants are correlated with the degree of unsaturation of phosphatidylglycerol (PG). Plants, such as squash which contain a low level of Δ12-*cis* fatty acids in PG are chilling-sensitive, whilst those typified by *Arabidopsis*, which have a higher level of Δ12-*cis* in PG, are more chilling-tolerant. The composition of these lipids will be regulated by both the specific fatty acid composition of the acyl pool and the selectivity of the acyltransferases.

Plants have two types of desaturases, a soluble chloroplast Δ9 desaturase and membrane-bound desaturases which are involved in Δ12 and Δ15 product synthesis. From a biochemical point of view the latter present a much greater difficulty in isolation, due to their membrane requirement.

The Δ9 (stearoyl) desaturase has recently been cloned from cucumber and castor seeds [104] using antibodies made against the purified avocado protein. The cDNA from safflower embryos has been cloned and includes a 33 amino acid transit peptide [139]. Modulation of the stearoyl-ACP desaturase level, using antisense technology, in transgenic rapeseed has resulted in a marked decrease in the level of oleic acid with an increase in the level of stearic acid [54]. This is a soluble ferrodoxin requiring enzyme and should not be confused with other desaturases which are membrane-bound, such as the Δ12 and Δ15 desaturases which respectively form linoleate (18:2) and linolenate (18:3). It is well documented that such desaturation occurs on phosphatidylcholine and monogalactosyldiacylglycerol backbones respectively [123, 136] and components of the electron transport chain have been identified [125]. Purification of these enzymes would present a formidable challenge even to the most experienced biochemists; Somerville and Browse, with some foresight, have created a series of mutants of lipid metabolism in *Arabidopsis* (Fig. 10) and experimentation is currently underway to isolate the genes by gene walking. This area is exceptionally important, but outside the scope of this review (for details, see [127]). Other approaches, including T-DNA tagging, are also expected to provide a significant way forward [22].

Cyanobacterial desaturases and temperature acclimatization

Desaturases are important in temperature adaptation, yet how does one obtain the enzymes and subsequently the genes involved in the production of unsaturated lipids if the proteins are membrane bound and rapidly lose activity on purification? Murata and colleagues of NIBB, such as Okasaki, have taken a lateral approach. They selected for mutants in the cyanobacterium *Synechocystis* PCC 6803 that were defective in the desaturation of fatty acids and were chilling-sensitive, having a lower growth rate than the wild type at 22 °C [145]. A mutant was shown, by lipid analysis, to be deficient in Δ12 desaturation. By complementation they were able to clone a gene which allowed growth at low temperatures and restored Δ12 desaturation. This gene could be either a structural or regulatory gene, as is the case for the desaturase mutants in *Arabidopsis*. Insertion of the gene into *Anacystis nidulans* R2-SPc, a species of *Cyanobacterium* completely lacking in Δ12 desaturation, resulted in the synthesis of Δ9,12 unsaturated fatty acids. This was at the expense Δ9 monounsaturated fatty acids and concomitant in-

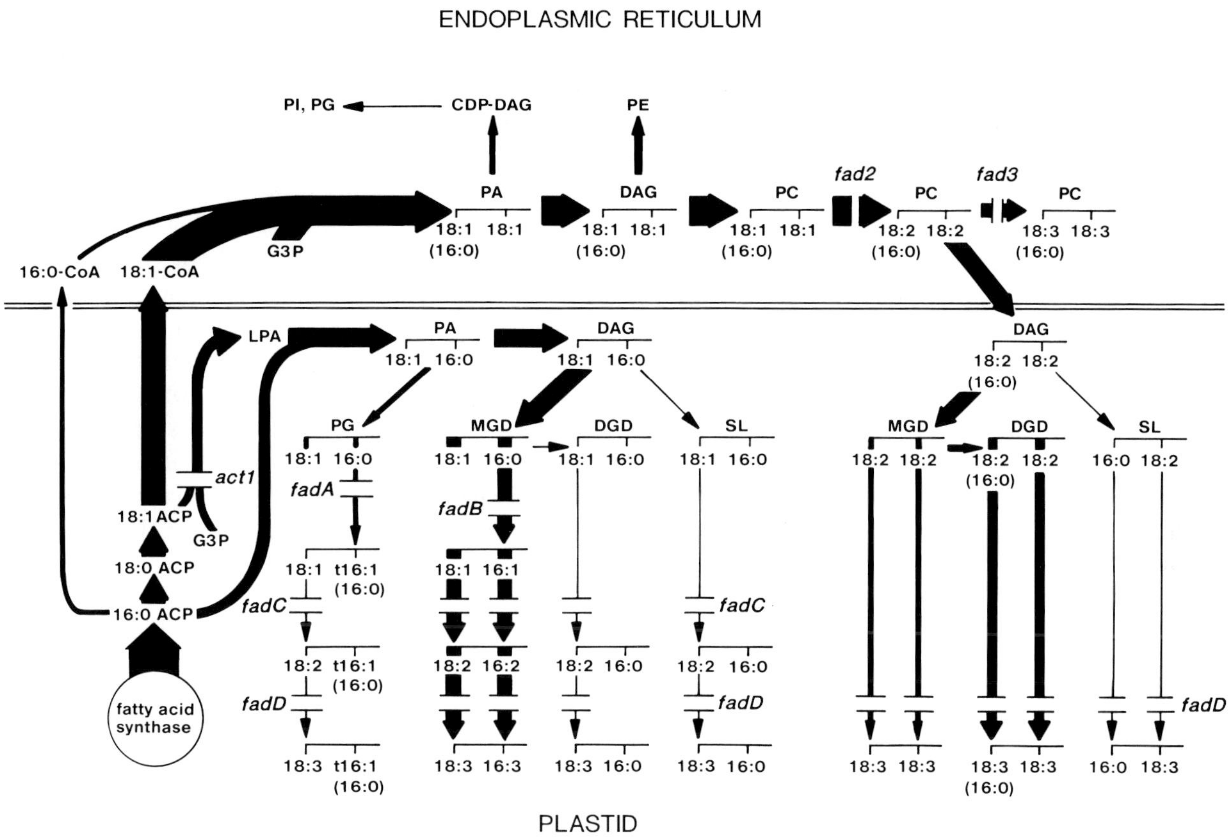

Fig. 10. Mutants in lipid biosynthesis of *Arabidopsis*, reprinted from Somerville and Browse, [Science 252: 80–87 (1991)] with permission (copyright 1991 by the AAAS). Seven classes of mutant in glycerolipid metabolism are shown. These were identified by measuring the fatty acid composition of approximately 10 000 individuals, by gas chromatography, from randomly chosen plants in a mutagenized population.

crease in the chilling tolerance of the transformant [146]. This represents the first example of the alteration of lipid composition by genetic engineering.

Experimentation on rye protoplasts, utilizing a 'membrane engineering approach' have demonstrated the importance of unsaturated lipids, specifically in phosphatidylcholine with respect to increasing the stability of plasma membrane during freezing-induced dehydration [130].

The incorporation of acyl groups into complex lipids and the genetic engineering of altered acyl lipid composition in plants

There is an enormous variety of complex lipids within cells such as sphingolipids, MGDG, DGDG, phosphatidylcholine, phosphatidylserine, phosphatidylinositol, phosphatidylethanolamine and triacylglycerols. The incorporation of fatty acids onto the backbone of these lipids requires the action of acyltransferases. Acyltransferases, responsible for the synthesis of plastid lipids, use ACP derivatives preferentially [23] whilst those involved in triacylglycerol biosynthesis use CoA substrates [25].

The soluble plastid glycerol-3-P acyltransferase (G-3-P-AT) has been purified from a number of plant sources: squash [76], pea [4, 19] and spinach [4]. Recently the cDNA for the squash [43], *Arabidopsis* [76] and pea G-3-P-AT have been cloned and sequenced.

In *E. coli*, the G-3-P-AT is membrane-bound, the enzyme was very labile during purification and was inactive until inserted into a biological membrane [24] – a case of a lipid synthesizing enzyme requiring a lipid environment! The protein has several membrane-spanning domains and insertion into the membrane was topicologically correct. This gives an optimistic view to reactivation of the plant membrane-bound enzymes upon insertion into the appropriate biological membrane. The use of chaperonins to refold denatured protein may also prove advantageous if applied during the isolation of membrane proteins.

The acyltransferases for position *sn*-2 and *sn*-3 in plants, like their *E. coli* equivalents, are membrane-bound; the 1-acylglycerol-3-phosphate acyltransferase of pea, which catalyses the acylation of the *sn*-2 position, has a marked selectivity for palmitate [23]. 1-acylglycerol-3-phosphate acyltransferases of seed tissue have a similar high selectivity for certain acyl chains and as such are seen as a potential target for genetic engineering. Since acyltransferases have different substrate specificities it should be possible to selectively alter the composition of plant lipids by manipulating the type and level of acyltransferases in plants. Murata introduced the squash and *Arabidopsis* G-3-P-AT gene into tobacco plants. This resulted in marked alteration in the lipid composition of the phosphatidylglycerol, verifying that *in vitro* selectivities are also seen *in vivo* under conditions where the acyl pool is the same [70].

Work is currently underway on the acyltransferases from plant mitochondria and other systems; they appear to have different substrate properties. It will be of interest to see if the mitochondrial enzyme also uses ACP as this has been reported in plant mitochondria [11]. A microsomal acyltransferase has recently been partially purified from etiolated shoots of pea and this will not use ACP substrates [31].

Fatty acid-binding proteins/membrane formation and targeting

Free fatty acids can be highly toxic within the cell, hence there are particular species of proteins which bind to them. Initially, in animals, it was believed that acyl-CoA- and fatty acid-binding proteins were one and the same. However, more recent studies have shown these functions reside in separate proteins [89]. In plants non-specific lipid-binding proteins have been well characterized [144, 151], appearing during germination. cDNAs have been isolated from maize [137], barley [2, 68] and spinach [3] for non-specific lipid transfer proteins. Amino acid sequences deduced from cDNA data show that the polypeptides are produced by cleavage of a preprotein and evidence suggests that processing is by the secretory pathway [3]. Originally it was believed that lipid-binding proteins were possibly involved in lipid transfer between organelles. Three recent observations would strongly question this conclusion: (1) maize lipid transfer protein has been immunolocalized and found predominantly in the external cellular layers and around the leaf veins [128]; (2) the EP2 lipid transfer protein in embryogenic carrot suspensions is secreted into the medium [131]; (3) the *Arabidopsis* lipid transfer protein has an extracellular location [138]. It is possible that lipid transfer proteins are involved in wax or cutin synthesis. Apart from the lipid transfer proteins referred to above, phospholipid-transfer proteins have been the subject of much recent research. These can exchange the lipid molecules between membranes without altering their own lipid content. There seem to be many different transfer proteins in eukaryotic cells which are capable *in vitro* of directing net lipid movement between organelles. Recent experiments in yeast systems [1] have demonstrated that the *SEC14* gene, which is essential for transport of proteins from the golgi complex, encodes a phosphoinositol/phosphatidylcholine transfer protein. These results establish an *in vivo* function for phospholipid transfer proteins, namely compartment-specific stimulation of protein secretion.

It is of interest to know if either fatty acid-binding proteins or lipid-binding proteins are present in seeds during lipid accumulation to act as carriers between cellular compartments.

Formation of oil bodies and oil body proteins

The mechanism of oil body formation has been described in detail elsewhere (see [71] for review). Early studies by Slack *et al.* [123] found that two major proteins co-purified with safflower and linseed oil bodies; these hydrophobic proteins are now known as oleosins. cDNA, gene and protein sequence data are available for two maize proteins [142, 88], carrot [32], rape and radish [72]. All of these proteins have similar amino acid sequences, sharing a central hydrophobic domain. It has been suggested that oleosins prevent coalescence of oil droplets during desiccation of the seed and may contain a lipase-binding site [72]. Clearly, there are other proteins associated with oil bodies. Herman *et al.* [35] have reported a hydrophilic oil body-associated protein from soybean, but as yet no function has been ascribed to this protein.

Differential gene expression and lipid synthesis

Perhaps the most advanced study in this area concerns the regulation of rape embryo ACP expression. The rape embryo clones were broadly divided into two classes based on differences in the transit sequence (group I and group II). It has been demonstrated that group II shows both temporal and tissue-specific expression [112]. Recent experiments from the Unilever laboratory have shown that a 1.4 kb 5′ flanking sequence of a rape seed ACP gene will confer temporal and seed specific expression on reporter genes which have this promoter (R. Safford, pers. comm.). Since napin, cruciferin and the oleosins [71] are deposited later during seed development than the onset of lipid deposition it is highly likely that promoter elements from these genes would not have the same temporal expression as the lipid synthesis genes. Other ACP genes exist which are apparently constitutively expressed [80]. In future it will be interesting to find the mechanism involved in such temporal and tissue-specific expression. Analysis of the 5′ sequences of several genes specifically expressed in seeds at the same time should provide information about common *cis* elements and hence lead to the identification of *trans*-acting factors which might regulate their expression.

Current directions and key questions

Despite an increased activity in the area of plant lipids, little is still known about the structure of many of the component enzymes of complex lipid biosynthesis (see review by Joyard *et al.* [47]). In part this is due to the problems of purifying membrane-bound enzymes. However this could be possible if a concerted scientific effort is made. Generating mutants of lipid metabolism in *Arabidopsis* and the ability to 'gene-walk' is one approach that aims to overcome this. Such mutants will arise from lesions in both structural and regulatory genes; mutations in *de novo* fatty acid synthesis up to $C_{18:0}$ the FAS genes are likely to be lethal. Since the fundamental reactions of fatty acid synthesis are conserved it is possible that antibody and gene probes derived from soluble proteins could be used to clone membrane-bound counterparts of central fatty acid synthesis that are involved in chain elongation.

The essential requirement of particular fatty acids in certain structural relationships in the cell can be tested by mutation and/or transformation studies. This has already been achieved for *trans* Δ3-hexadecenoic acid, by characterization of a mutation at a single nuclear locus (*fadA*), the carriers of which completely lack *trans*-16:1 [6]. This acyl chain was thought to play an important role in some aspect of photosynthesis, but its loss did not have a significant effect on chloroplast function. In the longer term mutational studies can also be extended to the importance of certain lipid classes, as is being undertaken in studies with *E. coli* [10].

With the availability of a number of cDNA and genomic clones for plant FAS components it will be possible to study the transcriptional regulation of these genes.

At the biochemical level, a number of questions remain unresolved. Are components of the FAS reaction associated *in vivo*? What is the exact nature of the number and stoichiometry of components of plant FAS? What are the physiological triggers stimulating the expression of genes for lipid synthesis in plants? What is the exact nature of the substrate which enters the plastid and the nature of the product which is exported?

The fruits of this in time will not only be a greater understanding of lipid biochemistry, which is scientifically satisfying, but will eventually lead to genetically engineered lipid compositions, both of membranes and storage lipids. The feasibility of this has already been demonstrated by Murata [70] and Kridl *et al.* [54]. It may well be an important consideration in the genetic engineering of the lipid content of seeds to alter the specificities of the lipases involved in the germination process. If newly produced lipids are not readily mobilized by the endogenous lipase, problems may occur on germination.

Alteration of the lipid content of plants will eventually supply key intermediates for the chemicals industry which will allow for sustainable agriculture, renewable raw resources and an environmentally cleaner route to chemical production.

Acknowledgements

A.R.S. would like to thank Professor Anthony T. James for introducing him to this area of biochemistry, Dr R. Safford for critically reading this manuscript and Dr Kieran Elborough and Bill Simon for kindly providing the art work.

References

1. Bankaitis VA, Aitken JR, Cleves AE, Dowhan W: An essential role for a phospholipid transfer protein in yeast golgi function. Nature 347: 561–562 (1990).
2. Bernhard WR, Somerville CR: Coidentity of putative amylase inhibitors from barley and finger millet with phospholipid transfer proteins infered amino acid sequence homology. Arch Biochem Biophys 269: 695–697 (1989).
3. Bernhard WR,Thoma S, Botella J, Somerville CR: Isolation of a cDNA clone for spinach lipid transfer protein and evidence that the protein is synthesized by the secretory pathway. Plant Physiol 95: 164–170 (1991).
4. Bertrams M, Heinz H: Positional specificity and fatty acid selectivity on purified *sn*-glycerol-3-phosphate acyltransferases from chloroplasts. Plant Physiol 68: 653–657 (1981).
5. Brody S, Mikolajczyk S: *Neurospora* mitochondria contain an acyl carrier protein. Eur J Biochem 173: 353–359 (1988).
6. Browse J, McCourt P, Somerville CR: A mutant of *Arabidopsis* lacking a chloroplast-specific lipid. Science 227: 763–765 (1985).
7. Cavin DT: The biosynthesis of long chain fatty acids in the developing castor bean. Can J Bot 43: 49–62 (1965).
8. Caughey I, Kekwick RGO: The characteristics of some components of the fatty acid synthetase system in the plastids from the mesocarp of avocado (*Persea americana*). Eur J Biochem 123: 553–561 (1982).
9. Chandler CS, Ballard FJ: Multiple biotin-containing proteins in 3T3-LI cells. Biochem J 237: 123–130 (1986).
10. de Chavigney A, Heacock PN, Dowhan W: Sequence and inactivation of the *pss* gene of *Escherichia coli*. Phosphatidylethanolamine may not be essential for cell viability. J Biol Chem 266: 5323–5332 (1991).
11. Chuman L, Brody S: Acyl carrier protein is present in mitochondria of plants and in eukaryotic microorganisms. Eur J Biochem 184: 643–649 (1989).
12. Cooper CL, Hsu L, Jackowski S, Rock CO: 2-acylglycerol phosphoethanolamine acyltransferase/acyl-acyl carrier protein synthetase is a membrane associated acyl carrier protein binding protein. J Biol Chem 264: 7384–7389 (1989).
13. Cortes J, Haydock SF, Roberts GA, Bevitt DJ, Leadlay PF: An unusually large multifunctional polypeptide in the erythromycin-producing polyketide synthase of *Saccharopolyspora erythraea*. Nature 348: 176–178 (1990).
14. Cottingham IR, Austin A, Sidebottom C, Slabas AR: Purified enoyl-[acyl-carrier-protein] reductase from rape seed (*Brassica napus*) contains two closely related polypeptides which differ by a six amino acid N-terminal extension. Biochim Biophys Acta 954: 201–207 (1988).
15. Cottingham IR, Austin AJ, Slabas AR: Inhibition and covalent modification of rape seed (*Brassica napus*) enoyl ACP reductase by phenylglyoxal. Biochim Biophys Acta 995: 273–278 (1989).
16. Cronan JE Jr, Rock CO: Biosynthesis of membrane lipids in *Escherichia coli* and *Salmonella typhimurium*. In: Neidhart FC (ed) Cellular and Molecular Biology. vol. 1, pp. 474–497. American Society for Microbiology.
17. de Duve C: Exploring cells with a centrifuge. Science 189: 186–194 (1975).
18. Danielli JF, Davson H: A contribution to the theory of permeability of thin films. J Cell Comp Physiol 5: 495–508 (1935).
19. Douady D, Dubacq J-P: Purification of acyl CoA: glycerol-3-phosphate acyltransferase from pea leaves. Biochim Biophys Acta 921: 615–619 (1987).

20. Dai DH, Pape ME, Lopez-Casillas F, Luo XC, Dixon JE, Kim KH: Molecular cloning of cDNA for acetyl-coenzyme A carboxylase. J Biol Chem 261: 12395–12399 (1986).
21. Egin-Buhler B, Ebel J: Improved purification and further characterisation of acetyl CoA carboxylase from cultured cells of parsley (*Petroselinum hortease*). Eur J Biochem 133: 335–339 (1983).
22. Feldmann KA: T-DNA insertion mutagenesis in *Arabidopsis*: mutational spectrum. Plant J 1: 71–82 (1991).
23. Frentzen M, Heinz E, McKeon A, Stumpf PK: Specificities and selectivities of glycerol-3-phosphate acyltransferase and monoacylglycerol-3-phosphate acyltransferase from pea and spinach chloroplasts. Eur J Biochem 129: 629–636 (1983).
24. Green PR, Bell RM: Asymmetric reconstitution of homogeneous *Escherichia coli sn*-glycerol-3-phosphate acyltransferase into phospholipid vesicles. J Biol Chem 259: 14688–14694 (1984).
25. Griffith G, Stymne S, Stobart K: Biosynthesis of triglycerides in plant storage tissue. In: Applewhite TM (ed) Proceedings World Conference on Biotechnology for the Fats and Oils Industry. American Oil Chemists Society, Glenview, IL (1988).
26. Guerra DJ, Ohlrogge JB: Partial purification and characterization of two forms of malonyl coenzyme A:acyl carrier protein transacylase from soybean leaf tissue. Arch Biochem Biophys 246: 274–285 (1986).
27. Guerra DJ, Ohlrogge JB, Frentzen M: Activities of acyl carrier protein isoforms in reactions of plant fatty acid metabolism. Plant Physiol 82: 448–453 (1986).
28. Guerra DJ, Dziewanowska K, Ohlrogge JB, Beremand PD: Purification and characterization of recombinant spinach acyl carrier protein I expressed in *Escherichia coli*. J Biol Chem 263: 4386–4391 (1988).
29. Hahlbrock K, Grisebach H: Enzymic controls in the biosynthesis of lignin and flavanoids. Annu Rev Plant Physiol 30: 105–130 (1979).
30. Harder ME, Ladenson RC, Schimel SD, Silbert DF: Mutants of *Escherichia coli* with temperature-sensitive malonyl coenzyme A:acyl carrier protein transacylase. J Biol Chem 349: 7468–7475 (1974).
31. Hares W, Frentzen M: Substrate specificities of membrane-bound and partly purified microsomal acyl CoA:1-acylglycerol-3-phosphate acyltransferase from etiolated shoots of *Pisum sativum* (L.). Planta 185: 124–131 (1991).
32. Hatzopoulos P, Franz G, Choy L, Sung RZ: Interaction of nuclear factors with upstream sequence of a lipid body membrane protein from carrot. Plant Cell 2: 457–467 (1990).
33. Hellyer A, Bambridge HE, Slabas AR: Plant acetyl-CoA carboxylase. Biochem Soc Trans 14: 565–568 (1986).
34. Hellyer A, Slabas AR: Acyl-[acyl-carrier protein] thioesterase from oil seed rape – purification and characterization. In: Quinn BJ, Harwood JL (eds) Plant Lipid Biochemistry, Structure and Utilization, pp. 157–159. Portland Press, London (1990).
35. Herman EM, Melroy DL, Buckhout TJ: Apparent processing of a soybean oil body protein accompanies the onset of oil mobilization. Plant Physiol 94: 341–349 (1990).
36. Hilditch TP, Williams PN: Chemical Constitution of Natural Lipids. Chapman and Hall, London (1964).
37. Hoffman NE, Pichersky E, Cashmore AR: A tomato cDNA encoding a biotin binding protein. Nucl Acid Res 15: 3928 (1987).
38. Hoj PB, Mikkelsen JD: Partial separation of individual enzyme activities of an ACP-dependent fatty acid synthetase from barley chloroplasts. Carlsberg Res Commun 47: 119–141 (1982).
39. Hoj PB, Svedson I: Barley acyl carrier protein: its amino acid sequence and assay using purified malonyl CoA:ACP transacylase. Carlsberg Res Commun 48: 285–305 (1983).
40. Holak TA, Kearsley SK, Kim Y, Prestegard JH: Three-dimensional structure of acyl carrier protein determined by NMR pseudoenergy and distance geometry calculations. Biochemistry 27: 6135–6142 (1988).
41. Hopwood DA, Sherman DH: Molecular genetics of polyketides and its comparison to fatty acid biosynthesis. Annu Rev Genet 24: 37–66 (1990).
42. Imai H, Nishida I, Murata N: Acyl-[acyl-carrier-protein] hydrolase of squash cotyledons: purification and characterization. In: Quinn BJ, Harwood JL (eds) Plant Lipid Biochemistry, Structure and Utilization, pp. 160–162. Portland Press (1990).
43. Ishizaki O, Nishida I, Agata K, Eguchi G, Murata N: Cloning and nucleotide sequence of cDNA for the plastid glycerol-3-phosphate acyltransferase from squash. FEBS Lett 238: 424–430 (1988).
44. Jackowski S, Rock CO: Acetoacetyl-acyl carrier protein synthease, a potential regulator of fatty acid biosynthesis in bacteria. J Biol Chem 262: 7927–7931 (1987).
45. Jaworski JG, Clough RC, Barnum SR: A cerulenin insensitive short chain 3-ketoacyl-acyl carrier protein synthease in *Spinacia oleracea* leaves. Plant Physiol 90: 41–44 (1989).
46. Jaworski JG, Clough RG, Barnum SR, Post-Beittenmiller D, Ohlrogge JB: Initial reactions of fatty acid synthesis and their regulation. In: Quinn BJ, Harwood JL (eds) Plant Lipid Biochemistry, Structure and Utilization, pp. 97–104. Portland Press, London (1990).
47. Joyard J, Block MA, Douce R: Molecular aspects of plastid envelope biochemistry. Eur J Biochem 199: 489–509 (1990).
48. Kater MM, Koningstein GM, Nijkamp HJJ, Stuitje AR: cDNA cloning and expression of *Brassica napus* enoyl-acyl carrier protein reductase in *Escherichia coli*. Plant Mol Biol 17: 895–909 (1991).
49. Kim Y, Prestegard JH: A dynamic model for the struc-

ture of acyl carrier protein in solution. Biochemistry 28: 8792–8797 (1989).

50. Kim Y, Ohlrogge JB, Presegard JH: Motional effects on NMR structural data-comparison of spinach and *Escherichia coli* acyl carrier proteins. Biochem Pharmacol 40: 7–13 (1990).
51. Knudsen J, Clark S, Dils R: Purification and some properties of a medium chain acyl-thioester hydrolase from lactating rabbit mammary gland which terminates chain elongation in fatty acid synthesis. Biochem J 160: 683–691 (1976).
52. Knudsen J, Grunnet I: Transacylation as a chain-termination mechanism in fatty acid synthesis by mammalian fatty acid synthetase. Biochem J 202: 139–143 (1982).
53. Kolattakudy PE: Lipid-derived defense polymers and waxes and their role in plant-microbe interaction. In: Stumpf PK, Conn EE (eds) The Biochemistry of Plants, vol. 9, pp. 291–314. Academic Press (1987).
54. Kridl JC, Knutzon DS, Johnson WB, Thompson GA, Radke SE, Turner JC, Knauf VC: Modulation of stearoyl-ACP desaturase levels in transgenic rapeseed. The International Society for Plant Molecular Biology, 3rd International Congress, Tucson, AZ, 6–11 October 1991, Poster 723.
55. Kuo TM, Ohlrogge JB: The primary sequence of spinach acyl carrier protein. Arch Biochem Biophys 234: 290–296 (1984).
56. Lal SK, Johnson S, Conway T, Kelly PM: Characterization of a maize cDNA that complements an enolase deficient mutant of *Escherichia coli*. Plant Mol Biol 16: 787–796 (1991).
57. Lamppa G, Jacks C: Analysis of two linked genes coding for the acyl carrier protein (ACP) from *Arabidopsis thaliana (columbia)*. Plant Mol Biol 16: 469–474 (1991).
58. Larson TJ, Lightner VA, Green PR, Modrich P, Bell RM: Membrane phospholipid synthesis in *Escherichia coli*. Identification of the *sn*-glycerol-3-phosphate acyltransferase polypeptide as the *plsB* gene product. J Biol Chem 255: 9421–9426 (1980).
59. Lerouge P, Roche P, Faucher C, Maillet F, Truchet G, Prome JC, Denarie J: Symbiotic host-specificity of *Rhizobium meliloti* is determined by a sulphated and acylated glucosamine oligosaccharide signal. Nature 344: 781–784 (1991).
60. Lessire R, Stumpf PK: Nature of the fatty acid synthetase systems in parenchymal and epidermal cells of *Allium porrum* L. leaves. Plant Physiol 73: 614–618 (1983).
61. Lowe PN, Rhodes S: Purification and characterization of [acyl carrier protein] acetyltransferase from *Escherichia coli*. Biochem J 250: 789–796 (1988).
62. Lynch DV, Thompson GA: Retailored lipid molecular species: a tactile mechanism for modulating membrane properties. Trends Biochem Sci 9: 442–445 (1984).
63. Mackintosh RW, Hardie DG, Slabas AR: A new assay procedure to study the induction of β-ketoacyl-ACP synthease I and II, and the complete purification of β-ketoacyl-ACP synthease I from developing seeds of oil seed rape (*Brassica napus*). Biochim Biophys Acta 1002; 114–124 (1989).
64. Masterson C, Wood C, Thomas DR: L-acetylcarnitine, a substrate for chloroplast fatty acid synthesis. Plant Cell Environ 13: 755–765 (1990).
65. McKeon TA, Stumpf PK: Purification and characterization of the stearoyl-acyl carrier protein desaturase and the acyl-acyl carrier protein thioesterase from maturing seeds of safflower. J Biol Chem 257: 12141–12147 (1982).
66. Mitchell P: Keilins respiratory chain concept and its chemiosmotic consequences. Science 206: 1148–1159 (1979).
67. Mudd JB, McManus TT: Metabolism of acetate by cell-free preparations from spinach leaves. J Biol Chem 237: 2057–2063 (1962).
68. Mundy J, Rogers JC: Selective expression of a probable amylase/protease inhibitor in barley aleurone cells: Comparison to the barley amylase/subtilisin inhibitor. Planta 196: 51–63 (1986).
69. Murata N: Molecular species composition of phosphatidylglycerols from chilling-sensitive and chilling-resistant plants. Plant Cell Physiol 24: 81–86 (1983).
70. Murata N: Gene-technological manipulation of fatty-acid unsaturation: desaturase and acyltransferase. The International Society for Plant Molecular Biology, 3rd International Congress, Tucson, AZ, 6–11 October 1991 (1991).
71. Murphy DJ: Storage lipid bodies in plants and other organisms. Plant Lipid Res 29: 299–324 (1990).
72. Murphy DJ, Keen JN, O'Sullivan JN, Au DMY, Edwards EW, Jackson PJ, Cummins I, Gibbons T, Shaw CH, Ryan AJ: A class of amphipathic proteins associated with lipid storage bodies in plants. Possible similarities with animal serum apolipoproteins. Biochim Biophys Acta 1088: 86–94 (1991).
73. Nikolau BJ, Wurtele ES: Molecular cloning and characterization of acetyl-CoA carboxylase and other biotin enzymes of plants. The International Society for Plant Molecular Biology, 3rd International Congress, Tucson, AZ, 6–11 October 1991, Poster 720 (1991).
74. Nishida I, Kawaguchi A, Yamada M: Selective inhibition of type II fatty acid synthetase by the antibiotic thiolactomycin. Plant Cell Physiol 25: 265–268 (1984).
75. Nishida I, Kawaguchi A, Yamada M: Effects of thiolactomycin on the individual enzyme of the fatty acid synthase system in *Escherichia coli*. J Biochem (Tokyo) 99: 1447–1454 (1986).
76. Nishida I, Frentzen M, Ishizaki O, Murata N: Purification of isomeric forms of acyl [acyl-carrier-protein]: glycerol 3-phosphate acyltransferase from greening squash cotyledons. Plant Cell Physiol 18: 1071–1079 (1987).

77. Nishida I, Tasaka Y, Shiraishi H, Okada K, Shimura Y, Beppu T, Murata N: The genomic gene and cDNA for the plastidial glycerol-3-phosphate acyltransferase of *Arabidopsis*. In: Quinn PJ, Harwood JL (eds) Plant Lipid Biochemistry, Structure and Utilization, pp. 462–464. Portland Press, London (1990).
78. Ohlrogge JB, Kuo TM: Control of lipid synthesis during soybean seed development: enzymic and immunochemical assay of acyl carrier protein. Plant Physiol 74: 622–625 (1984).
79. Ohlrogge JB, Kuo TM: Plants have isoforms of acyl carrier proteins that are expressed differently in different tissues. J Biol Chem 260: 8032–8037 (1985).
80. Ohlrogge JB, Browse J, Somerville CR: The genetics of plant lipids. Biochim Biophys Acta 1082: 1–26 (1991).
81. Ottey KA, Takhan S, Munday MR: Comparison of two cyclic-nucleotide-independent acetyl CoA carboxylase kinases from lactating rat mammary glands: identification of the kinase responsible for acetyl CoA inactivation *in vivo*. Biochem Soc Trans 17: 349-350 (1989).
82. Parker WB, Marshall LC, Burton JD, Somers DA, Wise DL, Gronwald JD, Gengenbach BG: Dominant mutations causing alterations in acetyl Coenzyme A carboylase confer tolerance to cyclohexanedione and aryloxyphenoxyproprionate herbicides in maize. Proc Natl Acad Sci USA 87: 7175–7179 (1990).
83. Pollard MR, Anderson L, Fan C, Hawkins DJ, Evans HM: A specific acyl-ACP-thioesterase implicated in medium chain fatty acid production in immature cotyledons of *Umbellularia california*. Arch Biochem Biophys 284: 306–312 (1991).
84. Post-Beittenmiller MA, Housek-Radojcic A, Ohlrogge JB: DNA sequence of a genomic clone encoding an *Arabidopsis* acyl carrier protein. Nucl Acids Res 17: 1777 (1989).
85. Post-Beittenmiller MA, Jaworski JG, Ohlrogge JB: *In vivo* pools of free and acylated acyl carrier protein in spinach: evidence for sites of regulation of fatty acid synthesis. J Biol Chem 266: 1858–1865 (1991).
86. Poulose AJ, Rogers L, Cheesbrough TM, Kolattukudy PE: Cloning and sequencing of the cDNA for S-acyl fatty acid synthase from the uropygial gland of mallard duck. J Biol Chem 260: 15953–15958 (1985).
87. Prescott DJ, Vagelos PR: Acyl carrier protein. Adv Enzymol 36: 269–311 (1972).
88. Qu R, Huang AHC: Oleosin KD18 on the surface of oil bodies in maize. J Biol Chem 265: 2238–2243 (1990).
89. Rasmussen JT, Borchers T, Knudsen J: Comparison of the binding affinities of acyl CoA binding protein and fatty acid binding protein for long chain acyl CoA esters. Biochem J 265: 849–855 (1990).
90. Ap-Rees T: The plasticity of plant metabolism. In: Proceedings 'Manipulating Plant Metabolism', Edinburgh, 12 September 1991, pp. 6–8.
91. Robert G, Leadlay PF: Use of [3H] tetrahydrocerulenin to assay condensing enzyme activities in *Streptomyces erythreus*. Biochem Soc Trans 12: 642–643 (1984).
92. Roughan G: On the control of fatty acid compositions of plant glycerolipid. In: Stumpf PK, Mudd JB, Nea WD (eds) The Metabolism, Structure and Function of Plant Lipids, pp. 247–254. Plenum, New York (1987).
93. Runswick MJ, Fearnley IM, Skehel JM, Walker JE: Presence of an acyl carrier protein in NADH-ubiquinone oxidoreductase from bovine heart-mitochondria. FEBS Lett 286: 121–124 (1991).
94. Safford R, de Silva J, Lucas C, Windust JHC, Shedden J, James CM, Sidebottom CM, Slabas AR, Tombs MP, Hughes SG: Molecular cloning and sequence analysis of complementary DNA encoding rat mammary gland medium-chain S-acyl fatty acid synthetase thio ester hydrolase. Biochemistry 26: 1358–1364 (1987).
95. Safford R, Windust JHC, Lucas C, De Silva J, James CM, Hellyer A, Smith CG, Slabas AR, Hughes SG: Plastid-localized seed acyl-carrier protein of *Brassica napus* is encoded by a distinct, nuclear multigene family. Eur J Biochem 174: 287–295 (1988).
96. Scherer DE, Knauf VC: Isolation of complementary DNA clone for the acyl carrier protein-I of spinach. Plant Mol Biol 9: 127–134 (1987).
97. Schroder S, Brown JWS, Schroder J: Molecular analysis of resveratrol synthase cDNA, genomic clones and relationship with chalcone synthase. Eur J Biochem 172: 161–169 (1988).
98. Schulz R, Ebel J, Hahlbrock K: Partial purification of β-ketoacyl acyl carrier protein synthetase from a higher plant. FEBS Lett 140: 207–209 (1982).
99. Schweizer M, Lebert C, Holtke J, Roberts LM, Schweizer E: Molecular cloning of the yeast fatty acid synthetase genes, FAS 1 and FAS 2: Illustrating the structure of the FAS 1 cluster gene by transcript mapping and transformation studies. Mol Gen Genet 194: 457–465 (1984).
100. Schweizer M, Takabayashi K, Laux T, Beck KF, Schreglmann R: Rat mammary gland fatty acid synthetase: localization of the constituent domains and two functional polyadenylation/termination signals in the cDNA. Nucl Acids Res 17: 567–586 (1989).
101. Shanklin J, Somerville C: Stearoyl-acyl-carrier protein desaturase from higher plants is structurally unrelated to the animal and fungal homologs. Proc Natl Acad Sci USA 88: 2510–2514 (1991).
102. Shearman CA, Rossen L, Johnston AWB, Downie JA: The *Rhizobium leguminosarum* nodulation gene *nodF* encodes a polypeptide similar to acyl carrier protein and is regulated by *nodD* plus a factor in pea roots. EMBO J 5: 647–652 (1986).
103. Sheldon PS: A study of plant plastid NADPH dependent β-ketoacyl-(acyl carrier protein) reductase. Ph.D. thesis, University of Birmingham, UK (1988).
104. Sheldon PS, Safford R, Slabas AR, Kekwick RGO: 2-Oxoacyl-[acyl carrier protein] reductase; a component

of plant fatty acid synthase. Biochem Soc Trans 16: 392–393 (1988).

105. Sheldon PS, Kekwick RGO, Sidebottom C, Smith CG, Slabas AR: 3-oxoacyl-(acyl-carrier protein) reductase from avocado (*Persea americana*) fruit mesocarp. Biochem J 271: 713–720 (1990).
106. Sheldon PS, Kekwick RGO, Smith CG, Sidebottom C, Slabas AR: 3-Oxoacyl-[ACP] reductase from oilseed rape (*Brassica napus*). Submitted for publication.
107. Shimakata T, Stumpf PK: The prokaryotic nature of the fatty acid synthetase of developing *Carthornus tinctorius* L. (safflower) seeds. Arch Biochem Biophys 217: 144–154 (1982).
108. Shimakata T, Stumpf PK: Fatty acid synthetase of *Spinacia oleracea* leaves. Plant Physiol 69: 1257–1262 (1982).
109. Shimakata T, Stumpf PK: Isolation and function of spinach leaf β-ketoacyl-[acyl-carrier-protein] synthetases. Proc Natl Acad Sci USA 79: 5808–5812 (1982).
110. Shimakata T, Stumpf PK: Purification and characteristics of β-keto-[acyl-carrier-protein] reductase, β-hydroxyacyl-[acyl-carrier-protein] dehydrase and enoyl-[acyl-carrier-protein] reductase from *Spinacia oleracea* leaves. Arch Biophys Biochem 218: 77–91 (1982).
111. Siggaard-Andersen M, Kauppinen S, von Wettstein-Knowles P: Primary structure of a cerulenin binding β-ketoacyl [acyl carrier protein] synthase from barley chloroplasts. Proc Natl Acad Sci USA 88: 4114–4118 (1991).
112. de Silva J, Loader NM, Jarman C, Windust JHC, Hughes SG, Safford R: The isolation and sequence analysis of two seed-expressed acyl carrier protein genes from *Brassica napus*. Plant Mol Biol 14: 537–548 (1990).
113. Simoni RD, Criddle RS, Stumpf PK: Fat metabolism in higher plants. 31. Purification and properties of plant and bacterial acyl carrier protein. J Biol Chem 242: 573–581 (1967).
114. Singer SJ, Nicholson GC: The fluid mosaic model of cell membranes. Science 175: 720–731 (1972).
115. Slabas AR, Hellyer A: Rapid purification of a high molecular weight subunit polypeptide form of rape seed acetyl CoA carboxylase. Plant Sci Lett 39: 177–182 (1985).
116. Slabas AR, Hellyer A, Bambridge HE: The basic polypeptide subunit of rape leaf acetyl-CoA carboxylase is a 220 kDa protein. Biochem Soc Trans 14: 714–715 (1986).
117. Slabas AR, Sidebottom CM, Hellyer A, Kessell RMJ, Tombs MP: Induction, purification and characterization of NADH-specific enoyl-acyl carrier protein reductase from developing seeds of oil seed rape, (*Brassica napus*). Biochim Biophys Acta 877: 271–280 (1986).
118. Slabas AR, Harding J, Hellyer A, Roberts P, Bambridge HE: Induction, purification and characterization of acyl carrier protein from developing seeds of oil seed rape (*Brassica napus*). Biochim Biophys Acta 921: 50–59 (1987).
119. Slabas AR, Hellyer A, Sidebottom C, Bambridge H, Cottingham IR, Kessell R, Smith CG, Sheldon P, Kekwick RGO, de Silva J, Windust J, James CM, Hughes SG, Safford R: Molecular structure of plant fatty acid synthesis enzymes. In: von Wettstein D, Chua NH (eds) Plant Molecular Biology, pp. 265–277. NATO ASI Series A, vol 140 (1987).
120. Slabas AR, Cottingham IR, Austin A, Hellyer A, Safford R, Smith CG: Immunological detection of NADH-specific enoyl-ACP reductase from rape seed (*Brassica napus*): induction relationship of α and β polypeptides, mRNA translation and interaction with ACP. Biochim Biophys Acta 1039: 181–188 (1990).
121. Slabas AR, Cottingham IR, Austin A, Fawcett T, Sidebottom CM: Amino acid sequence analysis of rape seed (*Brassica napus*) NADH-enoyl ACP reductase. Plant Mol Biol, in press.
122. Slabas AR, Swinhoe R, Slabas D, Simon W, Fawcett T: Biosynthesis and assembly of the enzymes involved in lipid metabolism in plants. In: Tubin AK (ed) Metabolic Interactions of Organelles in Photosynthetic Tissues. Society for Experimental Biology Seminar Series, in press.
123. Slack CR, Roughan PG, Balsingham N: Labelling studies *in vivo* on the metabolism of the acyl and glycerol moieties of the glycerolipids in the developing maize leaf. Biochem J 162: 289–296 (1977).
124. Slack CR, Bertaud WS, Shaw BD, Holland R, Browse J, Wright H: Some studies on the composition and surface properties of oil bodies from the seed cotyledons of safflower (*Carthanus tinctorius*) and linseed (*Linum ustatissimum*). Biochem J 190: 551–561 (1980).
125. Smith MA, Cross AR, Jones OTG, Griffiths WT, Stymne S, Stobart K: Electron-transport components of 1-acyl-2-oleoyl-*sn*-glycero-3-phosphocholine Δ12 desaturase (Δ12 desaturase) in microsomal preparations of developing safflower (*Carthanus tinctorius* L.) cotyledons. Biochem J 272: 23–29 (1990).
126. Smith S, Libertini CJ: Specificity and site of action of a mammary gland thioesterase which released acyl moieties from thioester linkages to the fatty acid synthetase. Arch Biochem Biophys 196: 92–92 (1979).
127. Somerville C, Browse J: Plant lipids: Metabolism, Mutants and Membrane. Science 252: 80–87 (1991).
128. Sossountzov L, Ruiz-Avila L, Vignols F, Jolliot A, Arondel V, Tchang F, Grosbois M, Miginiac E, Delseny M, Puigdomenech P, Kader J-C: Spatial and Temporal expression of a maize lipid transfer protein gene. Plant Cell 3: 923–933 (1991).
129. Stapelton SR, Jaworski JG: Characterization and purification of malonyl coenzyme A:[acyl carrier protein] transacylase from spinach and *Anabaenna variabilis*. Biochim Biophys Acta 794: 240–248 (1984).
130. Steponkus PL, Uemura M, Balsamo RA, Arvinte T,

Lynch DV: Transformation of the cryobehaviour of rye protoplasts by modification of the plasma membrane lipid composition. Proc Natl Acad Sci USA 85: 9026–9030 (1988).

131. Sterk R, Booij H, Schellekens GA, van Kammen A, de Vries SC: Cell specific expression of the carrot EP2 lipid transfer protein gene. Plant Cell 3: 907–921 (1991).
132. Stumpf PK: Biosynthesis of saturated and unsaturated fatty acids. In: Stumpf PK, Conn EE (eds) The biochemistry of plants, vol. 4 pp. 177–202. Academic Press (1980).
133. Stumpf PK: The biosynthesis of saturated fatty acids. In: Stumpf PK, Conn EE (eds) The Biochemistry of Plants, vol 9, pp. 121–136. Academic Press (1987).
134. Stumpf PK, Barber G: Fat metabolism in higher plants. IX. Enzymatic synthesis of long chain fatty acids by avocado particles. J Biol Chem 227: 407–417 (1957).
135. Stumpf PK, Shimakata T: Molecular structures and functions of plant fatty acid synthetase enzymes. In: Mudd JB, Gibbs M (eds) Biosynthesis and Fucntion of Plant Lipids, pp. 1–16. Waverly Press (1983).
136. Stymne S, Glad G: Acyl exchange between oleoyl-CoA and phosphatidylcholine in microsomes of developing soya bean cotyledons and its role in fatty acid desaturation. Lipids 16: 298–305 (1981).
137. Tchang F, This P, Steifel V, Arondel V, Morch MD, Pages M, Puigdomech P, Grellet F, Delseny M, Bouillon P, Huet JC, Guerbette F, Beavais-Cante F, Duranton H, Pernollet JC, Kader JC: Phospholipid transfer protein: full length cDNA and amino acid sequence in maize. J Biol Chem 263: 16849–16855 (1988).
138. Thoma S, Kaneko Y, Somerville S: An extracellular location of the supposed *Arabidopsis* lipid transfer protein. The International Society for Plant Molecular Biology, 3rd International Congress, Tucson, AZ, 6–11 October 1991, Poster 1044 (1991).
139. Thompson GA, Scherer DE, Foxall-Van Aken S, Kenny JW, Young HL, Shintani DK, Kridl JC, Knauf VC: Primary structures of the precursors and mature forms of stearoyl-acyl carrier protein desaturase from safflower embryos and requirement of ferredoxin for enzyme activity. Proc Natl Acad Sci USA 88: 2578–2582 (1991).
140. Truchet G, Roche P, Leronge P, Vasse J, Camet S, de Billy F, Prome J-C, Denarie J: Sulphated lipo-oligosaccharide signals of *Rhizobium meliloti* elicit root nodule organogenesis in alfalfa. Nature 351: 670–673 (1991).
141. Turnham E, Northcote DH: Changes in the activity of acetyl CoA carboxylase during rape-seed formation. Biochem J 212: 223–229 (1983).
142. Vance VB, Huang AHC: The major protein from lipid bodies of maize. J Biol Chem 262: 11275–11279 (1987).
143. Vanden Boom T, Cronan JE Jr: Genetics and regulation of bacterial lipid metabolism. Annu Rev Microbiol 43: 317–343 (1989).
144. Vergnolle C, Ostergaard J, Renard Kader J-C: Bifunctional lipid transfer/fatty acid binding proteins in oil-forming seeds. In: Quinn PS, Harwood JC (eds) Plant Lipid Biochemistry, Structure and Utilization pp. 272–274. Portland Press, London (1990).
145. Wada H, Murata N: *Synechocystis* PCC 6803 mutants defective in desaturation of fatty acids. Plant Cell Physiol 30: 971–978 (1989).
146. Wada H, Gombos Z, Murata N: Enhancement of chilling tolerance of a cyanobacterium by genetic manipulation of fatty acid desaturation. Nature 347: 200–203 (1990).
147. Walsh MC, Klopfenstein WE, Harwood JL: The short chain condensing enzyme has a widespread occurrence in the fatty acid synthetases from higher plants. Phytochemistry 29: 3797–3799 (1990).
148. Weaire PJ, Kekwick RGO: The synthesis of fatty acids in avocado mesocarp and cauliflower bud tissue. Biochem J 146: 425–437 (1975).
149. von Wettstein-Knowles P: Biosynthesis of epicuticular lipids as analysed with the aid of gene mutations in barley. In: Wintermans JFGM, Kuiper PJC (eds) Biochemistry and Metabolism of Plant Lipids, pp. 69–78. Elsevier, Amsterdam (1982).
150. Wurtele ES, Nikolan BJ: Plants contain multiple biotin enzymes: discovery of 3-methylcrotonyl CoA carboxylase, proprionyl-CoA carboxylase acid pyruvate carboxylase in the plant kingdom. Arch Biochem Biophys 278: 179–186 (1990).
151. Yamada M, Tsuboi S, Osafune T, Suga T, Takishima K: Multifractional properties of non-specific lipid transfer protein from higher plants. In: Quinn PJ, Harwood JC (eds) Plant Lipid Biochemistry, Structure and Utilization, pp. 278–280. Portland Press, London (1990).

Index